Urban Ecosystem Services IV

Urban Ecosystem Services IV

Guest Editors

Alessio Russo
Giuseppe T. Cirella

Basel • Beijing • Wuhan • Barcelona • Belgrade • Novi Sad • Cluj • Manchester

Guest Editors

Alessio Russo
School of Architecture and Built Environment
Faculty of Engineering
Queensland University of Technology
Brisbane
Australia

Giuseppe T. Cirella
Faculty of Economics
University of Gdansk
Sopot
Poland

Editorial Office
MDPI AG
Grosspeteranlage 5
4052 Basel, Switzerland

This is a reprint of the Special Issue, published open access by the journal *Land* (ISSN 2073-445X), freely accessible at: https://www.mdpi.com/journal/land/special_issues/PH0V38FA7W.

For citation purposes, cite each article independently as indicated on the article page online and as indicated below:

Lastname, A.A.; Lastname, B.B. Article Title. *Journal Name* **Year**, *Volume Number*, Page Range.

ISBN 978-3-7258-3219-4 (Hbk)
ISBN 978-3-7258-3220-0 (PDF)
https://doi.org/10.3390/books978-3-7258-3220-0

Cover image courtesy of Alessio Russo

© 2025 by the authors. Articles in this book are Open Access and distributed under the Creative Commons Attribution (CC BY) license. The book as a whole is distributed by MDPI under the terms and conditions of the Creative Commons Attribution-NonCommercial-NoDerivs (CC BY-NC-ND) license (https://creativecommons.org/licenses/by-nc-nd/4.0/).

Contents

About the Editors vii

Alessio Russo and Giuseppe T. Cirella
Urban Ecosystem Services in a Rapidly Urbanizing World: Scaling up Nature's Benefits from Single Trees to Thriving Urban Forests
Reprinted from: *Land* 2024, 13, 786, https://doi.org/10.3390/land13060786 1

Goran Krsnik, Sonia Reyes-Paecke, Keith M. Reynolds, Jordi Garcia-Gonzalo and José Ramón González Olabarria
Assessing Relativeness in the Provision of Urban Ecosystem Services: Better Comparison Methods for Improved Well-Being
Reprinted from: *Land* 2023, 12, 1088, https://doi.org/10.3390/land12051088 10

Atena-Ioana Gârjoabă, Cerasella Crăciun and Alexandru-Ionuț Petrișor
Natural Protected Areas within Cities: An International Legislative Comparison Focused on Romania
Reprinted from: *Land* 2023, 12, 1279, https://doi.org/10.3390/land12071279 26

Sheila K. Schueller, Zhelin Li, Zoe Bliss, Rachelle Roake and Beth Weiler
How Informed Design Can Make a Difference: Supporting Insect Pollinators in Cities
Reprinted from: *Land* 2023, 12, 1289, https://doi.org/10.3390/land12071289 53

Yi Le and Sheng-Yang Huang
Prediction of Urban Trees Planting Base on Guided Cellular Automata to Enhance the Connection of Green Infrastructure
Reprinted from: *Land* 2023, 12, 1479, https://doi.org/10.3390/land12081479 75

María de la Luz Espinosa Fuentes, Oscar Peralta, Rocío García, Eugenia González del Castillo, Rosa María Cerón Bretón, Julia Griselda Cerón Bretón, et al.
Soil Dynamics in an Urban Forest and Its Contribution as an Ecosystem Service
Reprinted from: *Land* 2023, 12, 2098, https://doi.org/10.3390/land12122098 93

Hyunsu Kim, Kyushik Oh and Ilsun Yoo
Analysis of Spatial Characteristics Contributing to Urban Cold Air Flow
Reprinted from: *Land* 2023, 12, 2165, https://doi.org/10.3390/land12122165 109

Neelesh Yadav, Shrey Rakholia and Reuven Yosef
Decision Support Systems in Forestry and Tree-Planting Practices and the Prioritization of Ecosystem Services: A Review
Reprinted from: *Land* 2024, 13, 230, https://doi.org/10.3390/land13020230 129

Ye Li, Junda Huang and Yuncai Wang
Identifying the Optimal Area Threshold of Mapping Units for Cultural Ecosystem Services in a River Basin
Reprinted from: *Land* 2024, 13, 346, https://doi.org/10.3390/land13030346 142

Giedrius Dabašinskas and Gintarė Sujetovienė
Spatial and Temporal Changes in Supply and Demand for Ecosystem Services in Response to Urbanization: A Case Study in Vilnius, Lithuania
Reprinted from: *Land* 2024, 13, 454, https://doi.org/10.3390/land13040454 158

Jan Łukaszkiewicz, Andrzej Długoński, Beata Fortuna-Antoszkiewicz and Jitka Fialová
The Ecological Potential of Poplars (*Populus* L.) for City Tree Planting and Management: A Preliminary Study of Central Poland (Warsaw) and Silesia (Chorzów)
Reprinted from: *Land* **2024**, *13*, 593, https://doi.org/10.3390/land13050593 . . . . **173**

About the Editors

Alessio Russo

Dr Alessio Russo is a Senior Lecturer in landscape architecture at the Queensland University of Technology. He holds a PhD in Arboriculture, Forestry, Ornamental and Landscape Agro-Ecosystems from the University of Bologna, Italy, and has extensive experience in both academia and professional practice. Dr Russo's research interests include the role of green infrastructure in providing ecosystem services, the relationship between urban green spaces and mental health, and nature-based solutions for resilient urban environments. He has published numerous peer-reviewed articles, books, and book chapters on these topics, significantly contributing to the field. Before joining the Queensland University of Technology, Dr Russo was a Senior Lecturer and Academic Course Leader at the University of Gloucestershire, UK, and an Associate Professor at RUDN University, Moscow, Russia. He also served as a Professor and Head of the Laboratory of Urban and Landscape Design at Far Eastern Federal University, Vladivostok, Russia. In addition to his academic achievements, Dr Russo has contributed to numerous international research projects and has been a keynote speaker at various conferences and seminars.

Giuseppe T. Cirella

Prof. Dr. Giuseppe T. Cirella is a Professor of Human Geography at the University of Gdansk, Poland, where he received a Doctor of Habilitation in Economics and Finance. He specializes in economic development, environmental social science, and sustainability. His background also includes socio-political research throughout Eastern Europe, Africa, and China. He is on the editorial board of a number of internationally renowned academic peer-reviewed journals.

Editorial

Urban Ecosystem Services in a Rapidly Urbanizing World: Scaling up Nature's Benefits from Single Trees to Thriving Urban Forests

Alessio Russo [1],* and Giuseppe T. Cirella [2]

1. School of Architecture and Built Environment, Faculty of Engineering, Queensland University of Technology, Brisbane 4000, Australia
2. Faculty of Economics, University of Gdansk, 81-824 Sopot, Poland; gt.cirella@ug.edu.pl
* Correspondence: alessio.russo@qut.edu.au

1. Introduction

The rapid advancement of urbanization, evident in the relentless expansion of concrete jungles, poses a significant threat to the delicate balance of ecosystem services. Throughout history, cities have endeavored to cultivate thriving human environments [1], a pursuit aligning with Elkington's [2] vision of full-cost accounting for nature and the imperative of sustainable urban development in the long term. Consequently, it becomes crucial to cultivate connections with nature for all urban dwellers while simultaneously mitigating the adverse effects of urbanization on these essential ecosystem services [3]. However, research reveals that only a paltry 13% of city residents live close enough to nature to enjoy its well-documented mental health benefits [1]. This stark reality underscores the challenge of reconciling the needs of a burgeoning urban populace with the imperative to maintain a healthy and sustainable built environment. Achieving sustainability in urban development requires a holistic approach that considers social, economic, and environmental factors. By integrating principles of sustainability into urban planning and design, cities can create resilient communities that thrive in the face of challenges. Measuring progress toward sustainability involves not only environmental metrics [4] but also indicators of social equity and economic prosperity [5], ensuring a balanced and inclusive approach to urban development.

Recent research highlights the potential of integrating nature into the urban landscape [6,7]. These intentionally designed or preserved nature-based solutions (NBSs), spanning from urban forests, tree-lined streetscapes, parks, and gardens to vegetation strategically embedded within the urban fabric, such as green roofs and walls, play a pivotal role in mitigating the adverse impacts of urbanization [8,9]. As a result, within fragmented urban landscapes, residual green patches can serve as vital habitat refuges [10]. These refuges offer crucial habitats for a diverse array of plant and animal species, thereby fostering and sustaining urban biodiversity [10,11]. This, in turn, promotes human health and wellbeing by providing opportunities for recreation, stress reduction, and communion with nature [12]. Moreover, they contribute to enhanced air and water quality by filtering pollutants and reducing rainwater runoff [13,14]. Additionally, green patches can assist in regulating urban temperatures and mitigating the adverse effects of urban heat islands [15,16], thereby creating a more pleasant and habitable environment for citizens.

Advancements in urban greening and the proliferation of tree cover across global cities represent encouraging progress [17,18]. Nevertheless, to effectively confront the biodiversity extinction crisis and make substantial contributions to human welfare, a more focused and inclusive strategy is needed. Local governments must explicitly incorporate initiatives within their urban planning frameworks that champion the preservation and enrichment of biodiversity alongside the correlated ecosystem services that bolster human wellbeing [17]. This involves not only conserving existing natural areas but also strategically implementing NBSs to enhance urban ecosystems altogether. Achieving these goals requires a nuanced understanding of the optimal design, placement, and management of NBSs within urban landscapes [19]. While progress has been made, there remains a critical knowledge gap in fully grasping how to maximize the benefits of these solutions across interconnected dimensions. This includes considerations of biodiversity conservation, ecosystem resilience, and human health and wellbeing.

Sustainability lies at the heart of this endeavor. Urban development must strive for sustainability by balancing economic growth, social equity, and environmental stewardship [9,20–22]. This entails fostering communities that are resilient to environmental challenges while promoting inclusivity and equitable access to green spaces and natural resources. Moreover, unlocking the measurability of sustainability requires developing robust metrics [23] that capture not only environmental indicators but also social and economic dimensions of urban wellbeing. By prioritizing sustainability principles in urban planning and policymaking, cities can pave the way toward a more resilient, equitable, and livable future for all inhabitants.

This Special Issue (SI) represents a significant attempt to bridge gaps in the existing urban ecology and sustainable city planning literature. Comprising eight articles and two reviews, it thoroughly explores various dimensions of ecosystem services within urban environments. Notably, this SI stands as the fourth edition in a comprehensive series dedicated to urban ecosystem services, building upon solid groundwork laid by preceding research [24–26]. It underscores a continued dedication to advancing our comprehension of urban ecology and sustainable urban development.

By synthesizing fresh perspectives and innovative insights, this edition strives to optimize the delivery of ecosystem services while addressing potential disservices inherent in urban areas. Covering a broad range of topics, the contributions within this SI offer a multifaceted exploration of ecosystem services, enriching our understanding and informing future urban planning strategies. Each contribution, whether exploring green infrastructure or mapping cultural ecosystem services (CESs) for urban conservation, offers valuable insights into the creation of more sustainable and resilient cities. Figure 1 visually presents the geographical distribution of the research and the authorship locations within this SI. This graphical representation offers a comprehensive overview of the diverse origins of the research showcased in this SI, emphasizing the global scope and collaborative ethos driving progress in this field.

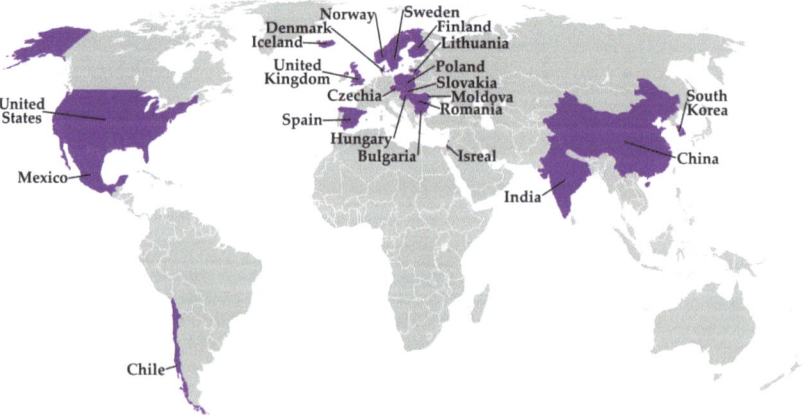

Figure 1. Dispersion of research contributions and authorship locations for this SI.

2. Synopsis of Contributions

In their recent study, Schueller et al. explored the potential of urban green spaces in addressing the concerning decline in pollinator populations. With pollinators playing a key role in plant reproduction and overall ecosystem health, their dwindling numbers due to habitat loss necessitate innovative interventions. The research explores the design of urban green spaces to effectively support diverse pollinator communities, aiming to provide habitats conducive to their survival and proliferation. Central to the study is the identification of key research questions aimed at optimizing the design of urban green spaces to meet the needs of pollinators. These questions encompass several crucial aspects, including the identification of target pollinator groups for conservation efforts, the determination of preferred plant species and optimal planting arrangements to support these pollinators, and an exploration of habitat requirements that extend beyond mere floral resources. Additionally, the study seeks to evaluate how surrounding landscapes influence the prioritization of creating new habitats within urban areas, considering factors such as connectivity, habitat fragmentation, and land use patterns.

By addressing these fundamental questions, Schueller et al. aim to provide valuable insights into the design and management of urban green spaces to support pollinator populations effectively. Ultimately, this research holds significant implications for urban planners, landscape architects, and policymakers involved in the development and maintenance of urban green infrastructure, offering guidance on how to enhance biodiversity and promote ecological resilience within cities. Through the implementation of evidence-based practices informed by this study, urban environments can become more hospitable to pollinators, contributing to the conservation of biodiversity and the sustainability of urban ecosystems.

Kim et al. investigated an innovative approach to counteracting urban heat islands during nocturnal hours: harnessing cold air emanating from nearby mountains and green spaces. Their study focused on optimizing city layouts to effectively channel this cool air flow (CAF) to achieve optimal temperature reduction. Through a careful investigation, they uncovered that taller buildings with ample wall surface area facilitate airflow, while densely packed, sprawling urban developments impede it. The research suggests a range of design considerations aimed at maximizing the benefits of CAF, providing invaluable insights for urban planners and designers striving to cultivate cooler and more sustainable urban environments. By shedding light on the intricate dynamics of airflow within urban settings and proposing actionable design strategies, this study contributes to the ongoing discourse on mitigating the adverse effects of urban heat islands and promoting urban sustainability.

Le and Huang conducted a groundbreaking study on utilizing cellular automata models to enhance urban tree planting strategies with the aim of creating interconnected green infrastructure networks that maximize their positive impact. By integrating CycleGAN models and cellular automata to replicate mycorrhizal networks, they predicted optimal urban tree layouts. This innovative approach leverages spatial data to simulate post-planting network connectivity, identifying priority planting locations that enhance ecological stability and climate resilience. The research addresses existing limitations in urban tree planting methods and offers valuable insights for sustainable urban planning and green infrastructure development. By optimizing tree placement to foster network connectivity, the study contributes to the creation of resilient urban ecosystems capable of mitigating environmental challenges such as heat island effects, air pollution, and biodiversity loss. The findings have significant implications for urban planners, policymakers, and landscape architects seeking evidence-based strategies to enhance urban greenery and promote environmental sustainability. Through the adoption of these innovative approaches, cities can foster healthier and more resilient environments, benefiting both human populations and the natural world.

The study conducted by Espinosa Fuentes et al. sheds light on the intricate dynamics of urban soils, which are crucial for ensuring the vitality and sustainability of urban forests. Focusing on the Bosque de Tlalpan Natural Protected Area (BT), the research examined the impact of conservation practices on soil quality. Through a comprehensive analysis of various soil properties such as exchangeable cations, heavy metal concentrations, soil carbon stock (SCS), and CO_2 effluxes, the authors examined four zones within the BT, each representing different levels of protection and public use across three climatic seasons. The investigation revealed that while concentrations of heavy metals generally adhered to Mexican regulations, mercury levels surpassed permissible limits. Significant variations in SCS and soil organic matter were observed among zones, with areas under stricter protection exhibiting higher values. Furthermore, CO_2 effluxes exhibited seasonal fluctuations, peaking during the rainy season. These findings underscore a positive correlation between conservation efforts and soil quality within the BT, with areas under stringent protection demonstrating enhanced carbon storage capacity and improved physicochemical properties. The study not only provides valuable insights into the current state of this urban forest ecosystem but also underscores the important role of conservation in preserving healthy soil conditions amidst urbanization pressures in densely populated areas.

In their review, Yadav et al. undertook a comprehensive evaluation of tree selection and plantation decision support systems (DSSs), analyzing their alignment with key objectives distilled from the existing literature. The review meticulously scrutinized the incorporation of multiple data sources and the usability of web interfaces within these DSSs. Five primary objectives for tree selection emerged from the analysis and were systematically compared across various existing systems: (a) climate resilience, (b) infrastructure/space optimization, (c) agroforestry, (d) ecosystem services, and (e) urban sustainability. Notably, the review highlighted a relative under-representation of climate resilience and urban sustainability considerations in current DSSs, indicating a potential gap in decision-making frameworks. The authors advocate for future DSS tools to adopt a more holistic approach by integrating these critical aspects into their frameworks.

Moreover, Yadav et al. proposed the utilization of deep neural networks (DNNs) to navigate the complexities inherent in achieving trade-offs between multiple objectives. By leveraging DNNs, decision-makers can address the intricate interplay between various goals, such as maximizing ecosystem services, ensuring the selection of climate-resilient tree species, and promoting agroforestry practices. In all, this review not only provides a comprehensive assessment of existing DSSs but also offers valuable insights into the potential enhancements needed to bolster their effectiveness in supporting informed tree-selection and plantation decisions. By advocating for a more inclusive approach as well as leveraging advanced technologies like DNNs, the study contributes to the ongoing discourse on sustainable urban forestry and resilient urban development.

Dabašinskas and Sujetovienė embarked on a comprehensive investigation into the dynamic interplay between the supply and demand for ecosystem services amidst urban expansion, stressing the critical need for ongoing monitoring and adaptability. Their research specifically investigated the spatiotemporal changes in ecosystem service provision, focusing on essential services such as food, carbon sequestration, and recreation, amidst the process of urbanization. The study rigorously quantified imbalances between the supply and demand for ecosystem services within the study area, with a particular emphasis on land use changes.

In particular, the most significant observed land use change entailed the conversion of agricultural land into forests and urban areas. Urban centers emerged as focal points, demonstrating the lowest supply and highest demand for all three ecosystem services investigated, thus underlining a stark negative correlation between the proportion of urban land and the provision of these vital services, especially in terms of food production. These findings underscore a pressing need for strategic interventions and policy measures aimed at mitigating the adverse impacts of urban expansion on ecosystem service provision. By highlighting these dynamics, the study provides valuable insights into the complex relationship between urbanization and ecosystem services, informing decision-making processes aimed at fostering sustainable urban development and bolstering the resilience of urban ecosystems.

In their innovative study, Krsnik et al. tackle the formidable challenge of comparing the value of various ecosystem services in urban settings. Their research aims to equip urban planners with the tools necessary to make informed decisions regarding urban development by prioritizing the services with the greatest potential benefits. The authors argue that existing methods for comparing the provision of ecosystem services in urban areas often produce misleading results. To address this issue, they propose a novel methodology that employs standardized thresholds to compare ecosystem service provision across different cities. This innovative approach promises to foster a fairer distribution of environmental benefits and wellbeing among urban populations, thereby contributing to more sustainable and resilient urban development strategies.

Although there is no consensus on the ideal level of ecosystem service provision, Krsnik et al. advocate for the adoption of standardized thresholds as a crucial step forward. By establishing a common framework for evaluating ecosystem service provision, their proposed methodology not only facilitates more accurate comparisons between cities but also lays the groundwork for future research and policy development in urban ecosystem management. In summary, this study represents a significant advancement in the field of urban ecosystem services assessment. It offers a systematic approach to evaluating and prioritizing ecosystem services in urban areas, potentially revolutionizing urban planning practices by promoting more informed decision-making and fostering greater equity in the distribution of environmental benefits across cities.

Gârjoabă et al., in their comprehensive study, explored the intricate planning strategies aimed at establishing resilient natural areas within urban environments. Their research focused on a comparative analysis between Nordic and Eastern European countries, recognizing the significance of their distinct political histories in shaping their approaches to this challenge. The researchers conducted a detailed examination of environmental laws and planning frameworks across these countries, seeking to uncover key elements that could offer valuable insights applicable to broader contexts within Europe. By identifying these crucial aspects, the study aimed to provide urban planners and policymakers with a robust foundation for effectively balancing urban development with the preservation of natural areas. This approach underscores the importance of considering local landscapes and urban layouts to ensure the sustainability and resilience of cities in the face of ongoing urbanization pressures. In all, by shedding light on the diverse approaches employed by Nordic and Eastern European countries, the study offers valuable insights that have the potential to inform and enhance urban planning practices across Europe, fostering the creation of more sustainable and resilient cities that harmonize with their surrounding natural environments.

In their research, Li et al. examine the critical importance of mapping CESs in river basins. This mapping is essential for identifying areas that require conservation efforts due to their significant contributions to CESs. However, existing studies often lack precise quantifications of the appropriate sizes for mapping units, which are fundamental for accurately assessing CESs. To bridge this gap, the study introduces the concept of the optimal area threshold of mapping units (OATMU). This approach involves the development of a multi-dimensional indicator framework and a validation methodology to determine mapping unit boundaries and suitable areas for CESs.

The multi-dimensional indicator framework integrates various indicators, including geo-hydrological, economic, and social management indicators. Each indicator's OATMU is calculated by identifying the inflection point in the second-order derivative of the power function. Through this process, the research defines the optimal size for mapping units, ensuring accurate CES assessment. The findings highlight the effectiveness of employing OATMU identification which, coupled with accessible basic data and simplified calculation methods, provides clear and universal technical support for optimizing CES mapping efforts. This approach enhances the precision and reliability of CES mapping, facilitating better-informed conservation and management decisions in river basin ecosystems.

Finally, Łukaszkiewicz et al. explored the potential of poplar trees (*Populus* L.) to enhance urban green spaces amidst the growing challenges of urbanization, population expansion, and climate change. The authors highlighted the rapid growth and adaptability of poplars across various environments, positioning them as a promising solution to address these complex urban issues. The study underscored the significant contributions of poplar trees to enhancing air quality and regulating microclimates within urban areas. However, the variable lifespans exhibited by different poplar cultivars present a notable challenge in their widespread adoption for urban greening initiatives. To address this issue, it has been recommended to employ strategic selection approaches that consider factors like growth rate and root system traits. It is proposed to integrate a diverse array of poplar species, each with different lifespans, to ensure the sustained and enduring presence of urban greenery efforts.

By exploring the potential of poplar trees in enhancing urban green spaces, Łukaszkiewicz et al. offer valuable insights that can inform urban planning and landscaping strategies. Their findings highlight the importance of selecting appropriate tree species tailored to the specific needs and challenges of urban environments, ultimately contributing to the creation of more sustainable and resilient cities in the face of ongoing urbanization and climate change pressures.

3. Conclusions

In an era of rapid urbanization, the importance of urban ecosystem services cannot be overstated. From mitigating climate change to enhancing public health and wellbeing, these services play vital roles in creating sustainable and resilient cities. At the forefront of this challenge is the need for strategic planning and decision-making. Urban planners and policymakers must recognize the value of urban green spaces and prioritize their conservation and enhancement. This involves not only preserving existing green spaces but also strategically integrating nature into the urban landscape. By doing so, cities can maximize the provision of ecosystem services while minimizing the negative impacts of urbanization. As noted by Łukaszkiewicz et al., poplar trees have shown promise in enhancing urban green spaces due to their rapid growth and adaptability. However, challenges such as variable lifespans will require strategic selection methods based on factors like growth rate and root system characteristics. By incorporating a diverse range of vegetation with varying lifespans, cities can ensure the continuity and longevity of urban greenery initiatives [27,28]. As such, this SI has uncovered compelling insights into the urgent task of scaling up nature's benefits within the context of rapid urbanization, transitioning from the scale of individual trees to the creation of vibrant and resilient urban forests.

In conclusion, prioritizing sustainability in the delivery of ecosystem services offers multifaceted benefits, including but not limited to enhancing the wellbeing of urban residents, safeguarding biodiversity, and mitigating environmental degradation. By embracing sustainable practices in urban planning and development, cities can significantly improve quality of life for their inhabitants. Sustainable urban landscapes not only provide essential ecosystem services such as clean air, water purification, and climate regulation but also offer opportunities for recreation, relaxation, and community engagement, enhancing the overall experience of the cityscape's look and feel [29]. Moreover, promoting sustainability fosters environmental conservation efforts, preserving natural habitats and protecting vulnerable species within urban ecosystems.

As a result, the integration of green infrastructure into a city enhances its resilience to extreme weather events, reduces urban heat island effects, and mitigates air and water pollution. The use of green spaces also serves as vital habitats for wildlife, contributes to urban biodiversity conservation, and promotes overall ecological health. Furthermore, sustainable urban development fosters social cohesion and equity by providing accessible green spaces for all residents, regardless of socioeconomic status.

In essence, the journey toward a greener and healthier future begins with a holistic understanding of the value of nature in urban settings. From the planting of single trees to the cultivation of thriving urban forests, each step toward sustainability in ecosystem services contributes to the creation of vibrant and resilient urban environments that benefit both present and future generations—an idea that has been underscored in the literature since the Brundtland Report [30]. Therefore, embracing sustainability as a guiding principle in urban development is not only essential but crucial for the wellbeing and prosperity of urban communities worldwide.

Author Contributions: Conceptualization, investigation, and writing—original draft preparation, A.R.; writing—review and editing, resources, software, and visualization, G.T.C. All authors have read and agreed to the published version of the manuscript.

Funding: This research received no external funding.

Data Availability Statement: The data presented in this study are available on request from the corresponding author.

Acknowledgments: We extend our sincere gratitude to the contributors to this Special Issue, as well as the dedicated team at *Land* for their invaluable assistance in facilitating the publication and editorial process.

Conflicts of Interest: The authors declare no conflicts of interest.

List of Contributions

1. Schueller, S.K.; Li, Z.; Bliss, Z.; Roake, R.; Weiler, B. How Informed Design Can Make a Difference: Supporting Insect Pollinators in Cities. *Land* **2023**, *12*, 1289. https://doi.org/10.3390/land12071289.
2. Kim, H.; Oh, K.; Yoo, I. Analysis of Spatial Characteristics Contributing to Urban Cold Air Flow. *Land* **2023**, *12*, 2165. https://doi.org/10.3390/land12122165.
3. Le, Y.; Huang, S.-Y. Prediction of Urban Trees Planting Base on Guided Cellular Automata to Enhance the Connection of Green Infrastructure. *Land* **2023**, *12*, 1479. https://doi.org/10.3390/land12081479.
4. Espinosa Fuentes, M. de la L.; Peralta, O.; García, R.; González del Castillo, E.; Cerón Bretón, R.M.; Cerón Bretón, J.G.; Tun Camal, E.; Zavala García, F. Soil Dynamics in an Urban Forest and Its Contribution as an Ecosystem Service. *Land* **2023**, *12*, 2098. https://doi.org/10.3390/land12122098.
5. Yadav, N.; Rakholia, S.; Yosef, R. Decision Support Systems in Forestry and Tree-Planting Practices and the Prioritization of Ecosystem Services: A Review. *Land* **2024**, *13*, 230. https://doi.org/10.3390/land13020230.
6. Dabašinskas, G.; Sujetovienė, G. Spatial and Temporal Changes in Supply and Demand for Ecosystem Services in Response to Urbanization: A Case Study in Vilnius, Lithuania. *Land* **2024**, *13*, 454. https://doi.org/10.3390/land13040454.
7. Krsnik, G.; Reyes-Paecke, S.; Reynolds, K.M.; Garcia-Gonzalo, J.; González Olabarria, J.R. Assessing Relativeness in the Provision of Urban Ecosystem Services: Better Comparison Methods for Improved Well-Being. *Land* **2023**, *12*, 1088. https://doi.org/10.3390/land12051088.
8. Gârjoabă, A.-I.; Crăciun, C.; Petrișor, A. I. Natural Protected Areas within Cities: An International Legislative Comparison Focused on Romania. *Land* **2023**, *12*, 1279. https://doi.org/10.3390/land12071279.
9. Li, Y.; Huang, J.; Wang, Y. Identifying the Optimal Area Threshold of Mapping Units for Cultural Ecosystem Services in a River Basin. *Land* **2024**, *13*, 346. https://doi.org/10.3390/land13030346.
10. Łukaszkiewicz, J.; Długoński, A.; Fortuna-Antoszkiewicz, B.; Fialová, J. The Ecological Potential of Poplars (Populus L.) for City Tree Planting and Management: A Preliminary Study of Central Poland (Warsaw) and Silesia (Chorzów). *Land* **2024**, *13*, 593. https://doi.org/10.3390/land13050593.

References

1. McDonald, R.I.; Beatley, T.; Elmqvist, T. The Green Soul of the Concrete Jungle: The Urban Century, the Urban Psychological Penalty, and the Role of Nature. *Sustain. Earth* **2018**, *1*, 3. [CrossRef]
2. Elkington, J. 25 Years Ago I Coined the Phrase "Triple Bottom Line". Here's Why It's Time to Rethink It. *Harv. Bus. Rev.* **2018**, *25*, 2–5.
3. Russo, A.; Escobedo, F.J.; Cirella, G.T.; Zerbe, S. Edible Green Infrastructure: An Approach and Review of Provisioning Ecosystem Services and Disservices in Urban Environments. *Agric. Ecosyst. Environ.* **2017**, *242*, 53–66. [CrossRef]
4. Cirella, G.T.; Zerbe, S. Quizzical Societies: A Closer Look at Sustainability and Principles of Unlocking Its Measurability. *Int. J. Sci. Soc.* **2014**, *5*, 29–45. [CrossRef]
5. Fritz, M.; Koch, M. Economic Development and Prosperity Patterns around the World: Structural Challenges for a Global Steady-State Economy. *Glob. Environ. Change* **2016**, *38*, 41–48. [CrossRef]
6. Ignatieva, M.; Dushkova, D.; Martin, D.J.; Mofrad, F.; Stewart, K.; Hughes, M. From One to Many Natures: Integrating Divergent Urban Nature Visions to Support Nature-Based Solutions in Australia and Europe. *Sustainability* **2023**, *15*, 4640. [CrossRef]
7. Vilanova, C.; Ferran, J.S.; Concepción, E.D. Integrating Landscape Ecology in Urban Green Infrastructure Planning: A Multi-Scale Approach for Sustainable Development. *Urban For. Urban Green.* **2024**, *94*, 128248. [CrossRef]
8. Buma, B.; Gordon, D.R.; Kleisner, K.M.; Bartuska, A.; Bidlack, A.; DeFries, R.; Ellis, P.; Friedlingstein, P.; Metzger, S.; Morgan, G.; et al. Expert Review of the Science Underlying Nature-Based Climate Solutions. *Nat. Clim. Change* **2024**, *14*, 402–406. [CrossRef]
9. Dunlop, T.; Khojasteh, D.; Cohen-Shacham, E.; Glamore, W.; Haghani, M.; van den Bosch, M.; Rizzi, D.; Greve, P.; Felder, S. The Evolution and Future of Research on Nature-Based Solutions to Address Societal Challenges. *Commun. Earth Environ.* **2024**, *5*, 132. [CrossRef]
10. Roberts, M.; Glenk, K.; McVittie, A. Urban Residents Value Multi-Functional Urban Greenspaces. *Urban For. Urban Green.* **2022**, *74*, 127681. [CrossRef]
11. Soares, R.M.V.; Lira, P.K.; Manes, S.; Vale, M.M. A Methodological Framework for Prioritizing Habitat Patches in Urban Ecosystems Based on Landscape Functional Connectivity. *Urban Ecosyst.* **2024**, *27*, 147–157. [CrossRef]
12. Marselle, M.R.; Hartig, T.; Cox, D.T.C.; de Bell, S.; Knapp, S.; Lindley, S.; Triguero-Mas, M.; Böhning-Gaese, K.; Braubach, M.; Cook, P.A.; et al. Pathways Linking Biodiversity to Human Health: A Conceptual Framework. *Environ. Int.* **2021**, *150*, 106420. [CrossRef]
13. Bazán, R.A.A.; Pórcel, R.A.D.; Rivera, G.L.D.; Vázquez, S.I.S. Enhancing Urban Water Quality with Green Infrastructure—A Study in Guadalupe, Nuevo Leon, Mexico. *Ecol. Eng. Environ. Technol.* **2023**, *24*, 216–224. [CrossRef]
14. Venter, Z.S.; Hassani, A.; Stange, E.; Schneider, P.; Castell, N. Reassessing the Role of Urban Green Space in Air Pollution Control. *Proc. Natl. Acad. Sci. USA* **2024**, *121*, e2306200121. [CrossRef]
15. Mokhtari, Z.; Barghjelveh, S.; Sayahnia, R.; Karami, P.; Qureshi, S.; Russo, A. Spatial Pattern of the Green Heat Sink Using Patch- and Network-Based Analysis: Implication for Urban Temperature Alleviation. *Sustain. Cities Soc.* **2022**, *83*, 103964. [CrossRef]
16. Ulpiani, G. On the Linkage Between Urban Heat Island and Urban Pollution Island: Three-Decade Literature Review towards a Conceptual Framework. *Sci. Total Environ.* **2021**, *751*, 141727. [CrossRef]
17. Oke, C.; Bekessy, S.A.; Frantzeskaki, N.; Bush, J.; Fitzsimons, J.A.; Garrard, G.E.; Grenfell, M.; Harrison, L.; Hartigan, M.; Callow, D.; et al. Cities Should Respond to the Biodiversity Extinction Crisis. *npj Urban Sustain.* **2021**, *1*, 11. [CrossRef]
18. Russo, A.; Chan, W.T.; Cirella, G.T. Estimating Air Pollution Removal and Monetary Value for Urban Green Infrastructure Strategies Using Web-Based Applications. *Land* **2021**, *10*, 788. [CrossRef]
19. Lafortezza, R.; Chen, J.; van den Bosch, C.K.; Randrup, T.B. Nature-Based Solutions for Resilient Landscapes and Cities. *Environ. Res.* **2018**, *165*, 431–441. [CrossRef]
20. Cirella, G.T.; Zerbe, S. Index of Sustainable Functionality: Application in Urat Front Banner. In *Sustainable Water Management and Wetland Restoration in Settlements of Continental-Arid Central Asia*; Cirella, G.T., Zerbe, S., Eds.; Bozen University Press: Bozen, Italy, 2014; pp. 137–155, ISBN 978-88-6046-069-1.
21. Kenter, J.O.; Raymond, C.M.; van Riper, C.J.; Azzopardi, E.; Brear, M.R.; Calcagni, F.; Christie, I.; Christie, M.; Fordham, A.; Gould, R.K.; et al. Loving the Mess: Navigating Diversity and Conflict in Social Values for Sustainability. *Sustain. Sci.* **2019**, *14*, 1439–1461. [CrossRef]
22. Weymouth, R.; Hartz-Karp, J. Principles for Integrating the Implementation of the Sustainable Development Goals in Cities. *Urban Sci.* **2018**, *2*, 77. [CrossRef]
23. Cirella, G.; Tao, L. The Index of Sustainable Functionality: An Application for Measuring Sustainability. *World Acad. Sci. Eng. Technol.* **2009**, *3*, 268–274. [CrossRef]
24. Russo, A.; Cirella, G.T. Urban Ecosystem Services: Current Knowledge, Gaps, and Future Research. *Land* **2021**, *10*, 811. [CrossRef]
25. Russo, A.; Cirella, G.T. Urban Ecosystem Services: New Findings for Landscape Architects, Urban Planners, and Policymakers. *Land* **2021**, *10*, 88. [CrossRef]
26. Russo, A.; Cirella, G.T. Urban Ecosystem Services: Advancements in Urban Green Development. *Land* **2023**, *12*, 522. [CrossRef]
27. Diekmann, M.; Andres, C.; Becker, T.; Bennie, J.; Blüml, V.; Bullock, J.M.; Culmsee, H.; Fanigliulo, M.; Hahn, A.; Heinken, T.; et al. Patterns of Long-Term Vegetation Change Vary between Different Types of Semi-Natural Grasslands in Western and Central Europe. *J. Veg. Sci.* **2019**, *30*, 187–202. [CrossRef]

28. Richards, D.R.; Belcher, R.N.; Carrasco, L.R.; Edwards, P.J.; Fatichi, S.; Hamel, P.; Masoudi, M.; McDonnell, M.J.; Peleg, N.; Stanley, M.C. Global Variation in Contributions to Human Well-Being from Urban Vegetation Ecosystem Services. *One Earth* **2022**, *5*, 522–533. [CrossRef]
29. Ananiadou-Tzimopoulou, M.; Bourlidou, A. Urban Landscape Architecture in the Reshaping of the Contemporary Cityscape. *IOP Conf. Ser. Mater. Sci. Eng.* **2017**, *245*, 042050. [CrossRef]
30. WCED World Commission on Environment and Development: Our Common Future—"Brundtland Report". Available online: https://www.are.admin.ch/are/en/home/medien-und-publikationen/publikationen/nachhaltige-entwicklung/brundtland-report.html (accessed on 8 May 2024).

Disclaimer/Publisher's Note: The statements, opinions and data contained in all publications are solely those of the individual author(s) and contributor(s) and not of MDPI and/or the editor(s). MDPI and/or the editor(s) disclaim responsibility for any injury to people or property resulting from any ideas, methods, instructions or products referred to in the content.

Article

Assessing Relativeness in the Provision of Urban Ecosystem Services: Better Comparison Methods for Improved Well-Being

Goran Krsnik [1,*], Sonia Reyes-Paecke [2], Keith M. Reynolds [3], Jordi Garcia-Gonzalo [1,4] and José Ramón González Olabarria [1,4]

[1] Forest Science and Technology Centre of Catalonia (CTFC), Ctra Sant Llorenç de Morunys, km 2, 25280 Solsona, Spain; j.garcia@ctfc.cat (J.G.-G.); jr.gonzalez@ctfc.cat (J.R.G.O.)
[2] Departamento de Ecosistemas y Medio Ambiente, Facultad de Agronomía e Ingeniería Forestal, Pontificia Universidad Católica de Chile, Av. Vicuña Mackenna, Santiago 4860, Chile; sreyes@mma.gob.cl
[3] US Department of Agriculture, Forest Service Research, 3200 SW Jefferson Way, Corvallis, OR 97331, USA; keith.reynolds2@usda.gov
[4] Joint Research Unit CTFC-AGROTECNIO, Ctra Sant Llorenç de Morunys, km 2, 25280 Solsona, Spain
* Correspondence: goran.krsnik@ctfc.cat; Tel.: +34-973-481-752

Abstract: In this study, we evaluated alternative methods for comparing the provision of ecosystem services among urban areas, stressing how the choice of comparison method affects the ability to compare the ecosystem service outcomes, in order to improve the management actions in urban green areas, reduce environmental inequality, and ensure satisfactory levels of human well-being. For the analysis, ten spatial indicators were quantified to assess the provision of urban ecosystem services in Barcelona, Spain, and Santiago, Chile. Two comparison methods were applied in both cities to evaluate the differences in their provision scores. The analysis was performed using the Ecosystem Management Decision Support (EMDS) system, a spatially enabled decision support framework for environmental management. The results depicted changes in the values of the provision of ecosystem services depending on the methodological approach applied. When the data were analysed separately for each city, both cities registered a wide range of provision values across the city districts, varying from very low to very high values. However, when the analysis was based on the data for both cities, the provision scores in Santiago decreased, while they increased in Barcelona, showing relativeness and a discrepancy in their provisions, hindering an appropriate planning definition. Our results emphasise the importance of the choice of comparison approach in the analyses of urban ecosystem services and the need for further studies on these comparison methods.

Keywords: urban ecosystem services; spatial modelling; urban green infrastructure; human well-being; urban planning

1. Introduction

Human well-being can be considerably increased by numerous services provided by ecosystems [1]. In urban areas, the demand for ecosystem services is significantly higher than that in rural environments, due to the limited natural recourses and high population concentrations in relatively small areas [2]. It is expected that the world's urban population will continue growing; therefore, an increased demand for urban ecosystem services (UES) in rapidly expanding urbanized areas can be expected, causing high pressures on urban green infrastructure [3,4]. In such an environment, urban green areas (UGA), including parks, urban forests, and street trees, are multifunctional sources of benefits, such as recreation, air purification, water drainage, or psychological relief [5,6]. The importance of their management is crucial, because they are heavily influenced by humans and can be modified relatively quickly according to their potential demand [7,8]. Therefore, the incorporation of ecosystem-service-based strategies into urban planning and management affords an opportunity to promote a more sustainable society and simultaneously enhance

human well-being [4]. Consequently, to promote the development of more sustainable cities, it is important to understand how UES are related to the structure of an urban landscape and how they spatially vary within a city [4,9].

The spatial management of UES supply helps to define the appropriate urban strategies for achieving ecologically sustainable cities [9]. In such a scenario, different planning methods can be applied based on the stage of urbanization that the city is passing through [10]. An analysis of current UES features can help to anticipate the future urban processes in some parts of the world [10] and enhance the practices relevant to maintaining or improving ecological [11,12], social [13], and climate mitigation outcomes [14]. Often, spatial comparison methods are used to achieve these goals, such as a comparison between different cities [3,15,16] or city districts within the same city [17]. These methods provide good feedback concerning UES assessments and address the needed management actions regarding UGAs.

The spatial patterns of UES supply are the result of both the physical and socioeconomic features of an urban environment, in which all the components are complexly intercorrelated [18]. Therefore, mapping and quantifying UES are powerful tools in the detection of the spatial heterogeneity in the provision of ecosystem services, and it is recommended as a first step towards a comprehensive management plan of green infrastructure, including comparison-based studies [6,19,20]. However, the lack of standardization in these comparison methods and comparable availability of spatial data significantly hinders ecosystem service modelling and quantification, due to the numerous dataset requirements [10,21]. Moreover, open-source, remotely sensed images considerably limit comprehensive analyses in urban areas, primarily due to their spatial resolution requirements [6]. These limitations substantially impede cartography-based comparisons and the detection of hotspots that could be used as examples of good or bad UES management. Namely, benefits are usually not equally provided within or between cities, due to unequal access to green infrastructure, causing environmental injustice in the distribution of environmental goods and well-being [22,23]. Differences in the distribution of environmental goods are mostly visible within cities in emerging countries, with an obvious socio-economical polarisation between city districts, or when comparing cities between more or less economically developed countries [24]. Due to continuing trends in urbanisation and, therefore, increased pressure on UGAs, reaching a desirable level of access to safe, inclusive, and accessible green spaces at the end of this decade is considered a global policy objective [25]. Nevertheless, due to different data scales, absolute values of provision are usually not directly comparable between cities, and the lack of a consensus on a possible reference scale that could define how high provision should be considered optimal deters improvements in the management of UES [26]. This raises the question of relativeness in the provision of UES, because the perception of satisfying provision can be significantly changed, depending on whether the values are analysed independently or compared to other urban areas. Therefore, an improvement in the methods for assessing and comparing UES is needed, allowing a comparison between different indicators and data sources though standardization processes, resulting in better and more comparable planning. Additionally, having information at the neighbourhood level and knowing the characteristics of all the relevant components implied can promote improved human well-being, caused by increase in the provision of UES and inspired by strategies applied in areas with similar geographical characteristics [27,28]. In such processes, spatial decision support tools are of great interest to assist in decision making and supporting relevant conclusions [29].

In this study, we apply the Ecosystem Management Decision Support (EMDS) system, a spatially enabled decision support framework for analysis and planning [30], to compare the provisions of UES in Barcelona, Spain, and Santiago, Chile. For this purpose, we follow the Millennium Ecosystem Assessment (MEA) framework for quantifying the regulating and cultural ecosystem services provided by green urban areas [31]. Our objective is to test the utility of EMDS in a comparison of urban environments, and make the first steps towards a more standardised assessment of comparing UES at the local level. We apply different comparison methods to detect differences in the results regarding the supply of benefits, pointing out the relativeness of UES provision.

2. Materials and Methods

2.1. Study Area

The study was conducted in the cities of Santiago, Chile, and Barcelona, Spain (Figure 1). Santiago and Barcelona are characterised by different socioeconomic and geospatial features. While Santiago is representative of rapid urbanisation processes, urban sprawl, and demographic transition, and reliably represents the socio–ecological–spatial patterns of Latin American cities, Barcelona is a dense, but planned, Mediterranean city with dominant post-transitional processes and limited space for expansion [32,33]. Ecosystem-service-based urban greening policies and sustainable strategies represent the main pillars of Barcelona's plan for its future development [34]. In contrast, the awareness of green infrastructure and its incorporation in urban planning have distinct applications in Santiago, depending on the commune (the administrative subdivisions of the city), although the needed sustainable policies for urban development have generally not been applied [35]. Both cities lack adequate green infrastructure within their city boundaries, but have continuous suburban forests in the cities' outskirts. In this study, we analyse the UES within the municipal limits of the city of Barcelona and the northern communes of the continuously urbanised part of the Santiago province. Only the northern communes in Santiago (the continuously urbanised parts of 20 communes) were chosen due to the large spatial extent of the city, which encompasses a large variety of urban morphological patterns. We used division on the statistical sections to conduct our UES analysis, because these divisions in both cities were detailed enough and comparable between our study areas. As a result, Santiago was divided into 179 districts and Barcelona was divided into 233 statistical areas. Regarding their populations, Barcelona has 1.6 million inhabitants and covers an area of 101.9 km^2 (its population density is about 16,000 people/km^2) [36]. The northern communes of Santiago have 3.1 million inhabitants and cover a total area of 256.2 km^2 (their population density is approximately 12,100 people/km^2) [37]. The climate of both cities is "Mediterranean hot summer climatic type" (CSa), but with a stronger maritime influence in Barcelona due to its coastal location [38].

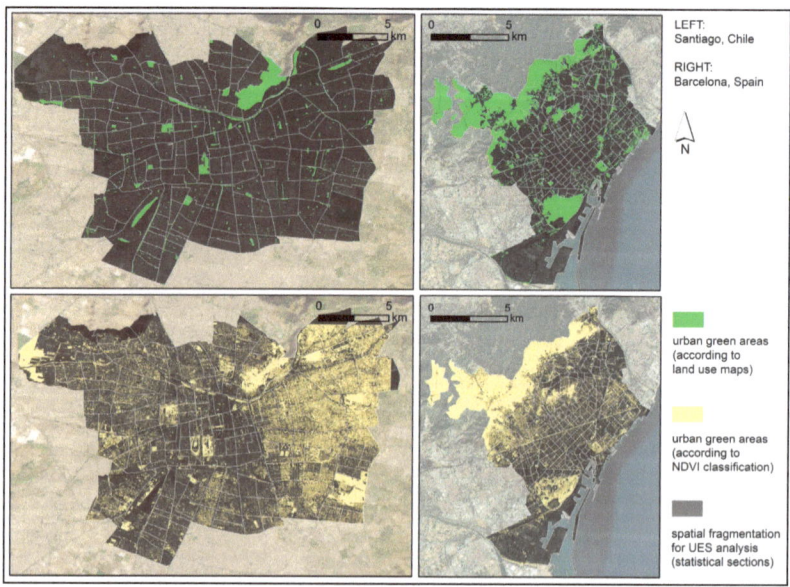

Figure 1. Study areas and relevant data layers.

2.2. Conceptual Design

The main objective of this study was to quantify and spatially assign the provision of urban ecosystem services at the district level in the cities of Barcelona and Santiago, and, via the use of different standardization methods, examine the applicability of the EMDS system in a comparison and quantification of these urban ecosystem service provisions at the local level. To meet this goal, the project was organized into four steps:

1. *Define the indicators for urban ecosystem services in terms of sets of metrics.* An analysis of data availability was performed in this first step to identify the metrics that could be used in the assessments of both cities for accurate comparisons. Metrics are variables that collectively quantify each ecosystem service indicator. In total, the provision of 10 urban ecosystem services was defined in each city, with each UES being represented by one metric.
2. *Analyse the provision of urban ecosystem services at the district level.* In this step, we designed the model to quantify the provision of each of the 10 UES defined in the first step. The same model, created in a logic-based geospatial modelling system, was applied in each city.
3. *Apply different normalization methods to compare the provision between Barcelona and Santiago.* In this step, we tested how different normalization methods affected the interpretation of the UES provision and comparison, and tested the utility of EMDS in this analysis. For this purpose, we used two different normalization methods. As a result, UES provision maps were obtained.
4. *Spatial aggregation and variation analysis.* In this final step, we analysed the differences in the spatial aggregation and variation in the provisioning between the results obtained by the two normalization methods.

2.2.1. Definition of Urban Ecosystem Services

The first step in the ecosystem services analysis was the definition of the UES indicators. An analysis of data availability was performed to identify the spatial data that could be used to define and quantify these UES indicators. Because the goal of the study was to compare the UES provisions in the two cities, it was necessary to use at least approximately similar metrics in both cities, and this requirement substantially reduced the data choices. Finally, one dataset of metrics was used to define each UES indicator in both cities. Therefore, 20 metrics in total were used in this study to define and quantify the 10 UES (Table 1). The geoprocessing operations in ArcMap 10.8 were applied to produce the desired metrics for each landscape unit of each city. Once calculated, all the metrics were attributed to the relevant spatial units. We used the MEA methodological framework [31] to model the data (Table 1) and, consequently, the categories of the regulating and cultural ecosystem services were used as the basis for the analysis.

Table 1. Metrics selected to evaluate each of the 10 UES data inputs.

UES Groups and UES Indicators	UES Metrics	Units	Format	Metrics References
REGULATING				
Micro-climate regulation	Intensity of urban heat island based on land surface temperature	°C	Raster	[39]
Air quality regulation	CO_2 storage by urban trees	kg/m^2	Raster	[40]
Drainage	Extension of impermeable surfaces or areas covered by vegetation	%	Raster	[41,42]
Noise reduction	Presence of green infrastructure along traffic axis	%	Polygon	[43,44]
Habitat provision	Continuity of green urban areas	m^2	Raster	[45]

Table 1. Cont.

UES Groups and UES Indicators	UES Metrics	Units	Format	Metrics References
CULTURAL				
Recreation	Distance to the closest green urban area suitable for recreational activities	m	Point	[46,47]
Social value	Quantity of sites within urban green areas serving as a meeting point with other citizens	num./km^2	Point	[48]
Psychological or health-related value	Abundance of urban green areas within neighbourhoods	m^2/inh.	Polygon	[49,50]
Cultural or historical value	Quantity of urban green sites relevant to local culture or history	num./km^2	Point	[51]
Aesthetics	Presence of green urban areas on the streets	%	Polygon	[52]

Regulating services are the benefits people obtain from the regulation of ecosystem processes [31]. We used five metrics to quantify the provision of five ecosystem services in each city (Table 1). To assess the micro-climate regulation, the urban heat island intensity was calculated. The calculation was performed using Landsat 8 imagery, band 10 from the TIRS sensor, and bands 4 and 5 from the OLI sensor. The list of images is shown in Table 2. All the selected images corresponded to the summer months, had minimum cloudiness in the scene, and were adjusted via an atmospheric correction process [53]. The Jiménez-Muñoz and Sobrino method [54] was used to calculate the land surface temperature and approximate urban heat island intensity. Emissivity values, which were needed for the land surface temperature calculations, were obtained using the Normalised Difference Vegetation Index (NDVI) thresholds approach for emissivity analyses [55], with the NDVI values modified according to the local imagery characteristics and established as shown in Table 3. Given the land surface temperature calculations, a mean temperature for each district was calculated.

Table 2. Landsat 8 images used in urban heat island calculation.

Barcelona		Santiago	
Date	Resolution	Date	Resolution
12 July 2013	30 m multispectral, 100 m thermal pixel	9 January 2014	30 m multispectral, 100 m thermal pixel
14 August 2016	30 m multispectral, 100 m thermal pixel	15 January 2016	30 m multispectral, 100 m thermal pixel
22 July 2019	30 m multispectral, 100 m thermal pixel	23 January 2019	30 m multispectral, 100 m thermal pixel

Table 3. NDVI thresholds applied in emissivity calculations.

Land Use Type	NDVI Thresholds	Emissivity Values
Vegetation	>0.4	0.99
Water	<0	0.98
Built-up areas	$0 \leq$ NDVI < 0.1	0.95
Bare ground	$0.1 \leq$ NDVI < 0.2	0.94
Mixed pixels	$0.2 \leq$ NDVI < 0.4	Equation by Valor and Caselles [55]

We used the CO_2 storage in urban trees to quantify the regulation of the air quality. A remote-sensing-based method using NDVI values was implemented, with the formula adjusted to our image resolution [40]. Rapid-Eye images from the Catholic University of Chile, from 11 October 2013, were used to calculate the NDVI values in Santiago. On the other hand, open-source NDVI data from 2017, provided by the City Council, were applied in the analysis in Barcelona [56]. Mean values were calculated for each city district. The same data sources were used in the assessment of drainage. We identified the pixels corresponding to impermeable surfaces and areas with vegetation cover and calculated the percentages these areas occupied within the city district area. Urban green continuity was used as a proxy for habitat provision [57,58]. Continuous areas of pixels with vegetation cover were detected and a mean value for each city district was calculated. Finally, the noise reduction was assessed as the share of green areas along the streets, using a buffer of 20 m on either side of the street.

Cultural services are the "nonmaterial benefits people obtain from ecosystems through spiritual enrichment, cognitive development, recreation, reflection, and aesthetic experiences" [31]. Due to their intangible characteristics, they are usually difficult to quantify, often being the subject of controversy because their definitions are typically vague and their indicators are not well established [48,59]. In this study, we used five metrics to quantify the provision of five cultural ecosystem services (Table 1). Recreation was assessed by the proximity of the UGA to an analysis unit suitable for recreational activities (UGA > 2 ha) [56,60]. A dot map, with a separation of 30 m between the dots, was created over the urbanised areas of both cities. The distance from each point to the nearest UGA was measured and a mean value was calculated for each district. Social value was measured by a quantification of the urban amenities for social activities located within a UGA, such as parks for children, open-air gyms, barbecue areas, and parks with registered social activities, etc. Their density per square kilometre was calculated in each city district. The same unit was used to quantify cultural or historical value. Here, only protected urban green areas, such as historical parks, monumental trees, and areas or trees of local interest, etc., were taken into account. Psychological and health-related value was assessed as the quantity of green areas within the city district (parks, urban forests, or green squares) per the number of inhabitants [56,60]. Finally, we analysed the presence of UGAs within the 20 m street buffers to assess aesthetics. The values of the share of a green area outside of a buffer zone were represented as a mean value at the city district level.

2.2.2. Analysis of the Provision of Urban Ecosystem Services

After quantifying the UES indicators with the metrics in the previous step, we proceeded to quantify their provision to the districts. A geospatially based logic model was built in the NetWeaver Developer [61], a component of the EMDS spatial decision support framework [30]. The provision of each metric in a district was quantified by the use of a specific measure of the strength of the evidence obtained from the model. The UES provision was quantified via the application of the unique rules applied to each metric that approximated the relations between the metrics and the UES. These rules defined type of relationship and interdependency between the metrics, as well as the degree of consideration of each metric in an indicator's quantification process.

The logic models in NetWeaver are built as networks of networks organised in a logical dependency structure. The strength of the evidence of dependent networks is logically derived from the evidence provided by antecedent networks [61]. Elementary networks, whose only antecedents are data (e.g., metrics), are located at the lowest level of the model structure and are the origin of the strength of the evidence measures. Each elementary network uses a fuzzy membership function to express the degree of support for a logical proposition provided by an observed data value. The evidence measures at the bottom of the network structure are propagated upward through the antecedent and dependent networks, connected by logic operators that specify how the evidence measures should be combined. In this study, we built two structurally equivalent logic models, one for each study area.

We used NetWeaver's graphical method of model design to build the model. The full model is documented in HTML (Archive 1) and we show a graphic representation of the regulating services in Barcelona in Figure 2. The evaluation of the regulating services considered five metrics (Table 1) and each were evaluated by a fuzzy membership function. In Figure 2, we show how the observed data relative to the climate regulation were converted into the strength of the evidence values. These ranged from −1 (meaning no evidence or no provision of UES) to 1 (full evidence or full provision of UES). Each of the other four metrics were similarly evaluated by a definition of the specific thresholds on the observed data used to define the strength of the evidence (Table 4). The U operator (Union in NetWeaver) specified that the measures of the strength of the evidence for all the metrics in our model were logically combined as an average, meaning that the lines of the evidence were additive and compensatory, so that low evidence values on one metric could be compensated by high values on others. Although NetWeaver allows for weighting the evidence of antecedent networks, our models used NetWeaver's default value of 1, so all the networks contributed equally to the conclusions of provisioning.

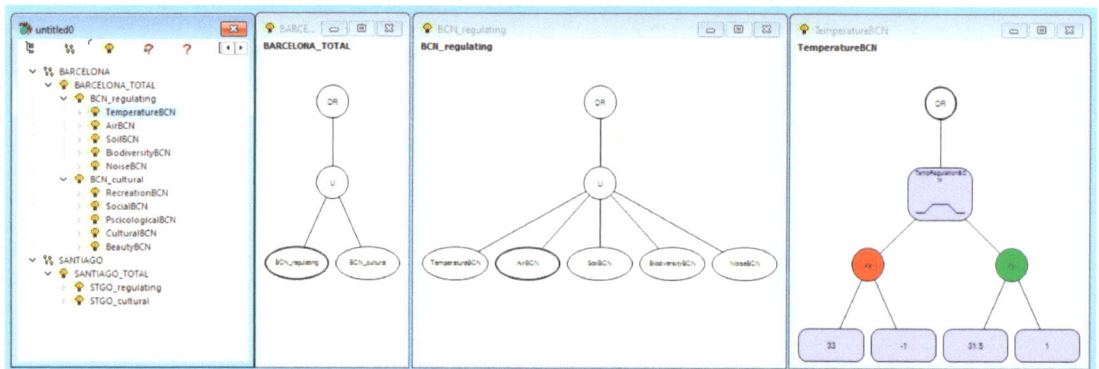

Figure 2. Graphic representation of NetWeaver model regarding the provision of regulating services in Barcelona.

Table 4. Observed values thresholds to define strength of evidence values.

UES Metrics	Metrics Units	Separated Thresholds Approach				Joint Thresholds Approach	
		BARCELONA		SANTIAGO			
		No Evidence	Full Evidence	No Evidence	Full Evidence	No Evidence	Full Evidence
Micro-climate regulation	°C	33	31.5	35	32	34	31.75
Air quality regulation	kg/m^2	1.25	1.7	1	1.45	1.35	1.57
Drainage	%	12	40	12	50	12	45
Habitat provision	m^2	900	80,000	2000	4,000,000	1450	2,000,000
Noise reduction	%	20	40	10	35	15	37.5
Recreation	m	700	150	1000	300	850	225
Social value	num./km^2	0	15	0	2	0	1.5
Psychological or health-related value	m^2/inh.	1	23	0.1	1.2	0.55	12.1
Cultural or historical value	num./km^2	0	10	0	1.3	0	5.65
Aesthetics	%	10	35	5	35	7.5	35

2.2.3. Comparison of Urban Ecosystem Services between the Two Cities

Given the model construction described in the previous section, we compared the provisions of the UES within the districts of the cities and between the cities. Because the metrics we implemented had different absolute ranges between the two cities, resulting in distinct scales for the provision of a metric, they were not directly comparable. Therefore, we tested two different methods to analyse the UES provision and, subsequently, compared the provisioning outcomes between Santiago and Barcelona. The difference between the two methods was based on the assignment of the thresholds used to define the fuzzy membership functions, as described in the previous section. In the first method, which we refer to as the separated thresholds approach, the observed data values were analysed independently for each city and unique thresholds were calculated separately for Santiago and Barcelona. The maximum and minimum threshold values for defining the fuzzy membership functions in each city were assigned based on the literature review for recreation and noise reduction, or based on the 15th and the 85th percentiles for the other UES indicators (Table 4). On the other hand, in the second method, the joint thresholds approach, the analysis was run with the unique threshold values for defining the fuzzy membership functions determined based on both study areas (Figure 3). In particular, the threshold values of the fuzzy membership functions in the joint threshold approach were calculated as the mean threshold values from the separated thresholds approach (Table 4).

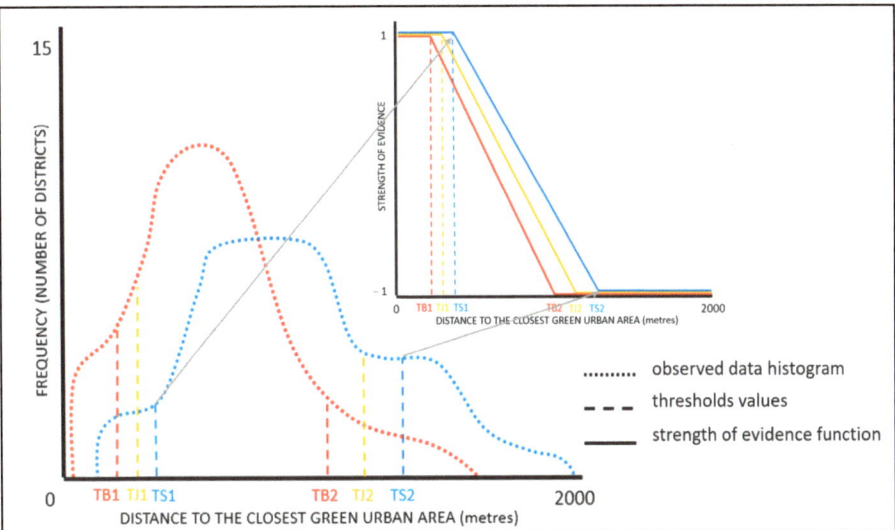

Figure 3. Schematic representation of threshold assignments for the strength of evidence function using recreation UES as an example. Red lines represent Barcelona (using separated thresholds approach), blue lines represent Santiago (using separated thresholds approach), and orange lines represent joint thresholds approach.

The spatial analysis was performed using the ArcGIS 10.8 software, as well as the EMDS 8.7 ArcMap Add-Inn. After running the NetWeaver model in EMDS, maps showing the provision of each UES indicator, the UES groups, and the total UES provision were generated. Lastly, we noted that the strength of the evidence measures computed in NetWeaver was a continuous variable, but the map values were classified into five categories using equal intervals, from very low to very high, for display purposes.

2.2.4. Spatial Aggregation and Variation Analysis

In this final step, the total provision scores resulting from the two different methods were compared and analysed. For this purpose, the strength of the evidence values (ranging from −1 to 1) were normalized to a (0–1) scale to simplify the interpretation. The normalized provision scores obtained with the separated thresholds approach were deducted from the value obtained using the joint thresholds approach. The results depict the degree (0–1 or 0–100%) and direction (positive or negative) of the changes in the provision resulting from the application of the two different methods. Additionally, the changes in the spatial aggregation of the provisions per district were analysed by applying global Moran's I statistics, a spatial autocorrelation tool that assesses both spatial locations and changes in the values of features [62]. Applying Moran's I, we aimed to assess the equality in the provision of the UES and, therefore, urban well-being. A lack of spatial correlation (negative I values) means a lower aggregation and greater equality in this provision, and vice versa [63]. We aimed to compare the results obtained by the two different methods to identify the gap between the provisions of the cities and study the relativeness regarding the optimal provision of UES.

3. Results

The spatial distribution of the provision of UES in Santiago, calculated using the separated thresholds approach, is shown in Figure 4. The provision of the cultural and regulating UES did not follow a common spatial pattern. The provision of cultural services was generally low within the entire study area, with the exception of a few specific districts where a higher provision could be noticed. On the other hand, the provision of regulating services showed a clear east–west spatial polarization. While a very low strength of evidence dominated the central and western districts of the city, with several low and medium values, the eastern districts displayed continuous areas of very high provision. With respect to the total provision of UES in Santiago, a more irregular spatial pattern was evident, maintaining the high strength of the evidence values in the east, but with less distinctive differences towards the west.

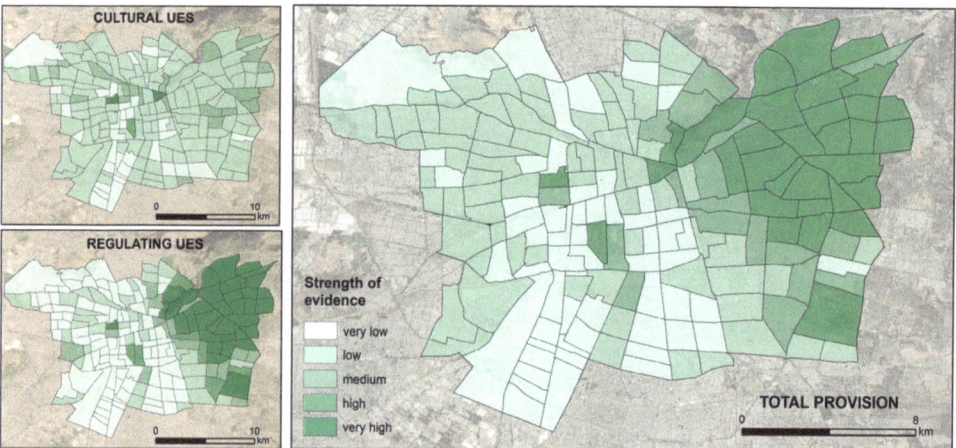

Figure 4. Maps of provision of UES in Santiago using separated thresholds approach.

Figure 5 shows the spatial distribution of the UES in Barcelona, calculated using the separated thresholds approach. In comparison to Santiago, the spatial distributions of both the cultural and regulating services were more irregular. Generally, higher values were observed in the marginal districts of the city, leaving the central districts characterised by a lower strength of evidence. The regulating services showed a greater polarisation in their values within the study area, while the score differences for the cultural services were smoother. The total provision of UES in Barcelona had an uneven spatial distribution. In general terms, very high provision values were observed in the districts where parks or urban forests were situated, located in the mountainous parts of the city; coastal districts had medium values, while an irregular representativeness of very low, low, and medium scores could be noticed in the central portion of the study area.

Figure 6 shows the total provision of the UES in Santiago and Barcelona when applying the joint threshold approach. The spatial distributions of the provisioning values followed a similar spatial pattern in both cities, as in the previous method. However, changes in these provision values could be noticed in both study areas. In Santiago, about 50% of the city districts showed an increase, while in the other 50% of the districts, lower provision values were observed. Nevertheless, negative variations were more frequent (the mean decrease value was −0.04, while the mean increase value was 0.02), indicated by the predominance of lighter shades in Figure 6. Most of the districts registered changes in their values passing from high to medium in the east, or low to very low in the western parts of the city. Only a few city districts had a significant increase in their provision after applying the joint thresholds approach compared to the separated thresholds approach (Figure 4). On the contrary, most districts of Barcelona showed an increase in their UES provision scores, about 80% of the total number. At the same time, the mean variation values were equal in both the positive and negative records. Most coastal districts passed from a medium to high strength of evidence. The provision increased from very low to low and low to medium in several central city areas, while increments from medium to high and high to very high were observed in the mountain districts. A significant decrease in provision was only registered in a few suburban areas, mostly passing from very high to high.

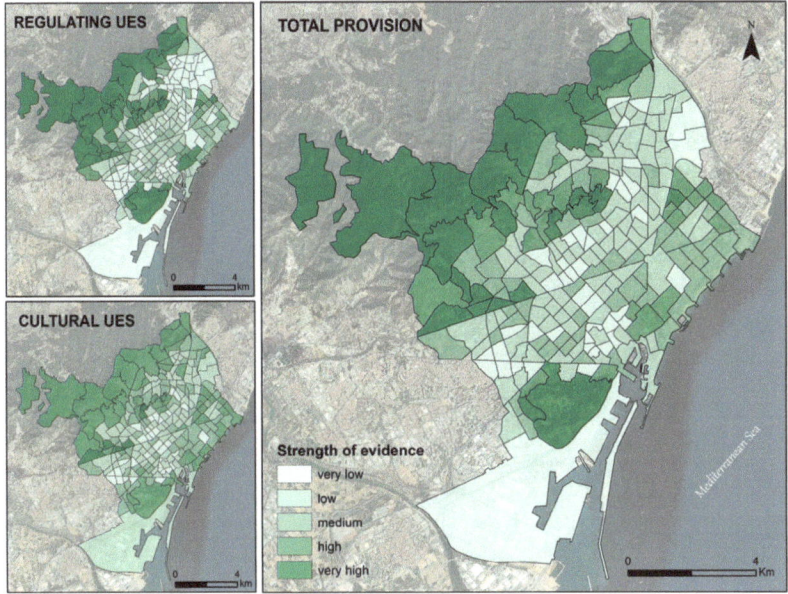

Figure 5. Maps of provision of UES in Barcelona using separated thresholds approach.

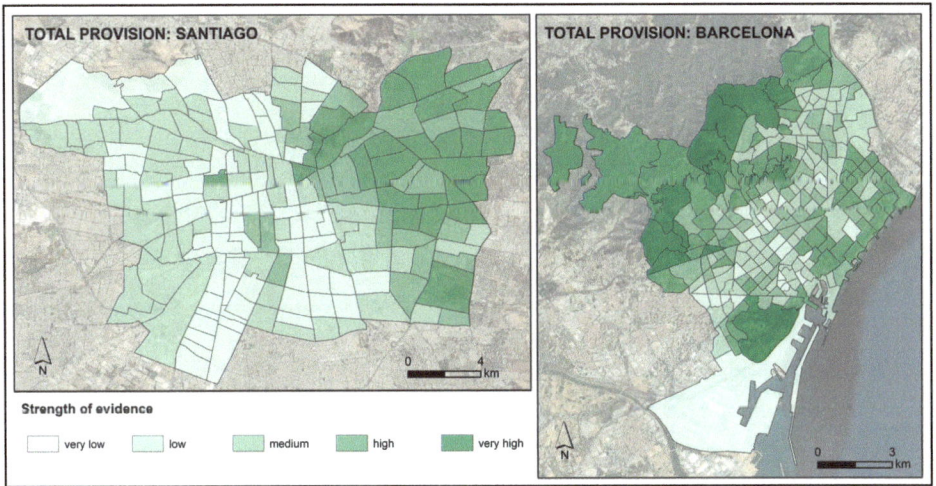

Figure 6. Maps of provision of UES in Santiago and Barcelona using joint thresholds approach.

The variations in the provisioning values between the two methods were much higher in Barcelona than Santiago, both in terms of positive and negative changes. After using the joint thresholds approach, the provision values in Barcelona predominantly increased, with scores of up to 20% higher. On the other hand, when the provision scores in Santiago were directly compared to the absolute provision values in Barcelona, smoother variations were noticed, but with a greater decreasing tendency, reaching up to −10%.

Regarding the spatial aggregation of the UES provision values, positive Moran's I values were obtained from both the separated and joint thresholds approaches in both cities, indicating some degree of aggregation. However, significant differences were observed between Santiago and Barcelona. First, the spatial aggregation of the total UES provision was higher in Santiago than Barcelona. The Moran's I in Santiago was 0.53, both for the separated thresholds and joint thresholds approaches, while the respective values in Barcelona were 0.19 and 0.15. Thus, the aggregation of the total provision varied more in Barcelona than Santiago when comparing the two threshold approaches. Differences in the spatial aggregation were also observed in the provision of regulating UES using the joint thresholds approach, in which Santiago had a high index value (0.63), whereas Barcelona had a low value (0.12).

4. Discussion

In this study, we conducted two different comparison methods for evaluating the provision of UES in the cities of Santiago and Barcelona. In both cases, an innovative spatial decision support framework, EMDS, was applied to allow the analysis of single and combined UES provisions. We used the same data in both approaches, only changing the threshold values to define low and high evidence of UES provision, which were primarily derived from evaluations of histograms of data distributions. After the application of the different threshold approaches, clearly distinct results were obtained for both cities, resulting in changed provision evidence values. Because the identification of the needed interventions to the green infrastructure and the specification of the actions required to improve urban green and human well-being are directly conditioned by the characteristics of UES supply, knowing their current state and managing them appropriately is crucial [64]. Our approach also attempted to develop standardised comparison methods for reducing ambiguities in the results, as well as provide a spatial solution for analysing UES to support urban planning policies that, in turn, provide a basis for a less ambiguous definition of a good urban strategy [65].

The provision of ecosystem services, including UES, primarily depends on the capacity of the ecosystem to deliver them [31]. While in rural environments it might be a challenging task to change the ecosystem capacity in the short term, urban environments are characterised by more dynamic geospatial features that are amenable to implementing changes. Namely, via land-use changes or interventions to urban facilities, environmental settings can be substantially changed in a relatively short time, enabling new scenarios for UES provision [66]. Ideally, these changes should be induced by prior analyses of the current UES characteristics, aimed at improving them [2]. The comparison analyses presented in this study can substantially help in defining the actions needed to initiate these changes and, therefore, act as a useful tool in UES management. These types of analyses can also improve the awareness of the need to continue improving urban ecosystems and increasing UES provision. Our results illustrate how the perceptions of provision values in a certain city can change by considering alternative comparison methods. Such perceptions can result in obtaining a wrong image of UES-related processes and their provisions and can lead to making inappropriate decisions regarding urban planning. For example, generally high provision values were observed in the eastern part of Santiago when the data were analysed independently, but after observing the results in the broader context of the joint comparison method, which included data from Barcelona, the provision in Santiago decreased, whereas the opposite outcome was observed in Barcelona. In other words, when using the separated thresholds approach, the high values in Santiago and

Barcelona were not equally high, which emphasized the relativeness in the provision of UES in this approach. This can easily cause difficulties with defining an appropriate urban planning strategy that attempts to improve the distribution of environmental goods and well-being. This also raises questions about the methodology for such comparison studies and which scores the provision should register to be considered as high enough. Until now, a UES comparison between different cities has only been conducted in a limited number of studies, with a clear lack of coherent comparison methods [3,16], and these have been based on what we refer to as the separated thresholds method, in which each dataset, before being compared, is analysed and normalised independently. As demonstrated in this study, such methods can provide misleading results, because their provision scales are based on different absolute values. While there is an objective at the global level focused on the urgent mitigation of the inequality of environmental goods, the development of research methodology does not follow the same path [25]. It is evident that each urban area has a unique geospatial reality defined by specific sets of features, including urban green infrastructure, and that the capacity of UES provision strongly depends on these characteristics, but effective improvements cannot be achieved at a broader global scale if each urban landscape is analysed independently. Thus, in this study, we emphasized the need to improve UES comparison methods, in order to obtain more comparable results, which would help to achieve a more equal distribution of urban well-being across cities by establishing more standardised comparison methods, such as the definition of UES thresholds that could be applicable over broad spatial extents.

Regarding provision values, the literature usually strives for an increase in UES supply, but there is no consensus on how high this provision should be to satisfactorily supply all the benefits. In rural environments, the goal is to achieve the maximum provision that the environment can provide according to its capacity, without putting it at environmental risk [67]. In urban environments, this capacity can easily be increased, but the environmental pressure on UGAs can also fluctuate drastically, depending on geographical circumstances [68]. The joint thresholds method that was demonstrated in this study can help to evaluate UES provision over broader spatial extents and provide a better perception of the comparability of UES characteristics (or lack thereof), but it cannot produce a complete solution for the actions needed to manage UGAs. At the same time, we recognize that, in some cases, the absolute values for the provision of UES can be so different that it may not make sense to adjust their interpretation to a common data scale. However, using the separated thresholds approach in the latter case would give even more problematic results, as discussed above. For this reason, the study also emphasizes the methodological constraints regarding UES comparisons, although it represents a first step towards more complete comparison methods, while also emphasising the need for more developed and elaborated methodological approaches.

The EMDS system that was used in this study enables the application of geospatial modelling to assess the complexity of the urban environment. Although EMDS had not previously been applied in UES-related studies, the system shows several strengths in resolving complex spatial problems. Apart from well-established terminology that facilitates the interpretation of results, a user-friendly interface enables the consideration of spatial complexities in a relatively simple way [69]. The latter features help to strengthen the collaboration between scientists and end-users, facilitating EMDS application in participatory planning. The possibility of the implementation of such methodology, with a combination of expert knowledge and scientific methods, is of great interest in UES-related decision making processes [70].

A spatial analysis of UES provision, as illustrated in this study, is a useful foundation for decision makers in setting policies and developing strategies for improving the provisioning of UES in urban landscapes, insofar as it spatially quantifies the current state of the urban environment with respect to its current status. However, to effectively support decision making in this context, additional decision tools are needed to: (1) identify which urban districts are the best targets for improvements in UES provision (e.g., strategic planning), and (2) identify what specific actions in those districts would produce the biggest gain in provisioning (e.g., tactical planning). While the spatial analysis of UES provisioning was relatively objective, the subsequent decision analyses were relatively subjective, but could be assisted by tools for a multi-criteria decision analysis (MCDA, [71]) that helps decision makers to organize the decision criteria into models, decide on the relative importance of these criteria, and document the decision models, in order to facilitate stakeholder participation. While the current study only addressed a foundational spatial analysis of UES provisioning, the EMDS system includes a variety of MCDA methods that can be applied to extend the current EMDS applications to the strategic and tactical phases of decision support for UES provisioning.

5. Conclusions

In this study, we assessed the provision of UES in Barcelona, Spain, and Santiago, Chile, implementing two different comparison methods. The EMDS spatial decision support framework was applied for the data modelling and results interpretation. The results demonstrated different levels of UES provisioning, depending on the methodological approach used, and reflected the relativeness in UES provision, which presents difficulties in developing effective strategic and tactical solutions for urban planning. Therefore, we suggested that UES comparison methods are useful tools for detecting environmental injustice in urban areas and supporting better UGA management. Still, it has to be considered that the standardization processes required for comparisons between urban entities may neglect the use of highly specific but relevant information.

Author Contributions: All authors participated on the conceptualization of the project. G.K. and S.R.-P. implemented the data preparation. G.K. did the analysis and wrote an initial draft. K.M.R. supervised the development and implementation of EMDS framework. G.K., J.R.G.O. and J.G.-G. analysed the results. All authors participated in in editing process of the manuscript. All authors have read and agreed to the published version of the manuscript.

Funding: This work was funded by the Catalan government predoctoral scholarship (AGAUR-FSE 2020 FI_B2 00147), SuFoRun Marie Sklodowska-Curie Research and Innovation Staff Exchange (RISE) Program (Grant agreement No: 691149) and AGAUR, as support to the "Forest Landscape Planning and Decision Making; towards resilient landscapes (ForPLADEM)", GRC (2021 SGR 01544).

Data Availability Statement: Archive 1, containing data used in the analysis, is elaborated and available upon request from the corresponding author.

Conflicts of Interest: The authors declare no conflict of interest. The use of trade or firm names in this publication is for reader information and does not imply endorsement by the U.S. Department of Agriculture of any product or service.

References

1. Bolund, P.; Hunhammar, S. Ecosystem services in urban areas. *Ecol. Econ.* **1999**, *29*, 293–301. [CrossRef]
2. Gómez-Baggethun, E.; Barton, D.N. Classifying and valuing ecosystem services for urban planning. *Ecol. Econ.* **2013**, *86*, 235–245. [CrossRef]
3. Kourdounouli, C.; Jönsson, A.M. Urban ecosystem conditions and ecosystem services–a comparison between large urban zones and city cores in the EU. *J. Environ. Plan. Manag.* **2020**, *63*, 798–817. [CrossRef]
4. Dobbs, C.; Kendal, D.; Nitschke, C.R. Multiple ecosystem services and disservices of the urban forest establishing their connections with landscape structure and sociodemographics. *Ecol. Indic.* **2014**, *43*, 44–55. [CrossRef]
5. Kraemer, R.; Kabisch, N. Parks in context: Advancing citywide spatial quality assessments of urban green spaces using fine-scaled indicators. *Ecol. Soc.* **2021**, *26*, 45. [CrossRef]
6. Derkzen, M.L.; van Teeffelen, A.J.A.; Verburg, P.H. REVIEW: Quantifying urban ecosystem services based on high-resolution data of urban green space: An assessment for Rotterdam, the Netherlands. *J. Appl. Ecol.* **2015**, *52*, 1020–1032. [CrossRef]
7. Wilkerson, M.L.; Mitchell, M.G.; Shanahan, D.; Wilson, K.; Ives, C.D.; Lovelock, C.; Rhodes, J. The role of socio-economic factors in planning and managing urban ecosystem services. *Ecosyst. Serv.* **2018**, *31*, 102–110. [CrossRef]
8. Niemelä, J.; Saarela, S.-R.; Söderman, T.; Kopperoinen, L.; Yli-Pelkonen, V.; Väre, S.; Kotze, D.J. Using the ecosystem services approach for better planning and conservation of urban green spaces: A Finland case study. *Biodivers. Conserv.* **2010**, *19*, 3225–3243. [CrossRef]

9. Holt, A.R.; Mears, M.; Maltby, L.; Warren, P. Understanding spatial patterns in the production of multiple urban ecosystem services. *Ecosyst. Serv.* **2015**, *16*, 33–46. [CrossRef]
10. Dobbs, C.; Hernández-Moreno, Á.; Reyes-Paecke, S.; Miranda, M.D. Exploring temporal dynamics of urban ecosystem services in Latin America: The case of Bogota (Colombia) and Santiago (Chile). *Ecol. Indic.* **2018**, *85*, 1068–1080. [CrossRef]
11. Dobbs, C.; Escobedo, F.J.; Zipperer, W.C. A framework for developing urban forest ecosystem services and goods indicators. *Landsc. Urban Plan.* **2011**, *99*, 196–206. [CrossRef]
12. Li, F.; Guo, S.; Li, D.; Li, X.; Li, J.; Xie, S. A multi-criteria spatial approach for mapping urban ecosystem services demand. *Ecol. Indic.* **2020**, *112*, 106119. [CrossRef]
13. Gold, S.M. Social benefits of trees in urban environmentst. *Int. J. Environ. Stud.* **1976**, *10*, 85–90. [CrossRef]
14. Parsa, V.A.; Salehi, E.; Yavari, A.R.; van Bodegom, P.M. Evaluating the potential contribution of urban ecosystem service to climate change mitigation. *Urban Ecosyst.* **2019**, *22*, 989–1006. [CrossRef]
15. Xu, C.; Jiang, W.; Huang, Q.; Wang, Y. Ecosystem services response to rural-urban transitions in coastal and island cities: A comparison between Shenzhen and Hong Kong, China. *J. Clean. Prod.* **2020**, *260*, 121033. [CrossRef]
16. Grêt-Regamey, A.; Galleguillos-Torres, M.; Dissegna, A.M.; Weibel, B. How urban densification influences ecosystem services—A comparison between a temperate and a tropical city. *Environ. Res. Lett.* **2020**, *15*, 075001. [CrossRef]
17. Subiza-Pérez, M.; Vozmediano, L.; Juan, C.S. Landscape and Urban Planning Green and blue settings as providers of mental health ecosystem services: Comparing urban beaches and parks and building a predictive model of psychological restoration. *Landsc. Urban Plan.* **2020**, *204*, 103926. [CrossRef]
18. Bing, Z.; Qiu, Y.; Huang, H.; Chen, T.; Zhong, W.; Jiang, H. Spatial distribution of cultural ecosystem services demand and supply in urban and suburban areas: A case study from Shanghai, China. *Ecol. Indic.* **2021**, *127*, 107720. [CrossRef]
19. Remme, R.; de Nijs, T.; Paulin, M. Natural Capital Model—Technical Documentation of the Quantification, Mapping and Monetary Valuation of Urban Ecosystem Services. RIVM Report 2017-0040. 2018, p. 76. Available online: www.rivm.nl/en (accessed on 28 April 2023).
20. Tavares, P.A.; Beltrão, N.; Guimarães, U.S.; Teodoro, A.; Gonçalves, P. Urban ecosystem services quantification through remote sensing approach: A systematic review. *Environments* **2019**, *6*, 51. [CrossRef]
21. Alemu, J.B.; Richards, D.R.; Gaw, L.Y.-F.; Masoudi, M.; Nathan, Y.; Friess, D.A. Identifying spatial patterns and interactions among multiple ecosystem services in an urban mangrove landscape. *Ecol. Indic.* **2021**, *121*, 107042. [CrossRef]
22. Shiraishi, K. The inequity of distribution of urban forest and ecosystem services in Cali, Colombia. *Urban For. Urban Green.* **2022**, *67*, 127446. [CrossRef]
23. Watkins, S.L.; Gerrish, E. The relationship between urban forests and race: A meta-analysis. *J. Environ. Manag.* **2018**, *209*, 152–168. [CrossRef] [PubMed]
24. Escobedo, F.J.; Clerici, N.; Staudhammer, C.L.; Corzo, G.T. Socio-ecological dynamics and inequality in Bogotá, Colombia's public urban forests and their ecosystem services. *Urban For. Urban Green.* **2015**, *14*, 1040–1053. [CrossRef]
25. Konijnendijk, C.C. Evidence-based guidelines for greener, healthier, more resilient neighbourhoods: Introducing the 3-30-300 rule. *J. For. Res.* **2022**, *34*, 821–830. [CrossRef] [PubMed]
26. Nahlik, A.M.; Kentula, M.E.; Fennessy, M.S.; Landers, D.H. Where is the consensus? A proposed foundation for moving ecosystem service concepts into practice. *Ecol. Econ.* **2012**, *77*, 27–35. [CrossRef]
27. Blanco, E.; Raskin, K.; Clergeau, P. Towards regenerative neighbourhoods: An international survey on urban strategies promoting the production of ecosystem services. *Sustain. Cities Soc.* **2022**, *80*, 103784. [CrossRef]
28. Haase, D.; Larondelle, N.; Andersson, E.; Artmann, M.; Borgström, S.; Breuste, J.; Gomez-Baggethun, E.; Gren, Å.; Hamstead, Z.; Hansen, R.; et al. A quantitative review of urban ecosystem service assessments: Concepts, models, and implementation. *Ambio* **2014**, *43*, 413–433. [CrossRef]
29. Grêt-Regamey, A.; Sirén, E.; Brunner, S.H.; Weibel, B. Review of decision support tools to operationalize the ecosystem services concept. *Ecosyst. Serv.* **2017**, *26*, 306–315. [CrossRef]
30. Reynolds, K.M.; Hessburg, P.F. An Overview of the Ecosystem Management Decision Support System. In *Making Transparent Environmental Management Decisions. Environmental Science and Engineering*; Springer: Berlin/Heidelberg, Germany, 2014; pp. 3–22. [CrossRef]
31. Millennium Ecosystem Assessment. *Ecosystems and Human Well-Being*; Island Press: Washington, DC, USA, 2005.
32. Montoya-Tangarife, C.; de la Barrera, F.; Salazar, A.; Inostroza, L. Monitoring the effects of land cover change on the supply of ecosystem services in an urban region: A study of Santiago-Valparaíso, Chile. *PLoS ONE* **2017**, *12*, e0188117. [CrossRef]
33. Zhang, S.; Ramírez, F.M. Assessing and mapping ecosystem services to support urban green infrastructure: The case of Barcelona, Spain. *Cities* **2019**, *92*, 59–70. [CrossRef]
34. Baró, F.; Palomo, I.; Zulian, G.; Vizcaino, P.; Haase, D.; Gómez-Baggethun, E. Mapping ecosystem service capacity, flow and demand for landscape and urban planning: A case study in the Barcelona metropolitan region. *Land Use Policy* **2016**, *57*, 405–417. [CrossRef]
35. Dobbs, C.; Escobedo, F.J.; Clerici, N.; De La Barrera, F.; Eleuterio, A.A.; MacGregor-Fors, I.; Reyes-Paecke, S.; Vásquez, A.; Camaño, J.D.Z.; Hernández, H.J. Urban ecosystem Services in Latin America: Mismatch between global concepts and regional realities? *Urban Ecosyst.* **2019**, *22*, 173–187. [CrossRef]

36. National Statistics Institute. Residents in Spain. 2022. Available online: https://www.ine.es/index.htm (accessed on 30 January 2023).
37. National Statistics Institute. Census. 2017. Available online: https://www.ine.gob.cl/censo (accessed on 30 January 2023).
38. Kourtzanidis, K.; Angelakoglou, K.; Giourka, P.; Tsarchopoulos, P.; Nikolopoulos, N.; Ioannidis, D.; Kantorovitch, J.; Formiga, J.; Verbeek, K.; de Vries, M.; et al. World map of the Köppen-Geiger climate classification updated. *Meteorol. Z.* **2006**, *15*, 259–263. [CrossRef]
39. Cui, F.; Hamdi, R.; Yuan, X.; He, H.; Yang, T.; Kuang, W.; Termonia, P.; De Maeyer, P. Quantifying the response of surface urban heat island to urban greening in global north megacities. *Sci. Total. Environ.* **2021**, *801*, 149553. [CrossRef]
40. Myeong, S.; Nowak, D.J.; Duggin, M.J. A temporal analysis of urban forest carbon storage using remote sensing. *Remote Sens. Environ.* **2006**, *101*, 277–282. [CrossRef]
41. Armson, D.; Stringer, P.; Ennos, A. The effect of street trees and amenity grass on urban surface water runoff in Manchester, UK. *Urban For. Urban Green.* **2013**, *12*, 282–286. [CrossRef]
42. Kuehler, E.; Hathaway, J.; Tirpak, A. Quantifying the benefits of urban forest systems as a component of the green infrastructure stormwater treatment network. *Ecohydrology* **2017**, *10*, e1813. [CrossRef]
43. Dzhambov, A.; Dimitrova, D. Urban green spaces' effectiveness as a psychological buffer for the negative health impact of noise pollution: A systematic review. *Noise Health* **2014**, *16*, 157–165. [CrossRef]
44. Gozalo, G.R.; Morillas, J.M.B.; González, D.M.; Moraga, P.A. Relationships among satisfaction, noise perception, and use of urban green spaces. *Sci. Total. Environ.* **2018**, *624*, 438–450. [CrossRef]
45. Rudd, H.; Vala, J.; Schaefer, V. Importance of backyard habitat in a comprehensive biodiversity conservation strategy: A connectivity analysis of urban green spaces. *Restor. Ecol.* **2002**, *10*, 368–375. [CrossRef]
46. Grunewald, K.; Richter, B.; Meinel, G.; Herold, H.; Syrbe, R.-U. Proposal of indicators regarding the provision and accessibility of green spaces for assessing the ecosystem service 'recreation in the city' in Germany. *Int. J. Biodivers. Sci. Ecosyst. Serv. Manag.* **2017**, *13*, 26–39. [CrossRef]
47. Zhang, H.; Chen, B.; Sun, Z.; Bao, Z. Landscape perception and recreation needs in urban green space in Fuyang, Hangzhou, China. *Urban For. Urban Green.* **2013**, *12*, 44–52. [CrossRef]
48. Plieninger, T.; Dijks, S.; Oteros-Rozas, E.; Bieling, C. Assessing, mapping, and quantifying cultural ecosystem services at community level. *Land Use Policy* **2013**, *33*, 118–129. [CrossRef]
49. Ulmer, J.M.; Wolf, K.L.; Backman, D.R.; Tretheway, R.L.; Blain, C.J.; O'neil-Dunne, J.P.; Frank, L.D. Multiple health benefits of urban tree canopy: The mounting evidence for a green prescription. *Health Place* **2016**, *42*, 54–62. [CrossRef]
50. Wolf, K.L.; Lam, S.T.; McKeen, J.K.; Richardson, G.R.; Bosch, M.v.d.; Bardekjian, A.C. Urban Trees and Human Health: A Scoping Review. *Int. J. Environ. Res. Public Health* **2020**, *17*, 4371. [CrossRef]
51. Riechers, M.; Barkmann, J.; Tscharntke, T. Perceptions of cultural ecosystem services from urban green. *Ecosyst. Serv.* **2016**, *17*, 33–39. [CrossRef]
52. Hu, T.; Wei, D.; Su, Y.; Wang, X.; Zhang, J.; Sun, X.; Liu, Y.; Guo, Q. Quantifying the shape of urban street trees and evaluating its influence on their aesthetic functions based mobile lidar data. *ISPRS J. Photogramm. Remote Sens.* **2022**, *184*, 203–214. [CrossRef]
53. Chuvieco, E. *Teledetección Ambiental*; Ariel: Barcelona, Spain, 2010.
54. Jime, J.C. A generalized single-channel method for retrieving land surface temperature from remote sensing data. *J. Geophys. Res. Atmos.* **2004**, *108*, D22. [CrossRef]
55. Valor, E.; Caselles, V. Mapping land surface emissivity from NDVI: Application to European, African, and South American areas. *Remote Sens. Environ.* **1996**, *57*, 167–184. [CrossRef]
56. Barcelona's City Hall Open Data Service. Available online: https://opendata-ajuntament.barcelona.cat/en (accessed on 31 January 2023).
57. Dormidontova, V.; Belkin, A. The Continuity of Open Greened Spaces-Basic Principle of Urboecology. *IOP Conf. Ser. Mater. Sci. Eng.* **2020**, *753*, 022048. [CrossRef]
58. Puigdollers, J. Barcelona Green Infrastructure and Biodiversity Plan 2020. Barcelona, Spain (in Spanish, English Summary). 2013, p. 111. Available online: http://scholar.google.com/scholar?hl=en&btnG=Search&q=intitle:Barcelona+green+infrastructure+and+biodiversity+plan+2020.#0 (accessed on 28 April 2023).
59. Hernández-Morcillo, M.; Plieninger, T.; Bieling, C. An empirical review of cultural ecosystem service indicators. *Ecol. Indic.* **2013**, *29*, 434–444. [CrossRef]
60. Geospatial Data Infrastructure. Available online: https://www.ide.cl/ (accessed on 31 January 2023).
61. Miller, B.J.; Saunders, M.C. *NetWeaver Reference Manual*; Pennsylvania State University: College Park, PA, USA, 2002; p. 127.
62. Getis, A.; Ord, J.K. The Analysis of Spatial Association by Use of Distance Statistics. *Geogr. Anal.* **1992**, *24*, 189–206. [CrossRef]
63. Goodchild, M.F.; Janelle, D.G. (Eds.) Spatially Integrated Social Science. In *Spatial Information Systems*; Oxford University Press: Oxford, UK, 2004.
64. Liu, L.; Wu, J. Scenario analysis in urban ecosystem services research: Progress, prospects, and implications for urban planning and management. *Landsc. Urban Plan.* **2022**, *224*, 104433. [CrossRef]
65. Romero-Duque, L.P.; Trilleras, J.; Castellarini, F.; Quijas, S. Ecosystem services in urban ecological infrastructure of Latin America and the Caribbean: How do they contribute to urban planning? *Sci. Total. Environ.* **2020**, *728*, 138780. [CrossRef] [PubMed]

66. McPhearson, T.; Andersson, E.; Elmqvist, T.; Frantzeskaki, N. Resilience of and through urban ecosystem services. *Ecosyst. Serv.* **2015**, *12*, 152–156. [CrossRef]
67. Polasky, S.; Lewis, D.J.; Plantinga, A.J.; Nelson, E. Implementing the optimal provision of ecosystem services. *Proc. Natl. Acad. Sci. USA* **2014**, *111*, 6248–6253. [CrossRef] [PubMed]
68. Spangenberg, J.H.; Görg, C.; Truong, D.T.; Tekken, V.; Bustamante, J.V.; Settele, J. Provision of ecosystem services is determined by human agency, not ecosystem functions. Four case studies. *Int. J. Biodivers. Sci. Ecosyst. Serv. Manag.* **2014**, *10*, 40–53. [CrossRef]
69. Reynolds, K.M. EMDS 3.0: A modeling framework for coping with complexity in environmental assessment and planning. *Sci. China Technol. Sci.* **2006**, *49* (Suppl. S1), 63–75. [CrossRef]
70. Elliot, T.; Bertrand, A.; Babí Almenar, J.; Petucco, C.; Proença, V.; Rugani, B. Spatial optimisation of urban ecosystem services through integrated participatory and multi-objective integer linear programming. *Ecol. Model.* **2019**, *409*, 108774. [CrossRef]
71. Huang, I.B.; Keisler, J.; Linkov, I. Multi-criteria decision analysis in environmental sciences: Ten years of applications and trends. *Sci. Total. Environ.* **2011**, *409*, 3578–3594. [CrossRef]

Disclaimer/Publisher's Note: The statements, opinions and data contained in all publications are solely those of the individual author(s) and contributor(s) and not of MDPI and/or the editor(s). MDPI and/or the editor(s) disclaim responsibility for any injury to people or property resulting from any ideas, methods, instructions or products referred to in the content.

Article

Natural Protected Areas within Cities: An International Legislative Comparison Focused on Romania

Atena-Ioana Gârjoabă [1], Cerasella Crăciun [1,2] and Alexandru-Ionut Petrisor [1,3,4,5,*]

1. Doctoral School of Urban Planning, Ion Mincu University of Architecture and Urbanism, 10014 Bucharest, Romania; atena.garjoaba@gmail.com (A.-I.G.); cerasella.craciun@gmail.com (C.C.)
2. "Urban Planning and Landscape" Department, Faculty of Urbanism, Doctoral School of Urban Planning, Ion Mincu University of Architecture and Urbanism, 10014 Bucharest, Romania
3. Department of Architecture, Faculty of Architecture and Urban Planning, Technical University of Moldova, 2004 Chisinau, Moldova
4. National Institute for Research and Development in Constructions, Urbanism and Sustainable Spatial Development URBAN-INCERC, 21652 Bucharest, Romania
5. National Institute for Research and Development in Tourism, 50741 Bucharest, Romania
* Correspondence: alexandru_petrisor@yahoo.com; Tel.: +40-213-077-3191

Abstract: Urbanization occurs now more rapidly than before, due to the development of compact cities or urban sprawl, threatening quasi-natural areas, especially those protected within/near built-up ones. Europe lacks laws dedicated to natural protected areas within built-up areas, which are subject to the same provisions as natural protected ones, or a legislative vacuum. This research aimed to find the best planning approach for resiliently conserving and developing these areas and establishing grounds for a new tool used for planning the proximity of natural areas within cities. The methodology involved selecting two groups of countries, Nordic and eastern European, and treating these areas differently. The choice was based on specific political history. The study analyzed the legislative and planning framework and compared the approaches of 11 analyzed countries to pinpoint the basic aspects accounted for and applied to other European territories, in order to preserve the characteristics of urban morpho-typology and the particularities of local landscapes. The comparison results suggest solutions such as adopting specific regulations for urban protected areas and their adjacent zones through legal documents, completing/detailing environmental legislation in Nordic countries, adopting laws dedicated to protected natural areas within and/or close to built areas, and changing the approach to protecting natural areas with urban planning or land use tools.

Keywords: urban protected areas; environmental legislation; urban planning; biodiversity conservation; Nordic countries; eastern European countries

Citation: Gârjoabă, A.-I.; Crăciun, C.; Petrisor, A.-I. Natural Protected Areas within Cities: An International Legislative Comparison Focused on Romania. Land 2023, 12, 1279. https://doi.org/10.3390/land12071279

Academic Editor: Shicheng Li

Received: 14 May 2023
Revised: 8 June 2023
Accepted: 20 June 2023
Published: 23 June 2023

Copyright: © 2023 by the authors. Licensee MDPI, Basel, Switzerland. This article is an open access article distributed under the terms and conditions of the Creative Commons Attribution (CC BY) license (https:// creativecommons.org/licenses/by/ 4.0/).

1. Introduction

The conflict between the morpho-typology of urban tissue and quasi- or semi-natural areas is becoming more and more acute. This conflict is stronger in old human settlements that have developed organically, and in the current context, in ever faster urban sprawl and the development of systems, relationships, and specific connections at the territorial level. The most frequent conflicts of this type are found in rural–urban areas, located at the confluence between the urban tissue and its neighboring agricultural lands, but also in areas placed between the urban tissue and green spaces inside the cities. The second category is more problematic, considering the generally insular morphological layout of green spaces in relation to urban fabric. Protected urban areas are sometimes called "protected islands" due to their isolation from their surrounding environments [1]. All the borders and separation areas between these two types of tissues are most often areas of conflict that require careful management from the point of view of urban planning.

Unplanned urban development failing to preserve local character, without a coherent eco-sustainable and resilient strategy, represents a real threat to the conservation of biodiversity worldwide [2]. This statement is valid in particular in areas valuable from this point of view, such as natural protected ones [3]. Biodiversity conservation can have a considerable impact in terms of increasing ecosystem services [4]. Currently, urbanization follows two main trends: increases in building density (to create compact cities) and the expansion of peripheral areas through urban sprawl [5]. In this context, the ability to support the persistence of species in natural areas within cities becomes a conclusive and, at the same time, a difficult objective for maintaining long-term conservation [6]. Additionally, the competition for occupying space that occurs between activities such as agriculture and nature protection is one of the most obvious human imprints [7]. A major problem from this standpoint, evident especially in Europe, is the fragmentation of ecosystems [8], a major threat to nature conservation [9].

The urban planning process should also take into account natural protected areas, and land use regulations should complement and strengthen these natural protected areas and even be a pillar of biodiversity conservation, especially for land without a protection regime, but representing a special landscape feature with a high conservation value [10]. A common misconception sometimes also addressed by urban planning is that the term "protected area" designates a wild area devoid of human influence [11], but in reality, ecological systems (especially urban ones) are in an intense interaction with urban and social ones, thus facilitating an interdisciplinary research and planning framework, with the aim of ensuring the maintenance of biodiversity in urban areas [12]. These are so-called socio-ecological systems (SES), complex systems that take into account social and ecological variables [13].

Urban planners and political decision makers have experienced solutions that take into account both social and economic concerns, as well as environmental concerns, interconnected in a complex trans-disciplinary sense [14] to reduce environmental impacts [15]. Ecosystem services are crucial, especially those of urban green infrastructure [16]. In order to safeguard the values and natural resources of their territory, municipalities are mandated by European urban planning laws to draft "municipal green infrastructure plans" [17]. Urban planners are challenged to understand, temporally and spatially, ecosystem services [18]. Unfortunately, they are often underestimated and difficult to quantify, considering the lack of a complex integration of systems for monitoring the biodiversity and values of ecosystem services in natural protected areas [19]. Therefore, urban planning in accordance with the augmentation of ecosystem services becomes very difficult, especially given the limited guidance on how ecosystem services should be used in the context of land use and environmental planning [20]. Additionally, very few of the many publications have provided a structured analysis of the contribution and use of this concept in urban planning [21].

Therefore, the literature review highlights limited knowledge of urban planning in terms of developing multi-disciplinary or even trans-disciplinary approaches with ecology. It is important to consider that the creativity in the urban landscape stems from the ability to sensitively perceive space and surrounding landscapes, influenced by the unique perspectives and perceptions of each specialist [22]. Moreover, the analysis of previous studies identified no urban planning tools with the role of valorizing ecosystem services from the viewpoint of spatial relations. No tools were identified even for just analyzing/quantifying the compatibility between built urban tissue and the quasi-natural one. Additionally, no clear and specific recommendations were identified regarding what exactly this tool should analyze.

This research originated from the problem of lacking specific theoretical information about areas adjacent to natural protected ones in cities, from the viewpoint of urban development. Against the background of this theoretical void, the absence of planning guidelines is also noticeable. These guidelines could direct the planning process to support the interdependence between quasi-natural fabric, specific to natural protected areas, and its adjacent built fabric.

The purpose of this study is to provide a set of recommendations for urban planning in accordance with the needs of urban areas close to natural protected ones. These recommendations refer to quasi-natural areas in urban environments, with the most urgent need for correlation with their proximities. Such recommendations can substantiate an urban planning instrument aimed at reducing urban pressures on natural protected areas by adopting appropriate planning methods for the areas adjacent to sensitive natural ones lacking a conservation value.

In this sense, a comparative analysis of some urban planning models, selected for being as different as possible, can pinpoint the different approaches to the urban planning of areas close to protected urban natural areas. Therefore, we compared two types of approaches to urban planning and legislation dedicated to natural protected areas in cities and their adjacent areas. The evaluated models are the approaches of Nordic and eastern European countries, which represent particular situations from a historical–evolutionary point of view, but also from a morpho-urban typology standpoint. The aim is to discover the optimal planning attitude for ensuring resilient conservation and the development of these areas, and create a new instrument used in the vicinity of natural areas within cities. The purpose of this study is not to create the tool itself, which can take different forms (guidelines, urban indicators, and framework structure of urban plans for these areas, etc.), but to phrase a set of recommendations substantiating the development of this tool.

2. Materials and Methods

Selection of case studies: For the analysis of urban planning related to natural protected areas, relevant planning instruments and legislative acts were analyzed for the following countries: Nordic countries—Denmark, Finland, Iceland, Norway, and Sweden, and eastern European countries—Bulgaria, Czech Republic, Hungary, Poland, Slovakia, and Romania (Figure 1).

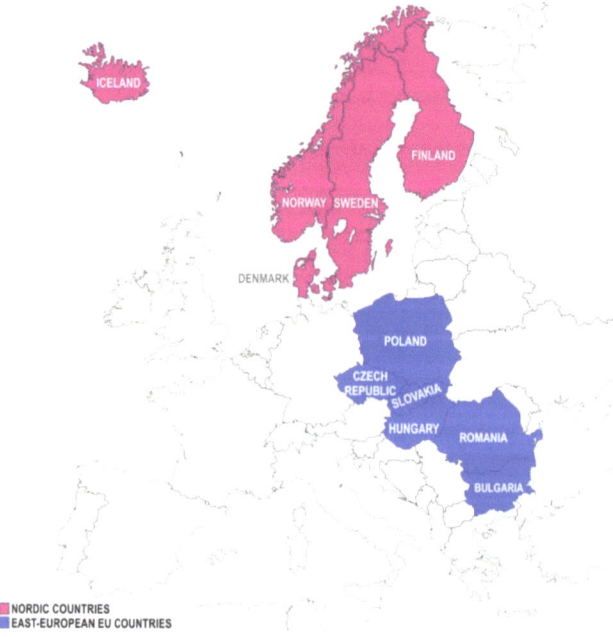

Figure 1. The countries selected for the analysis of legislative acts and planning instruments with incidence on natural protected areas.

These two groups of countries were selected for comparison due to their different evolutions from the viewpoint of urbanization and, implicitly, of urban planning. The main historical aspects differentiating the two groups over time are listed below.

- The emergence of cities: in Nordic countries, urbanization began to develop during the Middle Ages, with the emergence of trading cities and universities, while in eastern European countries, urbanization began later, in the 19th century, along with the industrialization process.
- Post-industrial urban evolution: in Nordic countries, post-industrial urban transformation has been characterized by the regeneration of abandoned industrial areas into modern and sustainable housing and business areas. In contrast, in eastern European countries, post-industrial urban transformation has often been slowed down or blocked by a lack of resources and economic problems [23].
- Urban size and density: Nordic countries generally have smaller and denser cities than eastern European countries. On the other hand, in eastern Europe, the tendency to decrease the area and density of cities, also known as "shrinking cities", is becoming more and more common, due to a decrease in population.
- The evolution of public policy: while Nordic countries had a stable democratic development in the 20th century, eastern European countries were under communist regimes for several decades, which had a significant impact on their economic and cultural development.

Establishing the analysis criteria: Taking into account the differences listed above, the study started from the premise that the two groups of countries also differ from the viewpoints of urban planning and its related legislation. Starting from this point, after identifying the differences at these two levels of the analysis, a set of recommendations were drawn up for each group of countries.

The two levels of analysis (legislation and urban planning instruments) were chosen in order to obtain an overview of how the areas adjacent to natural protected areas are treated from the point of view of urban planning. Another important aspect taken into account was the degree of attention given to them from a legislative point of view, especially considering the fact that not all planning instruments have legal value.

The legislation was selected to include legislative acts substantiating the urban–territorial planning systems specific to each country, those addressing quasi-natural areas in cities (for example, green spaces), natural protected areas and their adjacent areas, and the regulation of the relationship between natural protected areas and urban morphotypological tissue, in the specific context of each area. Considering the multitude of existing legislative acts in the 11 analyzed countries, the acts substantiating the basis of urban planning and environmental protection were chosen for analysis first.

An important starting point, in this sense, consists of the brochures produced for each state by the Organization for Economic Co-operation and Development (OECD), a forum where the governments of 37 democracies collaborate to develop policy standards for promoting sustainable economic growth. These sheets can be found on the OECD website in the "countries" section and provide an overview of the urban planning tools used in each country, as well as the main laws that regulate these urban planning aspects.

Having these sheets as a starting point, the following steps were taken for selecting the laws analyzed. First of all, general laws related to urban planning and the environment, which generally address the subject of these two fields, were selected. In order to identify the general environmental and urban planning laws of each country, the following sources of information were used: the official website of the respective country's government was accessed, and then its environment/urban planning/legislation sections were viewed, depending on the structure of the respective website. Online legislative databases (either at a national or European level), websites of environmental institutions, and those responsible for urban planning were also used.

Secondly, the legislative acts related to them, with a potential influence on the system of natural protected areas, were analyzed. Territorial planning instruments and their roles were also identified, especially with regard to the planning for areas adjacent to natural protected ones.

Comparison of case studies: After identifying the particular aspects of each country, a comparison was made between all the selected countries, depending on each analyzed level, but also on the territorial scale to which the respective instrument was applicable (legislative or planning). Other instruments considered relevant for the two fields were also taken into account—strategic instruments mentioned in the analyzed legislative acts.

Identification of principles used: Based on the comparison between the analyzed countries, we identified the main directions that each country follows with respect to environmental protection and urban planning. A parallel analysis of these issues facilitated the identification of the different aspects emphasized by each country/group.

Drawing up a set of recommendations for the two models: Comparing the two analyzed models—Nordic and eastern European—we noticed that the two adopted different instruments, in terms of territorial scale, required improvements, both from the urban planning and legislative viewpoints. For this purpose, a series of recommendations were made regarding the two levels of analysis, addressing the strategic approaches separately from the urban planning instruments, due to their lower legal value. The recommendations concerned the two groups, a single one, or a single country. Special attention was paid to Romania, first taking into account its specific natural heritage and environmental legislative proposals, differentiating it from the other analyzed countries.

Figure 2 presents a diagram of the methodology, indicating its main steps.

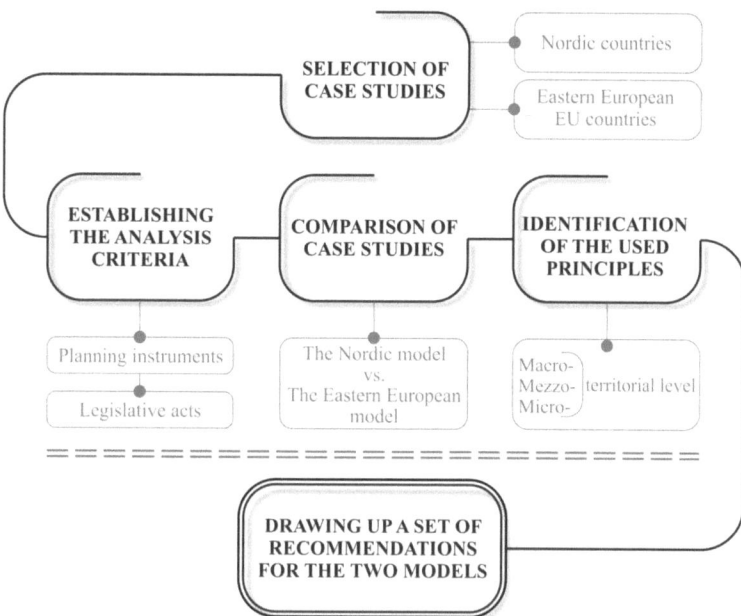

Figure 2. Phases of methodology used in the current study.

3. Results

3.1. Critical Analysis of Legislative Acts and Planning Instruments of Urban Protected Areas in the Analyzed EU Countries

3.1.1. Nordic Countries—Denmark, Finland, Iceland, Norway, and Sweden

Denmark

At the national level, the Danish Planning Act is the main guide for territorial planning. One of its objectives consists of "creating and conserving valuable buildings, settlements, urban environments and landscapes" [24]. The designation of protected areas is carried out through municipal plans, which can then be detailed through local development plans (also called detailed plans/Lokalplan). The Planning Act provides for the direct obligation of municipal plans to propose solutions for nature conservation—"The municipal plan must contain guidelines for safeguarding nature conservation interests, which are made up of natural areas with special nature protection interests, including existing Natura 2000 land and other protected nature areas, co-ecological links, potential natural areas, and potential ecological links"—§11a, Planning Act [24]. These municipal plans also mark/reserve component areas of The Green Denmark Map (Grønt Danmarkskort), a natural network that supports the planning of natural areas, with the aim of counteracting the loss of biodiversity.

Two other important legislative acts for territorial planning are the Nature Protection Law [25] and Land Registration Act [26]. The Nature Protection Law aims to protect nature, so that "social development can take place on a sustainable basis, both with respect for human living conditions and for the preservation of animal and plant life". The law contains provisions regarding protected areas from the viewpoint of protected species, public access, management, damages, and sanctions, but does not refer to the spatial relationship between them and the urban fabric.

The Land Registration Act contains provisions regarding property registration in the national real estate registration system. This act does not refer to protected natural areas.

No legislative act dedicated to protected areas located in urban areas was identified, thus concluding that they are subject to the same provisions as areas not directly spatially related to cities.

Finland

The Finnish planning system is mainly based on the Land Use and Building Act, but the Environmental Protection Act and Nature Conservation Act are also of particular importance, especially with regard to natural areas. The Land Use and Building Act has a chapter dedicated to national urban parks—chapter nine [27]. It mentions the way to establish one and the specific regulations for this. The establishment of such a protected natural area can be initiated upon the request of the local authorities, being completed by the institution made by the competent ministry. There are currently nine national urban parks established in Finland: Hämeenlinna, Heinola, Hanko, Kuopio, Pori, Forssa, Kotka, Turku, and Porvoo [28]. For each national urban park, a "maintenance and use scheme" is drawn up and approved by the responsible ministry; their regulations "must be taken into account in planning the areas of the park and in other planning and decision-making affecting the area" [27]. Therefore, urban planning and land use documentation must take this scheme into account. Plans are divided into the following categories, depending on the territorial scale at which they are designed: regional plans, local master plans, joint master plans (of several municipalities), and detailed local plans. The aspects that these plans must address include the "protection of the built environment, landscape and natural values". It is recommended for these plans to provide regulations regarding protected areas (natural or built). Regarding the content of detailed local plans, it is mentioned that "the built and the natural environment must be preserved and their special values must not be destroyed. There must be sufficient parks or other areas suitable for local recreation in the area covered by the plan or in its vicinity" (see Section 54 of the Land Use and Building Act).

The importance of urban planning in the conservation of biodiversity is also emphasized by the Finnish Biodiversity Action Plan 2013–2020. The Plan aims to "slow down the loss of biodiversity in urban and built-up areas by increasing knowledge of the subject and developing the related land-use planning, so as to take into account the conservation of biodiversity" [28]. Its actions consist of: encouraging the evaluation of unbuilt areas (important from the point of view of urban biodiversity), the promotion of important areas for conserving biodiversity in the urban environment, and the development of planning methods in the urban environment, in accordance with the conservation of biological diversity (see Section 2.12—urban and built areas of the Finnish Biodiversity Action Plan 2013–2020).

Iceland

A special aspect related to Iceland's legislation is the fact that its laws do not refer to green areas [29]. Planning is carried out at four territorial scales, through the following types of plans/documents: national planning strategies, regional plans, municipal plans, and detailed development plans. Considering its strategic nature, national documentation has the role of establishing the general directions that must be followed in the planning process. Regional plans are the main planning tool, to which the lower plans are subordinated from the point of view of the territorial scale. Municipal plans should respect and detail the provisions imposed by regional ones, and local plans should respect and detail the provisions imposed by municipal ones.

The main legislative act by which the documentation is guided is the Planning and Building Act [30]. Chapter III, "Preparation and implementation of development plans", Article 9, specifies that local conservation provisions are included in development plans, if there are valuable natural features in the area regulated by the plan considered to be desirable for preservation [30].

The main national law for environmental protection is the Nature Conservation Act [31]. According to Article 67, the Ministry of the Environment has the obligation to issue a comprehensive register of sites of natural interest once every 5 years at most, published in the Official Journal of Iceland. It can be completed at any time, if new areas are added to this register. The Nature Conservation Agency, in collaboration with the Icelandic Institute of Natural History, nature research centers, and nature conservation committees in question, is in charge of the preparation and collection of the data on register additions.

No legislative act adopted at the national level dedicated to natural protected areas located in urban areas was identified.

Norway

The law that governs the Norwegian planning system is the Planning and Building Act [32]. As mentioned in this act, Norway has the following planning instruments, different in terms of their legal power and territorial scale: national master plans (with a role in coordinating the planning process), regional planning strategies (determine where regional plans are needed), regional plans, municipal planning strategies (determine the areas where zonal plans are needed), municipal master plans (establish land use regulations), and zonal plans (detail the urban regulations for a given area; can be initiated both by local authorities and private actors). Additionally, at the national level, there are also sectoral plans, such as those for transport/mobility and the management of protected areas. An important mention included in the Planning and Building Act, regarding municipal master plans, is made in its Section 11-8—"Zones requiring special consideration". The Act specifies that, when protection regulations are adopted for a new protected area or when those in a management plan are revised, they can also be applied to the areas adjacent to national parks or protected landscapes. The aim is to prevent reducing the conservation value of these protected areas.

Regarding the content of zonal/detailed plans (see Section 12-5 "Land-use objectives in a zoning plan" of Norwegian Planning and Building Act), point three mentions "green structures" ("grønnstruktur"), consisting of nature areas, green corridors, outdoor recreation areas, and parks. These "green structures" are also provided for the content of municipal master plans, in the category of areas that require special attention. Therefore, this green system is established from the macro level through municipal master plans and then detailed at the mezzo level by zonal plans.

Another important legislative act for the conservation of natural protected areas is the Nature Diversity Act—Act of 19 June 2009 no. 100, relating to the management of biological, geological, and landscape diversity [33]. Its purpose consists of protecting biological, geological, and landscape diversity and ecological processes. Chapter V is dedicated to natural protected areas, but it does not include provisions dedicated to areas located inside or in the vicinity of urban areas.

The Act on natural areas in Oslo and its nearby municipalities [34] aims to "promote and facilitate outdoor life, nature experiences and sport". The law must also ensure the borders of the Marka—the area of forests and other quasi-natural lands around the city of Oslo, but the law does not deal with the protected areas that fall under the Nature Diversity Act and only refers to the creation of protected areas with special qualities for outdoor life—protected areas for recreation, which are established by the King.

In addition to these legislative acts that refer to natural protected areas, another two documents are available with roles in guiding/directing the planning process—Planning green structures in cities and towns [35]. As specified in this act, green structure planning should be based on the knowledge of animals, plants, their movement patterns, and their vulnerabilities, and should start from an overview of biotopes. An interesting aspect is the fact that the principles taken into account in the establishment of a green area, mentioned and explained in the guide, are similar to the principles for the establishment of a protected area, reported by John Wilson and Richard Primack [36] in their book, *Conservation Biology in Sub-Saharan Africa*. However, in this document, no special mentions are made regarding protected areas spatially related to urban fabric.

Sweden

The Swedish planning system is based on two legislative acts: the Planning and Building Act [37] and the Environmental Code [38]. As it appears from the Planning and Building Act, in Sweden, there is no planning at the national level, i.e., there is no national plan. However, the objectives of national interest are taken into account by documents at lower territorial scales. Additionally, at the regional level, planning is restricted, being currently regulated only for two regions—Skåne and Stockholm (through an addition to the act in 2019). At the level of each municipality, a municipal plan (comprehensive plan) must be adopted, containing guidelines for land use, but these plans are not legally binding. The next documentation at the territorial scale is a detailed regulation and development plan. This documentation has legal value and "can be" employed in certain areas for the urban regulation of land, through which the urban indicators for the studied area are established. For areas not covered by detailed regulatory plans, the municipality can adopt "zonal regulations", whose restrictive character is narrow, and which deal with aspects such as: land use, maximum buildable area, and buildable land sizes. The Planning and Building Act does not regulate natural protected areas.

The first mention of natural protected areas contained in the Environmental Code is made in its chapter three—"Basic provisions concerning the management of land and water areas", Section 2, which establishes that land/water surfaces that are not affected or are affected to a small extent by development projects must be protected from measures that can significantly affect their character. An interesting aspect appears in Section 15 of chapter 7—"Protection of areas" in the same Swedish Environmental Code, mentioning that "in special circumstances", the protection regime of a shore area (established at a distance between 100 and 300 m from the edge of the shore) may be suspended during the validity period of a detailed regulatory plan being adopted for the respective area.

Section 7 of the same chapter contains the first and only reference to an urban natural protected area, more precisely to the national urban park that covers the Ulriksdal-Haga-Brunnsviken-Djurgården area in Stockholm. It is mentioned here that interventions are allowed in national urban parks only if they can be implemented without affecting the park's landscapes or any natural or cultural assets.

3.1.2. Eastern European Countries—Bulgaria, Czech Republic, Hungary, Poland, Slovakia, and Romania

Bulgaria

The Bulgarian planning system includes four types of documentation, taking into account the territorial scale at which they are made: the National concept for spatial development, the Regional scheme for the spatial development of a region of level two, the Regional scheme for the spatial development of a district, and the Municipality concept for spatial development and detailed plans.

The National concept for spatial development 2013–2025 [39] includes six strategic objectives, one of which is dedicated to the protection of natural and cultural heritage—strategic objective four: "Well-preserved natural and cultural heritage". Subchapter 3.5.1. Natural values of the same Bulgarian act begins with the specification that "Protected nature areas account for a significant portion of the non-urbanized territories in the national space", and the document does not refer to protected areas in urban environments. The National concept for spatial development is not legally binding.

The following two types of documentation, made for NUTS-2- and NUTS-3-(/district)-type areas, generally include provisions regarding the functional and hierarchical structures of settlement networks, agglomeration areas, and their impact and development axes for infrastructure. The Regional scheme for the spatial development of a district should also include guidelines for plans made at the municipal level. Regional plans are not legally binding. The last two types of documentation—the Municipality concept for spatial development, the Municipality Spatial Development Concept, and the detailed plans have legal value and include specific functional regulations for targeted territories.

There are a series of legislative acts related to spatial planning and/or natural protected areas: the Spatial development act no. 1/2001 [40], The Regional Development Act no. 50/2008 [41], Environmental protection Act no. 91/2002 [42], Protected Areas Act no. 133/1998 [43], and Biodiversity Act no. 77/2002 [44].

According to the Spatial development act no. 1/2001, Art. 7 (1), the category of land designated as protected territories includes reserves, national parks, natural sites, maintained reserves, nature parks, protected areas, beaches, dunes, and water sources with sanitary protection zones, water areas, humid zones, and protected coastal strips. According to (2), the territories included in urbanized territories can be designated as protected territories. Article 62 (3) mentions that: "Existing green areas which are public ownership shall be developed and preserved as protected areas".

The Regional Development Act no. 50/2008 mentions natural and protected zones only as categories used in the functional zoning of space. There are no references to natural protected areas within cities.

The mentions regarding natural protected areas included in the Environmental Protection Act no. 91/2002 refer to the legislative act dedicated to them, which is analyzed below.

The Protected Areas Act no. 133/1998 regulates the categories of protected areas, the regimes for their protection and use, their designation, management, security, financing, and the penalties/fines in the case of non-compliance with the provisions of this act. In the legislative act, no references are made to natural protected areas spatially related to urban fabric.

The Biodiversity Act no. 77/2002 includes a chapter dedicated to the national ecological network. This chapter consists of: sites belonging to the Natura 2000 Network, protected areas of national importance, Corine sites, Ramsar wetlands, important areas for flora, and important areas for birds. Until 2005, the law included a section dedicated to the buffer zones of protected areas, whose articles were, however, repealed. The act does not refer to natural protected areas within or near urban environments.

Czech Republic

In the Czech Republic, there are three levels of government—levels at which spatial planning documents are drawn up: national, regional, and municipal. The national planning framework is represented by the National Development Policy [45]. This has the role of directing the national planning system, without having legal value. The protection of natural values is one of the priorities identified at the national level, together with the location of development projects that have the risk of significantly affecting the character of landscape. The importance of territorial systems with ecological stability, the protection of natural elements in the countryside, and the creation of conditions to ensure the permeability of landscapes for wild animals are also mentioned.

At the regional level, the Development Principles play a role similar to that of the National Development Policy at the national level. They highlight planning priorities at the regional level, but at the same time, coordinate planning at a municipal level, without having a legal character.

There are three types of plans drawn up at the municipal level—Local Territorial Plans, Regulatory Plans, and Planning Studies—out of which, only the first two have legal value. Local Territorial Plans cover the administrative area of an entire municipality and regulate land use. However, in many cases, these Plans leave a margin of appreciation for the Building Office, which is responsible for issuing building permits. Regulatory Plans are drawn up only for certain areas, where a need to detail the regulations from Local Territorial Plans is found. Planning Studies are documents that have the role of supporting the planning process by identifying possible solutions to certain problems. Territorial planning is regulated by The Building Act of 14 March 2006 [46].

Natural protected areas fall under the incidence of two legislative acts: the Act on the Environment [47] and the Act on Nature and Landscape Protection [48]. The first act has a general character and does not directly specify the term "natural protected area", speaking, in general, about environmental protection. The Act on Nature and Landscape Protection details how to protect each type of natural protected area. For example, for national parks and protected landscapes, it is mandatory to create both conservation and zoning plans (valid for 10 years in the case of national parks and between 10 and 15 years for protected landscapes). National parks are divided into three natural conservation zones, depending on their natural values, which differ in terms of the strictness of regulations. Regulations and zoning are mentioned in the legal act by which these protected areas are established. Similarly, protected landscapes are divided into three or four natural conservation zones.

Hungary

The Hungarian planning system includes three or four levels, depending on the region: National Spatial Plans, Spatial Plans for Special Regions, Spatial Plans for Counties, and Settlement Structural Plans. National Spatial Plans include a series of guidelines, strategic plans, and plans for a narrower scale of land use. These are documents with legal value and are renewed every seven years.

The following instruments, from a territorial scale viewpoint, are Spatial Plans for Special Regions. At the national level, there are two such plans—Budapest together with its adjacent urban agglomeration and the tourist area around Lake Balaton, which include both general provisions and strategic or land use plans and which have legal value. The main objectives of these two plans are to stimulate economic potential, support sustainable development, and protect natural and cultural heritage.

According to the Long-Term Development Concept of the City of Budapest 2030 [49], the main dangers for valuable natural areas are the "dynamic" mode of the spatial expansion of the city (sometimes insensitive to the limits of forests or borders between the natural and built environment), the occupation of land without a legal basis, and the lack of financial resources dedicated to the protection and management of these areas. For the preservation of natural values, the following "possible means of implementation" are proposed: the establishment of new green areas (to reduce the pressure on nature conservation areas—for example, by creating recreation areas), ensuring a balanced financing fund for nature conservation areas, the legal protection of areas significant from the point of view of nature but not yet protected, and environmental education. Within the plan, special attention is paid to the Danube, with its adjacent area being addressed in a dedicated chapter. The importance of the existing protected natural areas along the Danube, both from the Natura2000 network and in terms of national importance, is mentioned.

Within the Balaton Territorial Development Concept 2014–2030 [50], the significant set of natural protected areas that are components of the regulated area is mentioned, but no special mention is made about the natural protected areas near urban settlements.

Spatial Plans for Counties ensure the link between National Spatial Plans and local plans, detailing the regulations provided in national plans. Through them, regulations are established for the control of the development and protection of natural and built heritage.

The planning tool used at the local level, Settlement Structural Plans, includes both strategic planning and specific land use regulations. This document works together with local building regulations, which detail the approved types of use.

The framework legislation for the Hungarian spatial planning system is made up of three laws: Act XXI 1996 on Regional Development and Spatial Planning [51], Act LXXVIII 1997 on the Development and Protection of the Built Environment [52], and Act XXVI 2003 on the National Spatial Plan [53]. The first legislative act does not make specific references to natural protected areas. Act LXXVIII 1997 on the Development and Protection of the Built Environment refers only to green spaces important from the point of view of a settlement's structure, for which the provision of buffer zones is recommended. These areas are also specified in Act XXVI 2003 on the National Spatial Plan, where natural protected areas and the protection areas of natural protected areas are mentioned as "regional and county priority areas". Additionally, according to Chapter V, areas regarding national areas, Section 13, Article 4 of the same act, the natural ecological network is made up of natural protected areas, natural areas, and ecological corridor areas and must be included in regional- and county-level spatial plans. No specific references are made to natural protected areas in urban environments.

Poland

The legislative act substantiating spatial planning in Poland is the Spatial Planning and Development Act [54]. Planning is carried out on three levels, through the following types of documentation: national development plans, spatial development plans at the voivodship level, and local development/revitalization plans. The National Development Plan is valid for 20 years, and the one currently in force is The National Spatial Development Concept 2030—NSDC 2030 [55]. Its content highlights the importance of establishing ecological networks and conservation planning in protected areas. The plan does not refer to urban natural protected areas. NSDC is not legislative in nature.

The case of voivodeship plans is similar, as they are only indicative, have no legal value, and delimit, among others, protected areas. Additionally, at the regional level, it is mandatory to carry out landscape audits—territorial planning acts at the voivodeship level, through which the characteristics and value of local landscapes are determined.

Local development plans are legally binding documents (for more details, see Nowak et al., 2023 [56] and Blaszke et al., 2022 [57]). They regulate land use, urban indicators, and the protection of cultural assets. Local plans also estimate the costs of infrastructure projects and detail how the expropriation of the affected properties will be carried out following the implementation of these projects. No special specifications were identified for urban natural protected areas.

An important component of spatial planning in Poland is the system of environmental impact assessments—EIA [58], an auxiliary instrument for territorial development, the preparation of which is mandatory for all planning documents. The structure of the report that must be drawn up includes: a description of the project, a description of the natural elements that may be affected, a description of the historical monuments in the area (if applicable), a forecast in case the project is not carried out, a description of some development scenarios with the motivation of the most advantageous options, a description of the project monitoring, and a summary of the report. The legislative act contains a chapter (no. six) dedicated to the protection zoning of protected areas. It provides for the establishment of protection zones adjacent to protected natural areas, established by the same authority that established the respective natural protected area. If a protection zone is not provided, the area between the limit of the protected zone and a distance of 50 m from it is considered to be the protection zone. No special mentions were identified regarding natural protected areas located inside or in the vicinity of urban areas.

Slovakia

The Slovak Republic is characterized by three levels of government—national, regional, and municipal, and by four levels regarding the territorial planning system—the Slovak Spatial Development Perspective (at the national level), Regional Land Use Plans, and two types of local plans—Local Land Use Plans and Local Zoning Plans.

The Slovak Spatial Development Perspective [59] includes a chapter dedicated to landscape—Chapter 12, "Landscape structure of the Slovak Republic", which contains mentions about natural protected areas, but this refers to the Territorial System of Ecological Stability (TSES), a concept adopted by The Slovak Republic through the Decision of the Government no. 394 of 23 July 1991. The implementation of projects is carried out from the macro-territorial scale to the mezzo-/micro-territorial scale: the supraterritorial system of ecological stability (STSES) at the national scale, the regional territorial system of ecological stability (RTSES), and the local territorial system of ecological stability (LTSES). The local territorial system of ecological stability (LTSES) offers a landscape ecological basis for the elaboration of municipal plans [60]. The role of these planning tools is to conserve the biodiversity of landscapes, restore the connectivity of natural landscapes, and maintain or even improve ecosystem services [61] (for a more detailed discussion, see Popescu et al., 2022 [62]). It is important for the component elements of the TSES systems—bio-centres, bio-corridors, and interactive elements—to be included in the system of natural protected areas, as they are not fully legally protected.

Regional Land Use Plans establish strategic principles, include land use plans, establish the locations of major infrastructure and technical equipment, and establish indications for the protection of natural and cultural heritage. From a legal point of view, they are mandatory for local land use plans.

The first type of plan drawn up at the local level, Local Land Use Plans, must be adopted for settlements with over 2000 inhabitants. However, many settlements with less than 2000 inhabitants have adopted or are in the process of adopting a Local Land Use Plan, because they either have planned extensive developments or public buildings, or they have been obliged to by regional plans. Additionally, the preparation of a Local Land Use Plan is mandatory for settlements within which there are major infrastructure or public buildings.

The second type of local plan, Local Zoning Plans, describe the permitted use of land and are the most detailed ones, being usually drawn at a scale of 1:1000 or 1:500. Their preparation is necessary if required by Local Land Use Plans, or for planning public buildings.

Probably the most important three legislative acts for territorial planning are the Act on Land-use Planning and Building Order (Act 50/1976 Coll.), the Environmental Impact Assessment Act (Act 24/2005 Coll.), and the Act on Nature and Landscape Protection (Act 543/1994 Coll.).

According to the Act on Land-use Planning and Building Order [63], protected and protection areas are defined in land use planning documents. No special references are made to urban natural protected areas.

The Environmental Impact Assessment Act [64] regulates the professional and public assessment procedure for anticipated environmental impacts, the competence of the authorities responsible for this impact assessment, and the rights and obligations of the participants in the assessment process. If it is found that a strategic document has a negative impact on the system of protected areas, the only accepted reasons for it to be approved are related to human health, public order, national security, or other urgent reasons for which it has received the approval of the European Commission. No special references are made to urban natural protected areas.

The Act on Nature and Landscape Protection [65] regulates the territorial conservation of nature and landscapes at five levels of protection, the extent of its limitation increasing with the increase in the level of protection. In the legislative act, no references are made to urban areas and the spatial relationship between them and natural protected areas.

Romania

The basis of the planning system in Romania is Law 350/2001 on Territorial Development and Urban Planning [66]. According to Annex 1 of this law, the planning system includes three types of spatial planning documentation and three types of urban planning documentation. Territorial planning documents have a strategic character and are: national territorial planning plans (PATN), county territorial planning plans (PATJ), and zonal territorial planning plans (PATZ). PATNs include six sections, for six distinct levels—Section 3 is dedicated to protected areas, both natural and built, and constitutes the framework for identifying heritage areas, vulnerabilities, conservation priorities, and necessary financial support. PATJs and PATZs correlate with PATNs, and must indicate in their written part and graphic form the natural factors affected by human actions and unprotected or insufficiently protected natural heritage.

Urban plans have a regulatory character (normative character) and are divided into the following three categories: General Urban Plans (PUG), Zonal Urban Plans (PUZ), and Detailed Urban Plans (PUD). PUGs are drawn up at the level of an administrative territorial unit (municipality, city, or commune) and some of their objectives are to delimit areas with a special protection regime legally enforced and set urban growth boundaries. PUGs are most common land use policy instruments for managing urban land change, with a direct-impact on the land use conversion in cities [67]. PUZs are drawn up at the level of an area in an administrative territorial unit, for different purposes, among which is the protection of historical monuments. The development of a PUZ is mandatory for protected built-up areas. On the other hand, Law 350/2001 does not include special mentions regarding the preparation of PUZ-type documentation for green areas that have other characteristics than urban ones or for urban natural protected areas and/or their adjacent areas. PUDs are drawn up only for a single piece of land and their regulatory character is much more limited compared to PUZs, through PUDs not being able to change, for example, the regulated height regime for the respective land. For a more detailed presentation of these institutional actors and legal provisions, see Popescu and Petrisor, 2021 [68].

The written part of town planning plans, Local Town Planning Regulation (RLU), is based on the provisions of the General Town Planning Regulation (RGU), adopted by Government Decision no. 525/1996 [69]. RGU contains an article dedicated to areas with landscape value and natural protected areas (Article 8), from Section 1—"Rules regarding the preservation of the integrity of environment and protection of natural and built heritage", which mentions the prohibition of authorizing constructions/developments that depreciate the value of a landscape and the approval procedure for the execution of construction in natural areas protected by national interest. RGU does not contain mentions dedicated to natural protected areas spatially related to urban fabric, so has no correlation with the unique and particular urban morpho-typology of a territory.

Currently, there is a "Legislative proposal regarding the regime of urban protected natural areas and preservation of urban biodiversity" in preparation. This was adopted by the Romanian Senate and sent for debate in the Chamber of Deputies. According to the legislative proposal, natural urban protected areas and green corridors must be highlighted in urban planning and land development plans. This step could adapt the planning process toward the conservation needs of these areas (see also Popescu et al., 2022 [62]).

Regarding natural areas, especially those in cities, the most important legislative acts are Law no. 137 of 29 December 1995—the Environmental Protection Law [70], Emergency ordinance no. 195/2005 on environmental protection [71], and Law no. 24 of 15 January 2007, regarding the regulation and administration of green spaces in urban areas [72].

Law no. 137 of 29 December 1995—the Environmental Protection Law—regulates the regimes of natural protected areas and the activities subject to environmental impact assessment procedures within them.

Emergency ordinance no. 195/2005 on environmental protection prohibits changes in land use within and near natural protected areas, but "the vicinity of the natural protected area" is defined in vague terms: "the area outside the limit of a natural protected area, from which an impact on the natural protected area can be generated by a project or an activity, depending on their nature, size and/or location".

Law no. 24 of 15 January 2007, regarding the regulation and administration of green spaces in urban areas, does not refer to natural protected areas.

3.2. Comparison of the Existing Models and the Approach to the Problem of Conservation of Natural Protected Areas in Urban Areas in the Nordic Countries and in the Eastern European Countries

Figure 3 presents an outline of the Nordic countries, which is discussed in detail below.

Nordic countries do not attach particular importance to territorial planning carried out at the national level—Denmark, Finland, and Sweden do not have planning instruments at the national level. Additionally, the national plans for Iceland and Norway are of a strategic nature and presented in written form. For all Nordic countries, the actual planning starts at the regional or municipal level, through plans that regulate land use. Denmark is the only one of the countries for which no regional plans are provided, with planning starting at the municipal level. For the other countries, plans made at the mezzo-territorial level are drawn up at the regional and municipal levels. At the micro-territorial level, detailed plans are provided—for Finland and Iceland, zonal plans—for Norway or both types of plans—for Denmark and Sweden.

From a legislative point of view, each country has a law responsible for its planning system—The Planning and Building Act (in the case of Iceland, Norway, and Sweden), The Land Use and Building Act (Finland), and The Planning Act (Denmark). Regarding the mentions of natural protected areas contained in these acts, the Finnish law stands out in a chapter dedicated to national urban parks and the provision of a "maintenance and use scheme" for each of them. A basic law for environmental protection is also provided in each country. The Nature Conservation Act of Iceland mentions the obligation to create a register with areas that are of interest from the point of view of natural heritage. The Environmental Code of Sweden specifies the possibility of suspending the regime for the protection of coastal areas through a detailed regulatory plan during its validity period. Another interesting approach was identified in the Act on Nature Areas in Oslo and Nearby Municipalities. Here, attention is paid to the establishment of protected areas with special qualities for outdoor activities, especially with the aim of reducing the pressure on natural protected areas with special conservation value.

Among the strategic documents, even if they have no legal value, the following two mentions stood out: The Finnish Biodiversity Action Plan 2013–2020, which highlights the importance of developing planning methods in urban environments, in accordance with the conservation of biological diversity, and the Planning Green Structures in Cities and Towns document adopted for Norway, which draws attention to the planning of the system of green spaces based on an overview of the biotopes.

The planning systems adopted by eastern European countries are much more branched compared to those identified in the Nordic countries, at least from the point of view of the hierarchy of planning instruments. All the analyzed countries have adopted plans at the national level. Their analysis brought out the following mentions: The National Development Policy of the Czech Republic highlights the importance of locating development projects that endanger the natural heritage, The National Development Plan of Poland gives special importance to the creation of ecological networks, and The National Territory Development Plan of Romania contains a section dedicated to protected areas, both natural and built-up.

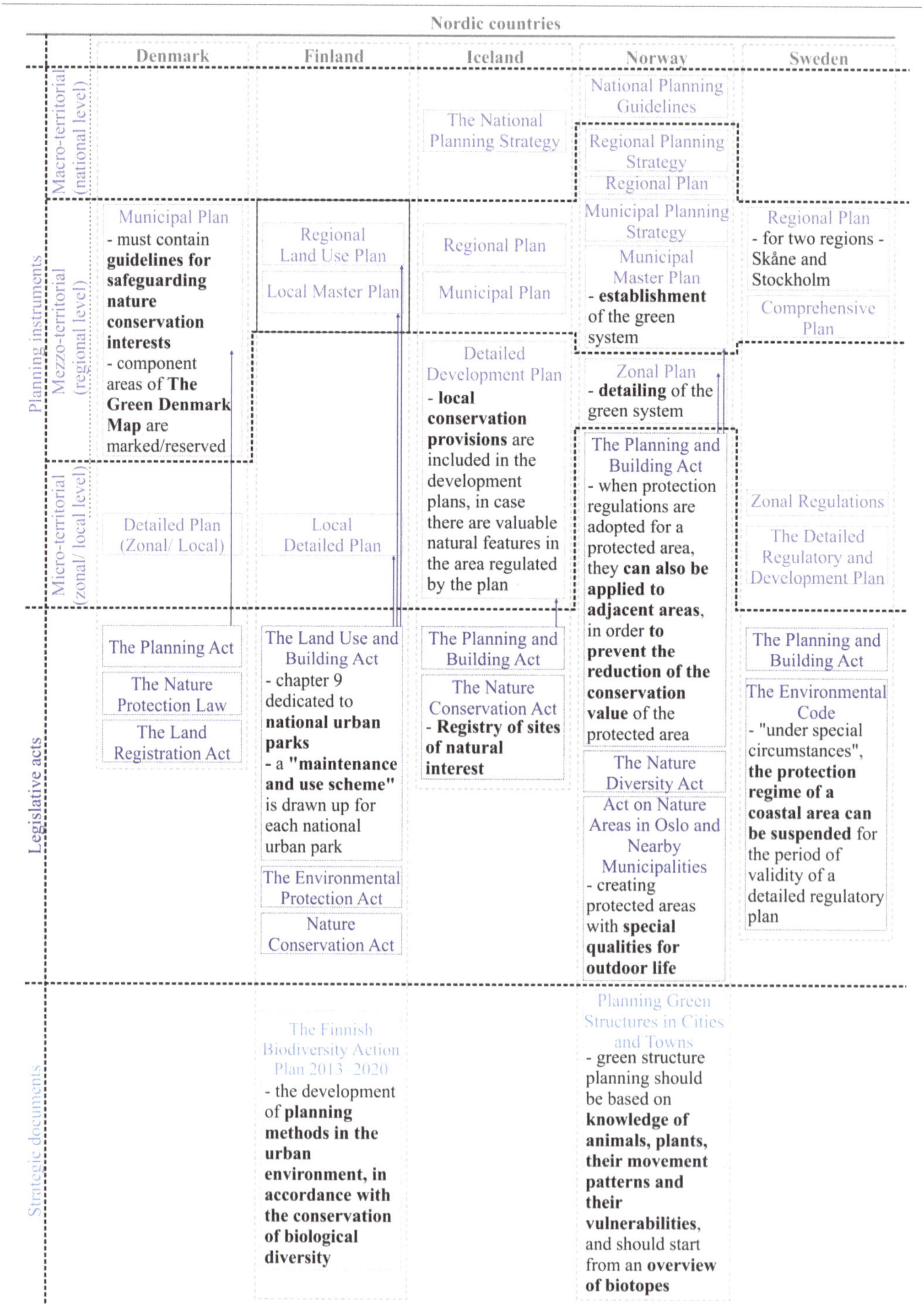

		Nordic countries				
		Denmark	Finland	Iceland	Norway	Sweden
Planning instruments	Macro-territorial (national level)			The National Planning Strategy	National Planning Guidelines / Regional Planning Strategy / Regional Plan	
	Mezzo-territorial (regional level)	Municipal Plan - must contain guidelines for safeguarding nature conservation interests - component areas of **The Green Denmark Map** are marked/reserved	Regional Land Use Plan / Local Master Plan	Regional Plan / Municipal Plan / Detailed Development Plan - **local conservation provisions** are included in the development plans, in case there are valuable natural features in the area regulated by the plan	Municipal Planning Strategy / Municipal Master Plan - **establishment** of the green system / Zonal Plan - **detailing** of the green system	Regional Plan - for two regions - Skåne and Stockholm / Comprehensive Plan
	Micro-territorial (zonal/local level)	Detailed Plan (Zonal/Local)	Local Detailed Plan		The Planning and Building Act - when protection regulations are adopted for a protected area, they **can also be applied to adjacent areas**, in order **to prevent the reduction of the conservation value** of the protected area	Zonal Regulations / The Detailed Regulatory and Development Plan
Legislative acts		The Planning Act / The Nature Protection Law / The Land Registration Act	The Land Use and Building Act - chapter 9 dedicated to **national urban parks** - a "maintenance and use scheme" is drawn up for each national urban park / The Environmental Protection Act / Nature Conservation Act	The Planning and Building Act / The Nature Conservation Act - **Registry of sites of natural interest**	The Nature Diversity Act / Act on Nature Areas in Oslo and Nearby Municipalities - creating protected areas with **special qualities for outdoor life**	The Planning and Building Act / The Environmental Code - "under special circumstances", **the protection regime of a coastal area can be suspended** for the period of validity of a detailed regulatory plan
Strategic documents			The Finnish Biodiversity Action Plan 2013–2020 - the development of **planning methods in the urban environment**, in accordance with the conservation of biological diversity		Planning Green Structures in Cities and Towns - green structure planning should be based on **knowledge of animals, plants, their movement patterns and their vulnerabilities**, and should start from an **overview of biotopes**	

Figure 3. Territorial planning instruments, legislative and strategic acts characteristic to the Nordic countries.

The next analyzed level is represented by urban planning plans at the mezzo-territorial level, which are considered in this case to be plans drawn up at regional, county, or municipal levels. Hungary has adopted two plans at the regional level—one for Budapest and its adjacent urban agglomeration and one for the tourist area near Lake Balaton, but these do not make special references to natural protected areas and their spatial relationship with the urban fabric. In parallel, a strategic concept was also adopted for these two areas. The one adopted for the Budapest area recommends the creation of new green spaces to reduce the pressure on valuable natural areas from a conservative point of view. Additionally, in the case of Hungary, the content of these plans made at a county level highlights the importance of establishing regulations for controlling development and protecting natural and built heritage.

At the next planning level, the micro-territorial level, the types of plans adopted at a zonal/detail level were analyzed. All the countries provide for the drafting of such plans, but no special mentions have been identified regarding natural protected areas and the planning of areas in their vicinity.

Regarding their legislative acts, each country has a basic law for its territorial planning system, with different names: The Spatial Development Act (Bulgaria), The Building Act (Czech Republic), the Act on Regional Development and Spatial Planning (Hungary), The Spatial Planning and Development Act (Poland), The Act on Land Use Planning and Building Order (Slovakia), and the Law on Land Use and Urban Planning (Romania). In Bulgaria, this act mentions the importance of public green spaces through a recommendation that they should be treated similarly to protected areas. It is very important to balance conservation requirements and the public use of an area [73]. Poland provides that the ecological network must be included in territorial and county plans. In Slovakia, the law mentions that protected areas and their areas of protection are defined in territorial planning documents.

Compared to the environmental legislation specific to the Nordic countries, eastern European countries seem to be more specific regarding the natural elements with conservative value. Figure 4 presents an outline of the eastern European countries, which is discussed in detail below.

Bulgaria adopted a law for the protection of biodiversity, which includes a chapter dedicated to the national ecological network. The Czech Republic has a law for the protection of nature and landscapes, which details the regulations and zoning for each type of natural protected area. In Hungary, it is recommended to provide a buffer zone for green spaces through its law on the development and protection of the built environment. In Poland, there is a law stating that all planning documents should contain an environmental impact assessment. Slovakia has paid special attention to planning a Territorial System of Ecological Stability (TSES), divided into three levels depending on the scale: supraterritorial, regional, and local, through a Government decision from 1991. At the same time, the nature protection laws and landscapes provide for five levels of protection, with the strictness of the regulations increasing with the level of protection.

In Romania, the General Urban Planning Regulation forbids the authorization of constructions that depreciate the value of a landscape. Despite this, the legislation in force is uncertain and too little focused on all the components of a landscape, allowing for changing the admissible function and indications of land use under certain conditions. It should be noted that these components of a landscape are important in constituting a planning and regulatory tool in the study, research, analysis, diagnosis, and prognosis of the resilience of human settlements, and with a significant role in increasing quality of life [74]. The law on environmental protection regulates the regimes of natural protected areas and the activities subject to environmental impact assessment procedures within them, but it does not clearly intervene with regard to green spaces with environmental potential but without historical and/or environmental heritage value (see also Stan, 2022 [75]). The Emergency Ordinance on Environmental Protection prohibits a change in land use near natural protected areas. In Romania, the adoption of a law dedicated to natural protected areas in urban environments is currently under debate.

Figure 4. Territorial planning instruments, legislative and strategic acts characteristic to Eastern European countries.

		Bulgaria	Czechia	Hungary	Poland	Slovakia	Romania
Planning instruments	Macro-territorial (national level)	The National Concept for Spatial Development	The National Development Policy - **Protecting natural values** and **locating development projects** that have the risk of significantly **affecting the character of the landscape**	The National Spatial Plan - legally **binding** document, renewed every seven years; Spatial Plans for Special Regions - Budapest and the adjacent urban agglomeration, the tourist area around Lake Balaton	The National Development Plan - the importance of **establishing ecological networks** and **conservation planning** in protected areas	The Slovak Spatial Development Perspective	The National Territory Development Plan - section 3 is dedicated to protected areas, both natural and built
	Mezzo-territorial (regional level)	The Regional Scheme for Spatial Development of a Region of Level 2; The Regional Scheme for Spatial Development of a District; The Municipality Concept for Spatial Development	Development Principles; Local Territorial Plan	Spatial Plan for Counties - establishing regulations for **the control of development** and for the **protection of natural and built heritage**	The Spatial Development Plan at The Voivodeship Level; Landscape Audit	Regional Land Use Plan; Local Land-Use Plan; Local Zoning Plan; The Act on Land-use Planning and Building Order - within the land use planning documents, **the protected and protection areas** are defined	The Development Plan of the County Territory; The Zonal Territory Development Plan; General Urban Plan; Zonal Urban Plan; Detailed Urban Plan; Law on Land Use and Urban Planning
	Micro-territorial (zonal, local level)	Detailed Plan	Regulatory Plan; Planning Studies	Settlement Structural Plan; Act on the National Spatial Plan - protected natural areas and protection zones are "**regional and county priority areas**" - the natural ecological network must be included **in regional and county spatial plans**.	The Local Development Revitalization Plan	The Environmental Impact Assessment Act; The Act on Nature and Landscape Protection - territorial conservation of nature and landscape on five levels of protection	General Urban Planning Regulation - prohibition of the authorization of constructions/developments that depreciate the value of the landscape
Legislative acts		The Spatial Development Act - Existing **green areas** which are public ownership shall be **developed and preserved as protected areas**; The Regional Development Act; The Environmental Protection Act; Protected Areas Act; Biodiversity Act - chapter dedicated to the national ecological network	The Building Act; Act on the Environment; Act on Nature and Landscape Protection - detailing the regulations and zoning for each type of natural protected area	Act on Regional Development and Spatial Planning; Act on the Development and Protection of the Built Environment - recommends the **provision of buffer zones for green spaces**	The Spatial Planning and Development Act; Act on the provision of information on the environment and its protection, public participation in environmental protection and environmental impact assessments - the system of **environmental impact assessments (EIA) - mandatory for all planning documents**	Decision of the Government no. 394 of 23 July 1991 - **Territorial System of Ecological Stability (TSES)** - supraterritorial, regional and local territorial systems - **a landscape-ecological basis** for the development of municipal plans - **preserving** the biodiversity of the landscape, **restoring** the connectivity of the natural landscape and **maintaining** or even **improving** ecosystem services	Environmental Protection Law - regulates the regime of protected natural areas and the activities that are subject to the environmental impact assessment procedure within them; Emergency Ordinance on Environmental Protection - prohibits the change of land use inside and in the vicinity of the protected natural area; Law regarding the regulation and administration of green spaces in the urban areas
Strategic documents				Concept of Long-Term Development of the City of Budapest 2030 - establishment of new green areas (to reduce the pressure on nature conservation areas; Balaton Territorial Development Concept 2014–2030			*Legislative proposal regarding the regime of urban protected natural areas and urban biodiversity conservation

Overall, although the anticipated results showed major differences in terms of the attention given to natural heritage, it can be seen that both Nordic and eastern European countries have adopted important and, at the same time, very varied provisions, both within territorial planning instruments and through legislative acts, at different levels of approach (Figure 5).

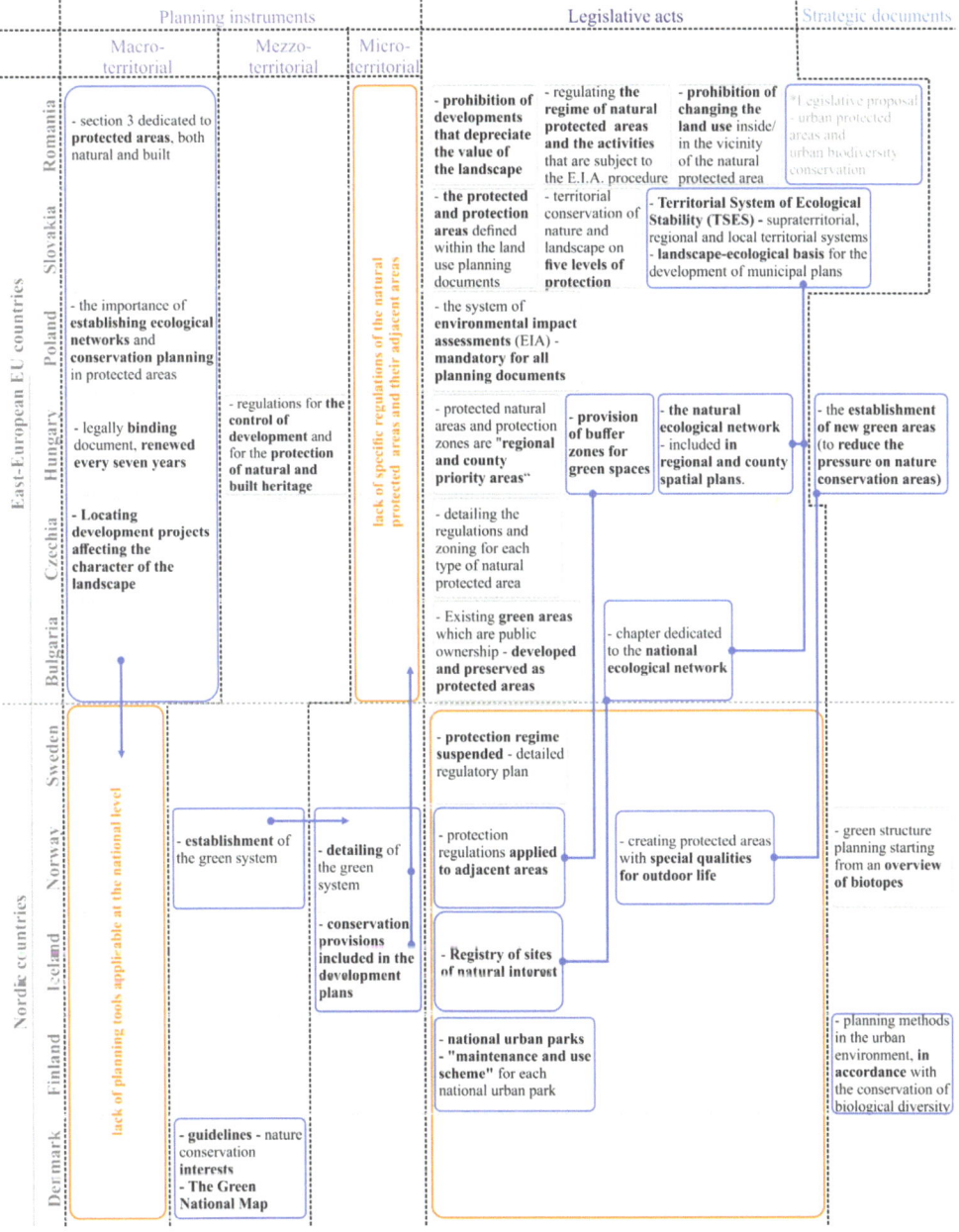

Figure 5. The parallel between the territorial planning instruments and the legislative and strategic acts characteristic of the Nordic and eastern European countries.

Very few dedicated provisions regarding natural protected areas within urban environments have been identified at the spatial level and with respect to urban morpho-typological structures, i.e., the regulation of national urban parks in Finland and the legislative act under debate for urban natural protected areas in Romania. It should be noted that these two areas are totally different from the viewpoints of the local natural and built environments or cultural particularity and specificity, and are in different situations of historical evolution, subject to the acute pressures of urban development, and, especially, are under different territorial statuses from a strategic viewpoint. In the case of territories with such a rich heritage, the need to create large natural protected areas is felt even more. Considering major natural elements, such as rivers, their protection requires the careful management of a larger area (at the scale of a hydrographic basin), which, in many cases, exceeds the limits of the natural protected area [76].

4. Discussion

4.1. The Significance of Results

The previous parallel made between the main conclusions extracted from the analysis of legislative documents and planning instruments, summarized in Figure 5, reveals that improvements can be made for the model of Nordic countries in terms of their approach to natural heritage and especially that directly related to their urban fabric, as well as in the case of eastern European countries.

This can only be achieved with the prospective, efficient, and, above all, ethical adaptation of possible models, only with a preservation of the specificity and particularity of local morpho-typologies and natural anthropic and cultural landscapes, as a priority in the strategy and planning processes for different typologies and concerted and correlated multiscale interventions, respectively, at different scales: urban, spatial, peripheral, rural–urban, territorial, national, cross-border, regional, or European.

The main differences in terms of the main aspects considered by the two groups of countries concern urban morpho-typologies and different historical, social, and cultural evolutions (Figure 5). These differences are highlighted in the case of their urban and territorial planning and strategy instruments. The Nordic countries do not attach importance to planning instruments drawn at the national level, with their actual planning starting from the mezzo-territorial level. Eastern European countries make important clarifications through national plans—the planning of protected areas, national ecological networks, or the locations of projects that depreciate the landscape character. Thus, the adoption of plans at the national/macro-territorial level has the potential to support these practices nationally.

The plans adopted at the mezzo-territorial level in Nordic countries do not make specific references to natural heritage. They were identified, however, in the cases of Denmark and Norway, and references are made to the system of green spaces. In eastern European countries, the only specific provision was identified in Hungary, which controls the development and protection of natural and built heritage. Planning at the mezzo-territorial level can be considerably improved at the levels of the two groups of countries by adopting specific regulations regarding the system of natural protected areas, including their protection/adjacent areas.

At the zonal/detailed territorial level, the plans drawn up for eastern European countries, perhaps due to the historical and political context of the communist regime that changed quite late in the 90s, do not make specific statements about natural heritage. Regarding the Nordic countries, Iceland provides conservation provisions included in its development plans, and in Norway, there is a detailing of the green system. Similar to the situation identified in the case of planning at the mezzo-territorial level, at this scale, the need to implement specific regulations, which are especially important for the planning of areas adjacent to natural protected areas, is noted.

From a legislative point of view, after the 90s, eastern European countries adopted more specific provisions for natural elements of conservative interest, an aspect highlighted in the diagram in Figure 5. However, similarities were also identified in the ways of treating certain problems: Norway provides protection regulations applied to areas near natural protected ones, while Hungary mentions the provision of some buffer zones, but for green spaces; Iceland has a national register of sites of conservation importance, while Bulgaria, Hungary, and Slovakia pay attention to the national ecological network. Eastern European countries bring important additional specifications regarding detailing the regulations and zoning for each type of natural protected area, the system of environmental impact assessments (EIA), protected and protection areas defined within land use planning documents, the levels of protection, and the prohibition of changing land use inside and near natural protected areas. From these points of view, important additions could be made to the environmental legislation of Nordic countries.

Norway signals the importance of creating protected areas from the viewpoint of the quality of outdoor activities through a legislative act. At the same time, Poland emphasizes the creation of new green spaces to reduce the pressure on natural protected areas, but through a strategic document without legal value. We note, therefore, the importance of the legal value of the acts through which regulations regarding natural protected areas are proposed, considering their value from a conservative point of view. However, a common element in all the countries was identified in terms of accessibility for people with disabilities and the provision of necessary visiting infrastructure. This element consists of emphasizing the importance of improving on and investing in accessible and inclusive tourism [77].

Additionally, taking into account the complex spatial conflict between natural protected areas and the urban fabric, a law dedicated to the regulation of natural protected areas in the urban environment would support the elimination of/reduction in pressure on these natural areas. Such a law was identified only in Romania, being currently, however, only at the level of a legislative proposal.

The approaches to protecting heritage and the natural landscape are very different from country to country, although similar aspects have been identified between the legislation of Nordic countries and that of eastern European countries. The main purpose of our study was the analysis of the legislation and planning instruments aimed at natural protected areas in inner cities. In the end, it was observed that these focused more on the generic, multiscale treatment of natural heritage, at different scales of approach. The reason for directing the analysis in this sense was the lack of specifications with reference to natural protected areas in urban environments. Very few mentions have been identified that address this subject, with the heritage and natural landscape in urban environments being subject to different urban-anthropic investment pressures [78], which are totally different in Nordic countries compared to Eastern countries, and with differences due to historical, social, cultural, or geo-strategic contexts.

From a legislative viewpoint, Finland is the only country that has provisions regarding the type of natural protected area in an urban environment—chapter nine of The Land Use and Building Act is dedicated to regulating national urban parks, providing a "maintenance and use scheme" for each national urban park in its content.

So far, no country has adopted a law dedicated to natural protected areas in urban environments, but in Romania, there is a legislative proposal for urban natural protected areas and urban biodiversity conservation, which was adopted by the Romanian Senate and sent for debate in The Chamber of Deputies. Through this legislative initiative, Romania's interest in natural protected areas in urban environments is noted, an interest that could be due to the large number/large areas belonging to these types of valuable areas. Approximately 59% of the cities in Romania overlap with at least one natural protected area and approximately 5% of the built areas of the urban settlements in Romania overlap with at least one natural protected area [79].

On the other hand, the territorial planning instruments provided by Romanian legislation may undergo advantageous changes in terms of a suitable approach possible for the natural heritage in urban areas. Urban territorial plans address natural protected areas to an insignificant extent, or even completely ignore them.

Additionally, considering the special natural heritage of Romania—the only country in Europe that has five biogeographical regions (continental, alpine, pannonic, pontic, and steppe)—considerable steps have been taken regarding areas of considerable conservation value, such as the adoption of the Framework Convention for the Protection and Sustainable Development of the Carpathians and the EU Strategy for the Danube Region, a major natural element that plays the role of a structuring axis at the European level.

The principles followed on the three analyzed levels—planning, legislation, and strategic approaches—were grouped according to the territorial scale of application in Figure 6.

Figure 6. The principles identified for planning, legislation, and strategic documents analyzed, grouped according to the territorial scale to which they are applicable.

The figure shows that, although planning instruments contain fewer and fewer regulations regarding natural protected areas and their adjacent areas as the territorial scale decreases from macro to micro, the situation is totally opposite from the viewpoint of laws and strategic documents. The provisions mentioned in the legislative and strategic acts are, in general, applicable on a micro scale, due to their specificity.

Planning instruments do not have generally valid models, nor do they generally provide for regulations at different scales of specific and particular approaches for areas adjacent to natural protected areas.

Thus, heritage and protected landscapes must be analyzed sensitively, treated as living organisms with their own changing internal metabolism and metamorphosis determining the urban metabolism [80], protected and conserved in an integrated correlation.

4.2. The Inner Validation of Results

The reason for choosing the two groups of countries was, as specified in the methodology, their different evolution from urban, political, and demographic viewpoints, starting from the premise that these different evolutions generated the creation of different models in terms of urban planning and its related specific legislation.

Although this analysis initially aimed to identify the way in which natural protected areas and their adjacent areas are approached in the documents mentioned above (legislative and related to urban planning), it was found that there are very few provisions strictly related to these areas, so all types of natural protected areas were included in the analysis, taking into account their possible treatment as any other natural protected area that does not necessarily have a direct spatial connection with an urban environment.

In the end, the hypothesis from the beginning was confirmed, with the analysis revealing two models with very different approaches, which created the possibility of making a comparison yielding later recommendations for all the countries, setting the grounds for an exhaustive urban planning approach for areas adjacent to natural protected areas.

4.3. The External Validation of Results

The natural heritage present in cities is not a new topic of research. There is a wide range of publications on this topic. However, as mentioned at the beginning of the paper, very few of them provide the guidelines needed in urban planning and territorial development to amplify the benefits generated by the presence of nature in a city, i.e., ecosystem services.

Despite the evolving role of cities as both consumers and producers of ecosystem services, challenges remain. These include the need for better methods and indicators for capturing urban ecosystem heterogeneity, a limited understanding of the link between urban ecosystem services and biodiversity, uncertainty regarding data transferability, and the lack of an analysis integrating people's preferences and values, particularly in assessing cultural ecosystem services [81]. Although there is still no tool for quantifying and locating ecosystem services, their presence is undeniable, even more so for natural protected areas, taking into account the increased diversity of species, including native ones, the value of landscapes, or any other resources that were the basis for establishing a protection regime for that area.

Marques et al. [23] concluded in their work, published in 2022, that urban planning needs to apply research/analysis methods related to urbanism, respectively, biophysical, socio-cultural, and economic methods, which can be combined and integrated to be able to analyze the urban–natural relationship and the benefits of ecosystem services from quantitative and qualitative points of view. More studies are needed to map the benefits of nature in order to be able to formulate appropriate solutions.

Although 13 of the Sustainable Development Goals adopted by the Aichi 2030 Agenda are relevant for the management of natural protected areas [82], none of them make specific references to those in urban environments.

No articles have been identified that dealt strictly with the subject of urban planning or legislative provisions regarding natural protected areas located within urban environments, even less with regard to a specific country. Although multiple sources of information are available for performing a comparative analysis between Nordic and eastern European countries in terms of environmental and urban planning legislation and its related instruments, until now, this analysis had not been carried out. Thus, our study is a premiere. Its objective was to fill in this gap by highlighting the specific challenges encountered by each country, but also the good practices and possible directions for improvement in each individual country. Therefore, this research can contribute to a deeper understanding of the specific context of each country and important lessons can be drawn for developing and improving the legislation and urban planning tools throughout the entire region and beyond, with an emphasis on the appropriate approach to natural protected areas.

There are many publications addressing the advantages of the presence of these areas in cities, but the information related to their capitalization from the viewpoint of planning for their adjacent areas is currently very limited.

4.4. The Importance of Results

The accentuated differences between the two groups of countries made it possible to phrase more recommendations due to the different approaches in this field. Considering that, currently, there are no in-depth studies on urban planning near natural protected areas, our results can constitute a starting point for developing an urban planning tool applicable for these areas. This tool could support the planning process in a bidirectional way, i.e., by approaching the natural protected area as a component of the urban fabric or by approaching urban natural protected areas as part of the network of natural protected areas.

4.5. Summary of the Study Limitations and Directions for Overcoming Them in the Future Research

This research was constrained by a series of practical limitations. The most obvious of these is the current very limited knowledge of the analyzed subject—that of the urban planning for areas adjacent to natural protected areas in urban environments. Another limitation was the very limited approach to this issue within the legislative and urban planning acts.

An elimination of these two limitations or their reduction would require the development of studies and the discovery of methods by which the ecosystem services generated by natural protected areas in urban environments could be quantified and located spatially. Subsequently, the creation of an urban planning tool could be adaptable according to the particularities of a respective area and could even be part of urban planning documentation.

5. Conclusions

The freedom to choose the tools for managing urban biodiversity conservation and planning the quasi-natural system in a city offers the possibility of developing plans adapted to the concerned territory/area [83]. However, upon comparing the two analyzed models—the Nordic and eastern European ones—it was found that the two have adopted different instruments in terms of the territorial scale of their application, but also that they both require improvements, from the viewpoints of both territorial planning and legislation.

As it emerged from the results of the analysis carried out individually for each of the 11 countries, and later in parallel (individually, but also based on the two groups), there was a major difference in the attention given to natural heritage. Although all the countries have adopted important and, at the same time, remarkably diverse provisions, only a few provisions dedicated to natural protected areas in urban environments have been identified (the existing provisions in the legislative acts of Finland and provisions contained in a draft law of Romania).

For this purpose, a series of recommendations were made regarding the two levels of analysis and the strategic approaches being approached separately from the urban planning instruments, due to a lower legal value. Certain recommendations target both groups of countries, such as the need to adopt specific regulations regarding the system of protected natural areas and their protection/adjacent areas in planning at the mezzo-territorial level and at the zonal and detailed levels. Others target only one of the two groups of countries or only one country in particular, such as the recommendation for the Nordic countries to bring specific additions to the regulation of each type of natural protected area or the need for Romania to draft provisions for approaching natural heritage in its urban planning documents.

Urban planning, together with its related legislation, must become a well-clarified process, in the sense of supplementing legislative acts with regulations as specific as possible for local heritage and, at the same time, appropriate to the particularities and context that define it.

The need for completion was identified at the two levels of analysis—urban planning legislation, in all the sensitive aspects and typologies of heritage and natural, anthropic, and cultural landscapes, in close development and connection with the historical evolution of unique and particular morpho-typological germs—this being the main particular characteristic at the EU level of all the countries, especially Romania.

Author Contributions: Conceptualization, C.C.; Methodology, A.-I.G., C.C. and A.-I.P.; Validation, A.-I.G., C.C. and A.-I.P.; Formal analysis, A.-I.G. and C.C.; Investigation, A.-I.G., C.C. and A.-I.P.; Resources, A.-I.P.; Data curation, A.-I.G. and C.C.; Writing—original draft, A.-I.G., C.C. and A.-I.P.; Writing—review & editing, A.-I.G., C.C. and A.-I.P.; Visualization, A.-I.G., C.C. and A.-I.P.; Supervision, A.-I.G. and C.C.; Project administration, A.-I.G.; Funding acquisition, A.-I.P. All authors have read and agreed to the published version of the manuscript.

Funding: This research received no external funding.

Institutional Review Board Statement: Not applicable.

Informed Consent Statement: Not applicable.

Data Availability Statement: Not applicable.

Conflicts of Interest: The authors declare no conflict of interest.

References

1. Iojă, C.; Breuste, J. Urban Protected Areas and Urban Biodiversity. In *Making Green Cities*; Breuste, J., Artmann, M., Iojă, C., Qureshi, S., Eds.; Springer International Publishing: Cham, Switzerland, 2020.
2. Petrișor, A.-I.; Sîrodoev, I.; Ianoș, I. Trends in the national and regional transitional dynamics of land cover and use changes in Romania. *Remote Sens.* **2020**, *12*, 230. [CrossRef]
3. Concepción, E.D. Urban sprawl into Natura 2000 network over Europe. *Conserv. Biol.* **2021**, *35*, 1063. [CrossRef] [PubMed]
4. Crăciun, C.; Gârjoabă, A.-I. Methods of Approaching Natural Protected Areas from the Towns of Europe. *Rev. Urbanism. Arhit. Construcții* **2022**, *13*, 11–28.
5. Popa, A.M.; Onose, D.A.; Sandric, I.C.; Dosiadis, E.A.; Petropoulos, G.P.; Gavrilidis, A.A.; Faka, A. Using GEOBIA and Vegetation Indices to Assess Small Urban Green Areas in Two Climatic Regions. *Remote Sens.* **2022**, *14*, 4888. [CrossRef]
6. Ivanova, I.; Cook, C. Public and private protected areas can work together to facilitate the long-term persistence of mammals. *Environ. Conserv.* **2022**, *50*, 1–9. [CrossRef]
7. Mouchet, M.A.; Rega, C.; Lasseur, R.; Georges, D.; Paracchini, M.-L.; Renaud, J.; Stürck, J.; Schulp, C.J.E.; Verburg, P.H.; Verkerk, P.J.; et al. Ecosystem service supply by European landscapes under alternative land-use and environmental policies. *Int. J. Biodivers. Sci. Ecosyst. Serv. Manag.* **2017**, *13*, 342–354. [CrossRef]
8. Kubacka, M.; Zywica, P.; Subiros, J.V.; Brodka, S.; Macias, A. How do the surrounding areas of national parks work in the context of landscape fragmentation? A case study of 159 protected areas selected in 11 EU countries. *Land Use Policy* **2022**, *113*, 105910. [CrossRef]
9. Santiago-Ramos, J.; Feria-Toribio, J.M. Assessing the effectiveness of protected areas against habitat fragmentation and loss: A long-term multi-scalar analysis in a mediterranean region. *J. Nat. Conserv.* **2021**, *64*, 126072. [CrossRef]
10. Lausche, B. *Guidelines for Protected Areas Legislation*; IUCN: Gland, Switzerland, 2011.
11. European Environment Agency. *Protected Areas in Europe—An Overview*; Publications Office of the European Union: Luxembourg, Luxembourg, 2012.
12. Yli-Pelkonen, V.; Niemela, J. Linking ecological and social systems in cities: Urban planning in Finland as a case. *Biodivers. Conservat.* **2005**, *14*, 1947–1967. [CrossRef]

13. Fedreheim, G.E.; Blanco, E. Co-management of protected areas to alleviate conservation conflicts: Experiences in Norway. *Int. J. Commons* **2017**, *11*, 756. [CrossRef]
14. Crăciun, C. Pluridisciplinarity, Interdisciplinarity and Transdisciplinarity—Methods of Researching the Metabolism of the Urban Landscape. In *Planning and Designing Sustainable and Resilient Landscapes*; Springer: Dordrecht, Holland, 2014; pp. 3–14.
15. Barton, M.A. Nature-Based Solutions in Urban Contexts: A Case Study of Malmö. Master's Thesis, Lund University, Lund, Sweden, 2016.
16. Petrișor, A.-I.; Mierzejewska, L.; Mitrea, A.; Drachal, K.; Tache, A.V. Dynamics of Open Green Areas in Polish and Romanian Cities During 2006–2018: Insights for Spatial Planners. *Remote Sens.* **2021**, *13*, 4041. [CrossRef]
17. Panagopoulos, T.; Jankovska, I.; Boștenaru Dan, M. Urban green infrastructure: The role of urban agriculture in city resilience. *Urbanism. Arhit. Construcții* **2018**, *9*, 58.
18. Petroni, M.L.; Siqueira-Gay, J.; Gallardo, A.L.C.F. Understanding land use change impacts on ecosystem services within urban protected areas. *Landsc. Urban Plan.* **2022**, *223*, 104404. [CrossRef]
19. Boștenaru, M.; Crăciun, C. Creativity and Spatial Urban and Landscape Perception in Architectural Imagination. In Proceedings of the 9th LUMEN International Scientific Conference Communicative Action & Transdisciplinarity in the Ethical Society, Lumen, Romania, 24–25 November 2017.
20. Blaszke, M.; Nowak, M.J. Objectives of spatial planning in selected Central and Eastern European countries. Analysis of selected case studies. *Ukr. Geogr. J.* **2022**, *4*, 57–62. [CrossRef]
21. Cocheci, R.M.; Ianoș, I.; Sârbu, C.N.; Sorensen, A.; Saghin, I.; Secăreanu, G. Assessing environmental fragility in a mining area for specific spatial planning purposes. *Morav. Geogr. Rep.* **2019**, *27*, 169–182.
22. Štrbac, S.; Veselinović, G.; Antić, N.; Stojadinović, S.; Stojić, N.; Živanović, N.; Kašanin-Grubin, M. Applicability of the PA-BAT+ in the evaluation of values of urban protected areas. *Front. Environ. Sci.* **2022**, *10*, 1–6. [CrossRef]
23. Marques, A.L.; Alvim, A.T.B.; Schröder, J. Ecosystem Services and Urban Planning: A Review of the Contribution of the Concept to Adaptation in Urban Areas. *Sustainability* **2022**, *14*, 2391. [CrossRef]
24. Danish Ministry of the Environment. *The Planning Act in Denmark Consolidated—Act No. 813 of 21 June 2007*; Danish Ministry of the Environment: Copenhagen, Denmark, 2007.
25. Danish Ministry of the Environment. *The Nature Protection Law—Act No. 951 of 3 June 2013*; Danish Ministry of the Environment: Copenhagen, Denmark, 2013.
26. Danish Ministry of Justice. *Land Registration Act—Act No. 622 of 15 September 1986*; Danish Ministry of Justice: Copenhagen, Denmark, 1986.
27. Finnish Ministry of the Environment. *The Land Use and Building Act No. 132 of 1999*; Finnish Ministry of the Environment: Helsinki, Finland, 1999.
28. Lidmo, J.; Bogason, A.; Turunen, E. *The Legal Framework and National Policies for Urban Greenery and Green Values in Urban Areas A Study of Legislation and Policy Documents in the Five Nordic Countries and Two European Outlooks*; Nordregio Report: Stockholm, Sweden, 2020.
29. Finnish Ministry of the Environment. *The Finnish Biodiversity Action Plan 2013–2020*; Finnish Ministry of the Environment: Helsinki, Finland, 2013.
30. Government of Iceland. *Planning and Building Act No. 73/1997, No. 135/1997 and No. 58/1999*; Government of Iceland: Reykjavík, Iceland, 1999.
31. Government of Iceland. *The Nature Conservation Act—Act No. 44 of 22 March 1999*; Government of Iceland: Reykjavík, Iceland, 22 March 1999.
32. Norwegian Ministry of Climate and Environment. *Act of 27 June 2008 No. 71 Relating to Planning and the Processing of Building Applications (the Planning and Building Act)*; Norwegian Ministry of Climate and Environment: Oslo, Norway, 27 June 2008.
33. Norwegian Ministry of Climate and Environment. *The Nature Diversity Act—Act of 19 June 2009 No.100 Relating to the Management of Biological, Geological and Landscape Diversity)*; Norwegian Ministry of Climate and Environment: Oslo, Norway, 19 June 2009.
34. Norwegian Ministry of Climate and Environment. *Act on Nature Areas in Oslo and Nearby Municipalities, 2009*; Norwegian Ministry of Climate and Environment: Oslo, Norway, 2009.
35. Norwegian Environment Agency. *Planning Green Structures in Cities and Towns, 2014*; Norwegian Environment Agency: Oslo Norway, 2014.
36. Wilson, J.; Primack, R. Designing Protected Areas. In *Conservation Biology in Sub-Saharan Africa*; Open Book Publishers: Cambridge, UK, 2019.
37. Swedish National Board of Housing, Building and Planning. *The Planning and Building Act, 2010*; Swedish National Board of Housing: Malmö, Sweden, 2010.
38. Swedish Ministry of the Environment. *The Environmental Code, 2000*; Swedish Ministry of the Environment: Malmö, Sweden, 2000.
39. National Centre for Regional Development—Bulgaria. *National Concept for Spatial Development 2013–2025, 2012*; National Centre for Regional Development—Bulgaria: Sofia, Bulgaria, 2012.
40. The Bulgarian Official Gazette. *Spatial Development Act No. 1/2001*; The Bulgarian Official Gazette: Sofia, Bulgaria, 2001.
41. The Bulgarian Official Gazette. *Regional Development Act No. 50/2008*; The Bulgarian Official Gazette: Sofia, Bulgaria, 2008.
42. The Bulgarian Official Gazette. *Environmental Protection Act No. 91/2002*; The Bulgarian Official Gazette: Sofia, Bulgaria, 2002.

43. The Bulgarian Official Gazette. *Protected Areas Act No. 133/1998*; The Bulgarian Official Gazette: Sofia, Bulgaria, 1998.
44. The Bulgarian Official Gazette. *Biodiversity Act No. 77/2002*; The Bulgarian Official Gazette: Sofia, Bulgaria, 2002.
45. The Czech Government. *Spatial Development Policy of the Czech Republic 2008, Approved by Government Resolution No. 929 of 20 July 2009*; The Czech Government: Prague, Czech Republic, 20 July 2009.
46. The Czech Parliament. *Act of 14th March 2006 on Town and Country Planning and Building Code (Building Act)*; The Czech Parliament: Prague, Czech Republic, 20 March 2006.
47. The Czech Government. *Act No. 17/1992 on the Environment*; The Czech Government: Prague, Czech Republic, 1992.
48. The Czech Government. *Act No. 114/1992 on Nature and Landscape Protection*; The Czech Government: Prague, Czech Republic, 1992.
49. The Urban Planning Department of Budapest City Hall. *The Long-Term Urban Development Concept of Budapest 2030 [Hosszú Távú Városfejlesztési Koncepció]*; The Urban Planning Department of Budapest City Hall: Budapest, Hungary, 2021.
50. Nowak, M.J.; Brelik, A.; Oleńczuk-Paszel, A.; Śpiewak-Szyjka, M.; Przedańska, J. Spatial Conflicts Concerning Wind Power Plants—A Case Study of Spatial Plans in Poland. *Energies* **2023**, *16*, 941. [CrossRef]
51. Blaszke, M.; Foryś, I.; Nowak, M.J.; Mickiewicz, B. Selected Characteristics of Municipalities as Determinants of Enactment in Municipal Spatial Plans for Renewable Energy Sources—The Case of Poland. *Energies* **2022**, *15*, 7274. [CrossRef]
52. Balaton Development Council. *Balaton Territorial Development Concept 2014—2030, 2013*; Balaton Development Council: Siófok Hungary, 2013.
53. The Parliament of Hungary. *Act XXI of 1996 on Regional Development and Regional Planning*; The Parliament of Hungary: Budapest, Hungary, 1996.
54. *The Parliament of Hungary, Act LXXVIII 1997 on the Development and Protection of the Built Environment*; The Parliament of Hungary: Budapest, Hungary, 1997.
55. The Parliament of Hungary. *Act XXVI 2003 on the National Spatial Plan*; The Parliament of Hungary: Budapest, Hungary, 2003.
56. Polish Ministry of Economic Development and Technology. *The Spatial Planning and Development Act of March 27, 2003*; Polish Ministry of Economic Development and Technology: Krakow, Poland, 2003.
57. The Government of Poland. *The National Spatial Development Concept 2030*; The Government of Poland: Krakow, Poland, 2010.
58. The Government of Poland. Act of 3 October 2008 on the provision of information on the environment and its protection, public participation in environmental protection and environmental impact assessments. *J. Laws* **2008**, *199*, 1227.
59. Ministry of Transport. *Construction and Regional Development of the Slovak Republic*; Ministry of Transport: Bratislava, Slovakia, 2011.
60. Izakovicova, Z.; Swiader, M. Building ecological networks in Slovakia and Poland. *Ekologia* **2017**, *36*, 303–322.
61. Kočická, E.; Diviaková, A.; Kočický, D.; Belaňová, E. Territorial system of ecological stability as a part of land consolidations (cadastral territory of Galanta—Hody, Slovak Republic). *Ekologia* **2018**, *37*, 164–182. [CrossRef]
62. Popescu, O.C.; Tache, A.V.; Petrişor, A.-I. Methodology for identifying the ecological corridors. Case study: Planning the brown bear corridors in the Romanian Carpathians. *Rev. Verde/Green J.* **2022**, *1*, 174–202.
63. Federal Assembly of The Czechoslovak Socialist Republic. *The Act on Land-Use Planning and Building Order (Act 50/1976 Coll.)*; Federal Assembly of The Czechoslovak Socialist Republic: Prague, Czechoslovak Republic, 1976.
64. The National Council of the Slovak Republic. *The Environmental Impact Assessment Act (Act 24/2005 Coll.)*; The National Council of the Slovak Republic: Bratislava, Slovakia, 2005.
65. The National Council of the Slovak Republic. *Act on Nature and Landscape Protection (Act 543/1994 Coll.)*; The National Council of the Slovak Republic: Bratislava, Slovakia, 1994.
66. The Parliament of Romania. *Law No. 350 of June 6, 2001 regarding Territorial Planning and Town Planning*; The Parliament of Romania: Bucharest, Romania, 2001.
67. Grădinaru, S.; Paraschiv, M.; Iojă, C.; Van Vliet, J. Conflicting interests between local governments and the European target of no net land take. *Environ. Sci. Policy* **2023**, *142*, 2. [CrossRef]
68. Popescu, O.C.; Petrişor, A.-I. Green infrastructure and spatial planning: A legal framework. *Ollen. Stud. Şi Comunicări Ştiinţele Nat.* **2021**, *37*, 217–224.
69. The Government of Romania. *Decision No. 525 of June 27, 1996 for the Approval of the General Urban Planning Regulation*; The Government of Romania: Bucharest, Romania, 27 June 1996.
70. The Parliament of Romania. *Law No. 137 of December 29, 1995—Environmental Protection Law*; The Parliament of Romania: Bucharest, Romania, 1995.
71. The Government of Romania. *Emergency Ordinance No. 195/2005 on Environmental Protection*; The Government of Romania: Bucharest, Romania, 2005.
72. The Parliament of Romania. *Law No. 24 of January 15, 2007 Regarding the Regulation and Administration of Green Spaces in the Urban Areas*; The Parliament of Romania: Bucharest, Romania, 2007.
73. Iojă, C.; Breuste, J.; Vânău, G.; Hossu, C.; Niţă, M.; Onose, D.; Slave, A. Bridging the People-Nature Divide using the Participatory Planning of Urban Protected Areas. In *Making Green Cities*; Breuste, J., Artmann, M., Iojă, C., Qureshi, S., Eds.; Springer International Publishing: Cham, Switzerland, 2020.
74. Kadar, M.; Benedek, I. The Branding and Promotion of Cultural Heritage. Case Study About the Development and Promotion of a Touristic Heritage Route in the Carpathian Basin. *J. Media Res.* **2017**, *10*, 80–102. [CrossRef]

75. Stan, M.-I. Are public administrations the only ones responsible for organizing the administration of green spaces within the localities? An assessment of the perception of the citizens of Constanța municipality in the context of sustainable development. *Technium Social Sci. J.* **2022**, *31*, 58–74. [CrossRef]
76. Opperman, J.J.; Shahbol, N.; Maynard, J.; Grill, G.; Higgins, J.; Tracey, D.; Thieme, M. Safeguarding Free-Flowing Rivers: The Global Extent of Free-Flowing Rivers in Protected Areas. *Sustainability* **2021**, *13*, 2805. [CrossRef]
77. Pasca, M.G.; Elmo, G.C.; Arcese, G.; Cappelletti, G.M.; Martucci, O. Accessible Tourism in Protected Natural Areas: An Empirical Study in the Lazio Region. *Sustainability* **2022**, *14*, 1736. [CrossRef]
78. Petrișor, A.-I.; Andrei, M. How efficient is the protection of biodiversity through natural protected areas in Romania? *Olten. Stud. Și Comunicări Științele Nat.* **2019**, *35*, 223–226.
79. Gârjoabă, A.-I.; Crăciun, C. Supporting the Process of Designing and Planning Heritage and Landscape by Spatializing Data on a Single Support Platform. Case Study: Romania. *Rev. Românească Pentru Educ. Multidimens.* **2022**, *14*, 54–68. [CrossRef]
80. Crăciun, C. *Urban Metabolism. An Unconventional Approach to the Urban Organism [Metabolismul Urban. O Abordare Neconvențională a Organismului Urban]*; "Ion Mincu" University Publisher: Bucharest, Romania, 2008; p. 48.
81. Geneletti, D.; Cortinovis, C.; Zardo, L.; Esmail, B.A. Planning for Ecosystem Services in Cities. In *Planning for Ecosystem Services in Cities*; Springer: Berlin/Heidelberg, Germany, 2020.
82. Dudley, N.; Ali, N.; Kettunen, M.; MacKinnon, K. Editorial Essay: Protected areas and the sustainable development goals. *Parks* **2017**, *23*, 9–12. [CrossRef]
83. Crăciun, C.; Gârjoabă, A.-I. Integration of Instruments for the Protection of Natural Protected Areas in Urban and Biodiversity Strategies and in Urban Planning Regulations. In Proceedings of the World LUMEN Congress, Iasi, Romania, 26–30 May 2021; Volume 17, pp. 140–158.

Disclaimer/Publisher's Note: The statements, opinions and data contained in all publications are solely those of the individual author(s) and contributor(s) and not of MDPI and/or the editor(s). MDPI and/or the editor(s) disclaim responsibility for any injury to people or property resulting from any ideas, methods, instructions or products referred to in the content.

Review

How Informed Design Can Make a Difference: Supporting Insect Pollinators in Cities

Sheila K. Schueller *, Zhelin Li, Zoe Bliss, Rachelle Roake and Beth Weiler

School for Environment and Sustainability, The University of Michigan, Ann Arbor, MI 48109, USA; lizhelin@umich.edu (Z.L.); zrbliss@umich.edu (Z.B.); rsterli@umich.edu (R.R.); blweiler@umich.edu (B.W.)
* Correspondence: schuel@umich.edu

Abstract: Pollinators are responsible for the reproduction of many plant and crop species and provide important diversity for food webs and cultural value. Despite the critical ecosystem services provided by pollinators, rapid pollinator declines are occurring in response to anthropogenic activities that cause the loss of suitable habitat. There is an opportunity for urban green space to support pollination ecosystem services locally and across the landscape. However, there is a lack of practical but evidence-based guidance on how urban green space can be designed effectively to provide floral resources and other habitat needs to a diverse assemblage of pollinators. We examine the existing pollinator research in this paper to address the following questions specific to insect pollinators in temperate urban settings: (1) Which pollinators can be the focus of efforts to increase pollinator ecosystem services in cities? (2) Which plants and what arrangements of plants are most attractive and supportive to urban pollinators? (3) What do urban pollinators need beyond floral resources? (4) How can the surrounding landscape inform where to prioritize new habitat creation within cities? Using these questions as a framework, we provide specific and informed management and planning recommendations that optimize pollinator ecosystem value in urban settings.

Keywords: pollinator; bees; pollination ecosystem services; urban green space; urban design; landscape typology; gardens; plant–pollinator interactions; urban biodiversity; research-practice gap

Citation: Schueller, S.K.; Li, Z.; Bliss, Z.; Roake, R.; Weiler, B. How Informed Design Can Make a Difference: Supporting Insect Pollinators in Cities. *Land* **2023**, *12*, 1289. https://doi.org/10.3390/land12071289

Academic Editors: Alessio Russo and Giuseppe T. Cirella

Received: 25 May 2023
Revised: 15 June 2023
Accepted: 20 June 2023
Published: 26 June 2023

Copyright: © 2023 by the authors. Licensee MDPI, Basel, Switzerland. This article is an open access article distributed under the terms and conditions of the Creative Commons Attribution (CC BY) license (https://creativecommons.org/licenses/by/4.0/).

1. Introduction

Pollinators are declining worldwide, and as pollinators are lost, so too are the ecosystem services they provide. Significant pollinator loss is occurring in response to several threats, including pesticides, pathogens, and habitat loss and fragmentation [1–4]. Loss of pollinator species and networks has been noted at the global scale [1,5], and through direct long-term studies in specific regions [6–9], which has significant implications for humans and the ecosystems on which they depend. An ecosystem service is a benefit, often economic, that nature provides to humans [10], and the service of animal-assisted reproduction of flowering plants, or pollination, is a recognized critical "regulating" ecosystem service [11]. Many pollinator species increase or ensure the reproduction of plants in both native plant communities and agricultural systems. Almost 90% of all wild flowering plant species depend, at least partially, on animal pollination services for their reproduction [12], and nearly 35% of global food crops rely on pollinators to reproduce [5,13], giving pollinators an estimated food provisioning service value in the USD billions [14]. The impact of pollinator declines is already measurable in agricultural systems [15]. In addition, the ecosystem service of value of pollinators extends beyond their role in food production and the persistence of natural plant communities to include supporting and even cultural ecosystem service value. Insect pollinators, in particular, are at the base of many food webs, providing rich, abundant resources for other species, which in turn support other ecosystem services [5]. The diversity of animals that serve as pollinators—from hummingbirds to metallic green bees—also provide aesthetic and cultural ecosystem service value, especially in cities where species richness enhances the psychological benefits of green space [16].

While urbanization has contributed to pollinator species loss, well-designed urban landscapes provide an opportunity not only to curb the pollination crisis but also to support the beneficial ecosystem services of pollinators, both within the city and beyond. Many of the drivers of pollinator loss are associated with urbanization, including the loss and fragmentation of pollinator habitat through increased land use intensity and impervious surfaces and threats such as non-native species, environmental contaminants, and urban warming [17–19]. While some studies have found pollinator declines along increasing urbanization gradients [20,21], the effect of urbanization on insects generally can vary greatly from negative to no impact to even positive effects [22]. In fact, cities can contain surprisingly abundant pollinators compared to rural or agricultural landscapes [19,23–25]. Hall et al. [24] suggest that cities with a variety of forage and nesting sites can serve as important refugia for pollinators compared to increasingly less hospitable rural and suburban landscapes that surround urban areas. The relatively small spatial and temporal scales of insect pollinators in terms of functional ecology (for example, habitat range, life cycle, and nesting behavior compared with larger mammals) offer opportunities for small actions to yield large benefits for pollinators [24]. This can support the ecosystem services of pollinators within the city itself, including the productivity of urban food gardens [26], and support the spillover of pollination ecosystem service benefits to agricultural and natural areas outside of the city [27,28].

Interest in urbanization as both a potential threat and opportunity for pollinators has accelerated research in pollinator urban ecology, but the actual practice of supporting pollinator ecosystem services in cities has not been fully realized. Our ability to inform urban design has been improved by systematic reviews of urban pollinators and pollination ecology [18,19,25,29], and general recommendations on ways to support pollinators in urban settings [19,30–32] and even assessments of pollinator ecosystem service value of green space in specific cities (from Grenada, Spain to Chicago, USA; [33,34]). However, there is inadequate specific and practical planning and management guidance directly linked to the research that supports it [19]. Gaps between research and practice are a common problem in conservation [35,36], highlighting the need for approaches that stimulate the connection between research and feasible implementation [37].

The purpose of this paper is to compile and organize relevant literature on pollinators in cities within a framework of practice-based questions relevant for urban designers, planners, and dwellers. The evidence-based and socially contextualized answers to these questions can guide ecologically informed design that maximizes pollinator ecosystem services in cities in the face of increasing urbanization. Specifically, we address the following questions:

- Which pollinators can be the focus of efforts to increase pollinator ecosystem services in cities?
- Which plants and what arrangements of plants are most attractive and supportive to pollinators in cities?
- What do urban pollinators need beyond floral resources?
- How can the surrounding landscape inform where to prioritize new habitat creation within cities?

We first discuss the relevant literature for each of these questions and also address the question of urban management practices that would support pollinator ecosystem services. Then, we provide a synthesis of recommendations for the local garden scale, including informed planting typologies for urban gardens and then, from an urban planning perspective, to guide landscape-scale management and design. Given the burgeoning field of urban pollination, we do not attempt to review all available evidence, but rather to contribute to the "research-implementation space" [37] in the form of some key actionable recommendations directly based on current knowledge and needs.

Although many urban design recommendations might apply to different pollinators and regions, we aimed to narrow our focus especially to *insect* pollinators for *temperate* zones, especially those native to North America. Our aim is to provide more species- and research-specific recommendations rather than make generic conclusions or falsely extrapolate to other systems. That is, recommended plant arrangements and typologies are likely to be quite different for temperate bees vs. tropical hummingbirds. We agree that the bias in the urban pollination literature to the global north is problematic [29,38], but specific recommendations for tropical cities (including bird and bat pollinators) are emerging as these systems are increasingly studied [38–40], and so parallel recommendations specific to other systems can become available over time.

2. Pollinators: Which Pollinators Can Be the Focus of Efforts to Increase Pollinator Ecosystem Services in Cities?

"Pollinator" is a large category of organisms defined by their function. Although bees and butterflies may be the most well known, pollinators span a range of taxonomic categories including not only wasps and other members of the *Apoidea* superfamily but also moths, which make up nearly 90% of species in the *Lepidoptera* order, as well as "hoverflies" in the *Syrphidae* family of the *Diptera* order (true flies). Even the term "bee" is a common name that can apply to a diversity of species within the *Anthophila* clade and includes not only honeybees and native (to North America) bumblebees but also mason, carpenter, and sweat bees. Birds and bats are also critical pollinators in many areas of the world, though insects remain the most frequently recorded pollinators in urban areas worldwide [29]. Specifically, across studies of urban areas globally, Hymenoptera, especially honeybees and bumblebees, were by far the most frequently recorded insects, followed by Lepidoptera and Diptera [29].

Public perceptions of pollinators or of "Save the Bees" campaigns often focus on the European honeybee (*Apis mellifera*), known for their honey production and pollination of certain crops [41]. In fact, wild bees are equally if not more critical for staple crop pollination (e.g., apples and blueberries [42]), and *A. mellifera* is non-native in most areas of the world and known to compete with native bees for limited floral resources [43]. While *A. mellifera* are the most frequently recorded and often the most abundant species in urban areas worldwide [29], an increase in urban beekeeping (hives specifically for *A. mellifera*) can negatively impact more diverse native bee assemblages in cities [44]. Thus, urban efforts to increase ecosystem services of pollination need to recognize the value of pollinator diversity to maintain services, as well as to consider the critical difference between managing habitat for pollinator services and preserving overall pollinator biodiversity [45]. Intense focus on honeybees spreads misinformation regarding pollinator biodiversity and its value, so it should be established that the term "pollinator" encompasses many species with different needs.

While pollinator responses to urbanization are quite varied and are trait- and scale-dependent, urbanization tends to lead to an increase in the abundance and dominance of generalist and social species and a higher rarity of specialist species [18,25,29]. For example, urban areas tend to support generalist, short-tongued bee species and not specialist bee species [46]. It may be possible to attract more specialist species with different plant selections (see below), but even if only generalist pollinator taxa are common in cities, these still span a diverse assemblage (Table 1) and so can support ecosystem services.

Table 1. A list of generalist insect pollinator families and genera (italicized) based on ref. [47].

Category	Species
Bees	Many (but not all): bumblebees (*Bombus* spp.), sweat bees (*Halictus* sp. and *Lasioglossum* sp.), leaf-cutter bees (*Megachile* sp.), carpenter bees (*Xylocopa* sp.)
Wasps	Paper wasps (*Polistes* sp.), yellowjacket (*Vespula*), bald-faced hornet (*Dolichovespula*)
Butterflies and moths	Hummingbird moths (*Hemaris* sp.), sulphurs (*Colias* sp.), swallowtails (*Papilio* sp.), fritillaries (*Speyeria* sp.)
Flies	Families: bee flies (*Bombyliidae*), Syrphid (hover) flies (*Syrphidae*), Tachinid flies (*Tachinidae*), thick-headed flies (*Conopidae*)
Beetles	Families: soldier beetles (*Cantharidae*), long-horned beetles (*Cerambycidae*), leaf beetles (*Chrysomelidae*), snout beetles (*Curculionidae*)

Understanding the finer-scale differential effects of urbanization on pollinators can allow urban designers to focus on providing habitat to the pollinator species most likely to be present or to take informed approaches to improve habitat suitability for missing taxa. For example, ground-nesting bee species richness decreases with an increase in the impervious surface of an area [48] as ground-nesting species require patches of bare dirt in which to nest. Conversely, cavity-nesting and aboveground-nesting bee species are more abundant in more densely built urban areas, given that they are able to build their nests in pre-existing cavities of urban structures [18,48]. Research also shows that even just the presence of permanent grassland can increase the number of rare bee species supported [48]. Thus, design efforts that increase bare soil and permanent grassland habitat in areas of highly impervious surfaces can diversify the pollinator assemblage. As we further discuss below, habitat patch size can also affect pollinator composition. Suburban and urban sites containing smaller habitat patches show an increases in small bee, social bee, and solitary cavity-nesting bee species [49]. These urban habitat patches also favor species in the family *Halictidae* (sweat bees) over *Apidae* (honey, bumble, carpenter, and cuckoo bees). Study results such as these can better direct plant selection and habitat construction in urban areas based on the floral preferences of the more favored species; for example, sweat bees prefer flowers from the plant families *Asteraceae* (asters and daisies) and *Lamiaceae* (mints) [47].

Managing habitats for diverse pollinator assemblages in cities is feasible, because although they may have different habitat and floral resource needs, there are many overlaps in needs within temperate insect pollinator communities. This overlap provides opportunities to create beneficial habitat for many pollinator groups at once [47]. For example, 30% of the 4000 bees native to the United States are cavity-nesting species, using dead wood and plant stems as shelter for developing larvae [50]. These cavity-nesting bees include mason, leafcutter, and carpenter bees, which all belong to different genera. Similarly, well-chosen plants, as we discuss below, can attract pollinators from completely different insect orders, e.g., butterflies and moths (*Lepidoptera*), bees (*Hymenoptera*), beetles (*Coleoptera*), and flies (*Diptera*).

3. Plants: Which Plants and What Arrangements of Plants Are Most Attractive and Supportive to Pollinators in Cities?

While more green space in urban areas is associated with increased pollinator species richness [25], the characteristics of that green space strongly influences the pollinators present [51] add [18,29] and therefore the ecosystem services they can provide. Urban green space spans a huge variety of types, from individual flowerbeds to private or community gardens planted with flowering and/or edible plants and to lawns, green roofs, recreational parks, remnant natural areas, vacant lots, and even urban cemeteries, golf courses, and university campuses [29,52]. These spaces differ in landscape management intensity, vegetation composition, vertical structure, microclimate, and patch size, all of which can have effects on pollinator populations [18,53]. Broadly speaking, the presence of floral foraging resources is consistently found to influence the presence of pollinators in urban areas [19,24]. More specifically, urban areas with higher floral diversity [54,55] and floral abundance [56–58] tend to have a higher pollinator abundance and/or diversity. This pattern holds even for a vertically isolated habitat in a highly urbanized context—green roofs in Chicago planted with native plants and with the highest plant diversity have the highest bee species richness and abundance [59]. These studies support the idea that it is not necessarily the degree of "urban-ness," but instead the specific characteristics of each urban green space that ultimately influence pollinator abundance and diversity. In this section on plant choice, we provide specific information to guide decisions at this *local* scale of a green space [18], including factors such as the origin, floral features, and arrangement of vegetation within a green space, noting how these factors influence the ability of an urban green space to support insect pollinators.

3.1. Native vs. Non-Native Plants

In recent years, gardeners and researchers alike have been debating the role of native and non-native ornamental plants in supporting local biodiversity and ecosystem function. Given the long history of coevolution between native plants and their insects, the widespread use of non-native plants across urban and non-urban landscapes has likely contributed to global insect declines through the loss of suitable food and habitat [60]. Although pollinators are often attracted to various non-native ornamental plants in urban landscapes [56,61], research increasingly demonstrates that overall pollinator abundance and diversity are greatest in landscapes with native plants [55,59,62]. Pardee and Philpott [56], for example, found that for city backyard gardens, both bee richness and abundance are higher in gardens that contain more native plants and thus have more floral abundance, taller vegetation, more cover, and more potential nesting sites. Similarly, Rollings and Goulson [63] found a significantly higher diversity of pollinators attracted to native plants as opposed to ornamental plants in backyard gardens. Additionally, there may be a native plant "threshold" of eight or more species of native plants within a landscape to increase the abundance and diversity of native bees [64,65]. These findings suggest that incorporating native plants into the urban landscape is critically important for supporting abundant, diverse pollinator populations. The use of native plants is especially important for supporting specialist species, for which non-host plant pollen is not only non-preferred but toxic [66], so cities without sufficient quantities of host plant pollen would have an overall lower bee species richness. While rare specialist pollinator species may not contribute as significantly to crop pollination ecosystem services [67], in an urban context, specialist pollinators and their associated plants can provide cultural and aesthetic ecosystem service value while contributing to overall diversity.

Beyond the value of native plants for pollinator diversity and ecosystem services, they should be favored in urban green spaces due to the risks posed by non-native plants. The flowers of non-native plants, especially when abundant, can draw pollinators away from native plants and thereby decrease their reproductive success, especially when the non-native flowers are similar to the native ones [68]. Thus, the use of non-native plants in city plantings can undermine the ecosystem service value of pollination for native plant communities. Non-native plantings also increase the risk of invasive species spread, which can further degrade natural ecosystems. Urban and suburban gardens are not only a key entry point for many invasive plant species but also a source for secondary escape and spread into natural area remnants within cities or along the urban–wildland interface surrounding cities [69,70].

Some non-native ornamental plants do offer important social value, and so if their risks of spread or competition with natives are low, they can play a valuable role as part of a mixed planting of both native and non-native species to support pollinator ecosystem services. Landscape aesthetics are a powerful cultural driver of a green space's success and long-term sustainability [71]. Non-native, ornamental plants have been human-selected for their beauty, scent, reliability, or cultural relevance among other socially determined reasons. It follows that these plants are familiar, broadly socially accepted, and legible to the general public. These spaces inherently feel orderly and cared for, which exemplifies Joan Nassauer's "Cues to Care" [72]—the social importance of cues to care cannot be underestimated, but it can be signaled in other intentional ways in plantings that are less ubiquitous or orderly. Regardless, the social legibility of culturally beloved, ornamental plants such as peonies, roses, or boxwoods can be leveraged to make less-familiar, more "wild" native plantings more socially acceptable and therefore more likely to be cared for and persist over the long-term, which is what pollinators depend on. Additionally, not all gardeners want to plant an all-native garden but want to help pollinators. By not vilifying ornamentals and instead encouraging a mix of ornamental and native plants, we may offer a steppingstone and olive branch for anyone who wants to garden for pollinators.

Beyond their potential social value, some non-native ornamental plants do directly provide resources for pollinators. For example, many popular, easy-to-grow annuals and perennials, as well as non-native flowering trees, can be very attractive to pollinators and provide abundant floral resource availability [61,73,74]. However, there is growing evidence that non-native plants attract a subset of pollinators, and frequently these pollinators are more generalist in their foraging preferences [57,61]. This support for generalists is valuable, especially in urban environments, which may be resource-poor, but it is important to acknowledge this limitation and provide a variety of foraging resources for both specialist and generalist species. The value of non-native plants is important to consider in the face of climate change. Many non-native plants are selected for longer blooming periods and blooming periods for the "shoulder" seasons of early spring and late fall, all of which can extend foraging resource availability temporally for pollinators [75], which may be especially important with climate change. Thus, in making decisions about which plant species to include in urban green spaces, it is important to not only consider their origin [76] but also their risks, social value, which pollinators they support, and their value for urban climate adaptation.

The growing number of "nativars," or cultivated varieties of native plants, in horticulture has raised questions about how the horticultural modification of plant traits affects the ecological function of native plants. To date, this research has produced variable answers. Cultivated plants are often selected for desirable growth traits (disease resistance, longer bloom time, tidy form, etc.) or aesthetic traits (double blooms, larger blooms, bloom color variety, variegated leaf/stem color, etc.). Either for growth form or aesthetics, alterations that change the physical traits of a plant may alter its attractiveness or resource availability to insects and thus change their ecological function. Robust garden trials from Mt. Cuba Center in Delaware are actively testing pollinator use (ecological function) alongside horticultural performance (social function). Though only a few genera have been tested so far (*Baptisia* spp., *Coreopsis* spp., *Echinacea* spp., *Monarda* spp.) results vary widely, with some cultivated varieties actually attracting more pollinators than the "straight" native plant. For example, within Monarda varieties, moths and butterflies were more attracted to selections that offered the largest abundance of 2–3" wide flowers, with *M. fistulosa* 'Claire Grace' attracting substantially more pollinator visits than the straight native *M. fistulosa* [77]. Importantly, the Mt. Cuba researchers note that this artificially resource-rich "buffet" of dense *Monarda* plantings may reveal pollinator preferences given unlimited options, but any of these *Monarda* varieties on their own in a landscape setting may perform equally well. Although the current research is limited regarding which specific alterations may affect ecological function, Tallamy et al. [60] found that native varieties that had leaves that altered from green to red, blue, or purple were eaten significantly less by herbivorous insects. The same study found no effect of altered plant habit, fruit size, disease resistance, or fall color, but leaf variegation seemed to increase insect herbivory. These limited studies suggest that some "nativars" still provide resources for pollinators and herbivorous insects [78], although more research is still needed to test cultivated varieties of more plant species and to investigate alterations to pollen and nectar quantity and quality and their potential invasiveness risk.

3.2. Floral Features

Beyond plant origin (native or non-native), it is known that floral features such as fragrance, color, shape, and nutritional quantity and quality influence pollinator attractiveness [47,61,79]. The broadest diversity of pollinator groups can be supported by representing a diversity of their associated flower traits (Table 2) within the garden or green space.

Table 2. Flower color and trait preferences of different insect pollinators.

Pollinator	Flower Trait Association
Bees	Primary color: pink, purple, or blue Secondary: white or yellow.
Wasps	Shallow corollas for nectar; white.
Butterflies and day-foraging moths	Flat-topped flowers or a structure to grasp while nectaring. Prefer composite flowers.
Night-foraging moths	White or cream with a strong fragrance.
Flies	Flat or bowl shapes, umbels. White or cream color. Musty fragrance.

Pollinator response to floral color varies across plant genera [61], but there are certain color–species associations based upon pollinators' species-specific ability to perceive a certain range of colors and some species' preferences. Red, for example, is perceived as black to bees [47], and so the color red is not attractive to bees; however, red is particularly attractive to hummingbirds, which are also essential pollinators. Bees and many other insect pollinators perceive ultraviolet light cues that often serve as nectar guides but are imperceptible to humans. White, cream, or green-colored flowers are attractive to the less popular but no less important pollinators: wasps, beetles, flies, and night-foraging moths. While these general color associations are helpful for attracting some pollinators, many other floral traits may be more important.

Flower shape, structure, and size also influence the type of pollinator that can access the floral resources, i.e., nectar and pollen. First, pollinators vary in their pollen- and nectar-gathering strategies and have different physical structures to gather, consume, and store these resources. Some physical structures that alter a pollinator's foraging strategy include tongue length, the presence of external storage structures (e.g., hairs), body size, and body weight. Long-tongued bees, butterflies, and moths can easily access tubular flowers, but short-tongued bees, beetles, and wasps can only access shallow, open nectaries (unless they "steal" resources by cutting into the flower base). For large, closed flowers such as *Baptisia* spp. or *Lobelia siphilitica*, resource access is limited to strong, large-bodied bumblebees that can push apart the heavy petals or very tiny bees that can slip in through the gaps. Flat-topped, umbel-shaped flowers including *Pycnanthemum* spp., *Eutrochium* spp., and *Zizia* spp. offer shallow nectaries that attract a greater diversity of pollinators with both short and long tongues. Floret-dense, open, composite flowers of the Asteraceae family including *Symphyotrichum* spp. (asters), *Echinacea* spp. (coneflowers), and *Solidago* spp. (goldenrods) also feature shallow nectaries accessible to a variety of pollinators at a high density, offering a large quantity of resources and a suitable landing pad for large-bodied butterflies. Given that flower shape, size, and structure influence resource availability to a certain subset of pollinators, it follows that by providing a diversity of floral morphologies, a diverse garden will attract a more diverse assemblage of pollinators.

Putting all of these features of certain flowers known to attract pollinators together, there are many existing lists of recommended pollinator-friendly plant species for urban green space designers to consider. While this can be exciting for those looking to improve their pollinator habitat, such recommendation lists have limitations. Garbuzov and Ratnieks [71] reviewed fifteen plant recommendation lists from various sources including pamphlets, websites, books, and botanical garden information stands/leaflets. They found that while these lists are useful communication tools for a general audience—and a good starting point for future research—they often contain poor recommendations, omit what would be good recommendations, lack overlap even when considering the same geographical regions, and are based on author experience rather than empirical evidence. They note "a list is only as good as the data that went into it," and the lists they reviewed "almost never refer to the empirical sources on which they are based." (p. 1019). Thus, traits and field observations may be more reliable to inform plant choice.

3.3. Floral Arrangement in Time

Providing not just a diversity of floral traits but also flowers that bloom throughout the growing season is key to supporting a diverse and abundant population of pollinators [47]. Phenology, or the timing of seasonal biological events, frequently orchestrates a tightly evolved relationship between plants and their pollinators. Different plant species bloom at different times for many reasons, including taking advantage of water availability, to have a competitive advantage over other plants and to attract specific pollinators. Many pollinators, especially native bees, live short lives and emerge, feed, and reproduce over a period of weeks or months. Mason bees (*Osmia* spp.), for example, are active only from spring to early summer, and overwintering queen bumblebees are some of the first pollinators to emerge during spring thaw. These early spring pollinators require pollen and nectar from early-blooming, often spring ephemeral flowers such as *Geranium maculatum* (wild geranium). In fall, many pollinators are preparing to overwinter or migrate (such as the monarch butterfly) and rely on late-blooming plants including *Solidago* spp. (goldenrods) and *Symphyotrichum* spp. (American asters). By providing a diversity of blooms from spring to fall, green spaces can support a greater diversity of pollinators throughout their life cycles. Flowering trees can also play a critical role in providing abundant floral resources over a whole season and can fill gaps in floral resource availability [52,73].

Additionally, the mowing management of lawns—a potentially critical green space in an urban area—can have significant effects on the timing of available floral resources for pollinators. Lawns are iconic to Western cities; they occupy a large proportion of urban areas and are culturally and aesthetically valued [80,81]. Many municipalities enforce 'weed laws' to ensure the conformity of the lawn ideal by restricting grass height (e.g., a Chicago ordinance prohibits lawn vegetation from exceeding 24.4 cm; Municipal Code of Chicago: §7-28-120). Households mow to conform to societal expectations, city ordinances, and the personal satisfaction of a neat and tidy yard [80]. Intensive lawn management requires time and financial commitments and is often driven by aesthetics and social norms to adhere to ideals of orderly, weed-free, lush carpets of green grass [72,82–84]. However, frequent mowing restricts plant diversity in urban lawns to only a few species that are able to tolerate repeated defoliation and soil disturbance, i.e., *Bellis perennis, Glechoma hederacea, Lolium perenne, Plantago major, Prunella vulgaris,* and *Trifolium repens* [85]. The homogeneity of plant species, together with herbicide application, deplete lawns of floral resources for pollinators [86]. Adjusting mowing frequency can lead to increases in pollinator value. Lerman et al. [87] found that the lawn with lowest mowing frequency has the highest floral abundance but not the highest bee abundance. They suggested that taller grass in less-mown yards might have prohibited access to the flowers and rendered the floral-abundant lawns less attractive to pollinators. Alternatively, the lawn flowers might lack the performance traits necessary for competing with the tall grass, leading to pollen limitation and hence less attractive habitats for bees [88]. Thus, an appropriate mowing frequency and the addition of floral diversity are both needed for urban lawns to support more pollinators throughout their active season.

3.4. Floral Arrangement within a Green Space

In addition to the origin, floral features, and flowering timing of the plants within a green space, another important local-scale consideration to support urban pollinators is the arrangement of floral resources *within* the green space or garden, which may be more important even than its size. For example, some research suggests that while garden size does not influence invertebrate communities [89], floral density (the number of blooms per unit area) frequently has a positive effect on pollinator abundance and diversity [90]. Keasar [91] found that clustering flowers, even if these clusters included resource-sterile plants, increased native bee visitation rates. Small community gardens that are densely packed with a variety of floral resources are highly attractive to pollinators [52,92], and increasing such plantings within larger recreational parks could increase the parks' pollinator value.

The spatial arrangement of flowers can strongly affect not only how many but which pollinator species visit a garden. Plascencia and Philpott [93] found that the honeybee was more abundant in sites with patchy floral resources (larger nearest-neighbor ratio of quadrats with ≥ 15 flowers), likely because it is a generalist species, and its medium size permits it to forage large distances. By contrast, native bee species were most abundant in gardens with clustered floral resources (smaller nearest-neighbor ratio). Thus, the local and small-scale arrangement of flowering plants can favor certain pollinators over others, and so heterogeneous arrangements are likely to support the greatest diversity of pollinators. In Section 5, we also discuss landscape-scale spatial considerations for supporting pollinators, such as the variety of green space sizes and their placement in an urban setting.

4. Nesting Habitat: What Do Urban Pollinators Need beyond Floral Resources?

Quality pollinator habitat includes more than just floral and foraging resources. Within the same local scale as floral resources, native pollinators, especially wild bees, must have habitat for nesting and overwintering. Variability in insect pollinator composition among urban green spaces can depend less on floral density or cover and more on available nesting habitat [53]. Many native bees nest in plant material such as hollow plant stems and leaf litter, while others require bare ground to nest. For example, across Californian urban gardens, Quistberg et al. [94] found that the number of cavity-nesting bees depended on the percent of leaf litter cover and the number of ground bees increased with an increased percentage of bare ground and decreased with wood chip mulching. Landscaping practices such as the widespread use of wood mulch can thus be detrimental to ground-nesting bees. Tree and shrub pruning/leaf removal, especially as "fall/spring clean ups," can also greatly diminish the nesting habitat for pollinators [32]. Leaving such materials can pose an aesthetic challenge, especially in urban gardens and campuses where the city or school may have yard care guidelines. Campaigns such as "Leave the Leaves!" are spreading thanks to the Xerces Society and the National Wildlife Federation. Both organizations have published materials explaining the ecological value to pollinators of leaf litter, plant stems, and other dead and decaying plant debris. Butterfly larva such as that of the great spangled fritillary (*Speyeria cybele*) overwinter in piles of leaves, and other species such as the red-banded hairstreak (*Calycopis cecrops*) lay their eggs on fallen oak leaves, which the hatched caterpillars will eat in the spring. Bumblebees (*Bombus* spp.) are a well-known pollinator group that depend on leaf litter for protection over the winter. These are just a few of many examples, demonstrating that planting pollinator-friendly plants is only one part of creating a quality pollinator habitat.

Bees can be organized into three guilds based on their nesting habits: ground-nesting, above-ground-nesting, and cleptoparasitic [95]. The ground-nesting guild is dominated by the families *Andrenidae, Melittidae, Halictidae,* and *Colletidae*, while the above-ground-nesting guild includes mostly Megachilidae and Apidae species [96]. Nesting sites for cavity-nesting bees in the form of "bee hotels" are increasingly being promoted as a way to aid pollinator conservation. Bee hotels vary in size and are typically constructed from wood and contain different-sized cavities and a variety of materials to be used for nesting, such as bamboo tubes and bricks with holes. Unfortunately, installing bee hotels can be counterproductive because most of North America's native bee species (and 70% of the ~20,000 bee species worldwide) nest under—not above—the ground [97]. A large majority of bees either nest underground or parasitize other bees' nests, which limits the value of above-ground-created habitats such as bee hotels.

Bee hotels are widely touted as a positive addition to any pollinator garden, but numerous studies have documented increased parasitization of native bees nesting inside such hotels [97]. Additionally, non-native and non-pollinating bee and wasp species have been demonstrated to use bee hotels more often than native, pollinating bee species, thus outcompeting native bees for nearby resources [97]. Geslin et al. [98] found that 40% of all individuals recorded using the 96 bee hotels they installed were *Megachile sculpturalis*, a leafcutting bee native to Japan and China. They also found a negative correlation between the presence of *M. sculpturalis* and native bees in the hotels. MacIvor and Packer [97] coined the term "bee-washing" (a form of green-washing) to warn promoters and users of bee hotels against spreading potentially misleading information. They note that, much like pamphlets of pollinator-friendly plants, bee hotels are useful tools for engaging the public in citizen science and pollinator conservation outreach but that their potential pitfalls must be thoroughly researched before they are recommended as "pollinator friendly".

Although bee hotels may not necessarily be the best choice for bee nesting sites, bees still do need habitat in which to nest and overwinter—nesting requirements are as important to consider as floral resources [99]. Bumblebees (*Bombus* spp.) often overwinter beneath the base of clumped grasses, while many ground-nesting bees will utilize bare patches of soil for nesting. Bare patches in particular pose an aesthetic challenge in an urban setting; those who do not know the purpose of the bare soil may find it less aesthetically pleasing than a patch of foliage or flowers. Such challenges could be overcome by signage that explains the purpose of the bare patches and their value to native pollinators. Alternatively, soil squares—smaller patches of bare soil that form a 0.5 m deep hole in the ground—could be constructed to provide nesting sites for cavity-nesting bees [100]. An interesting study by Cane [101] shows that native species of *Halictus* prefer to nest beneath decorative landscaping pebbles instead of bare soil patches. Much like providing an array of floral resources will tend to attract the most diverse pollinator assemblage, providing a variety of nesting materials and sites that can be maintained (rid of harmful parasites if necessary) is most beneficial to urban pollinators.

5. Landscape-Level Planning and Connectivity—How Can the Surrounding Landscape Inform Where to Prioritize New Habitat Creation within Cities?

Thus far, we have discussed local-scale factors that influence the ability of a green space to support pollinators effectively, but designing for pollinator ecosystem services in cities requires considering habitat at different scales, from the quality and size of the green space itself to the arrangement of potential habitat within the landscape and the quality of the urban context (Figure 1). Urbanization essentially fragments pollinator habitat, a process that can reduce pollinator and plant pollen movement among fragments and reduce pollinator abundance and diversity, especially for species sensitive to fragmentation [17]. Green space fragments that contain more floral resources and a higher diversity of flowering plants (e.g., community gardens, residential gardens) have significantly more abundant and diverse pollinator communities [102]. The patchwork of other urban habitat types (e.g., open parks, lawns, paved areas, buildings) surrounding these floral "hotspots" represent a resource-poor "matrix" that may be unusable and even impermeable to pollinators.

Pollinator habitat planning at a city-wide or landscape scale thus must also consider the spatial distribution of existing pollinator habitat and the land cover in between. While local, garden-scale characteristics are often stronger predictors of pollinator abundance and richness [56,103], the surrounding landscape matrix may influence the accessibility or quality of any individual garden. For isolated garden patches, finding opportunities to create "corridors" of even marginal pollinator habitat to connect patches of higher-quality habitat can facilitate movement across a landscape, encouraging opportunities for short- or long-distance migration, shifting life cycle habitat requirements, and genetic exchange. In contrast, some urban cover types or features such as buildings or busy roads may present barriers to pollinator movement across an urban landscape. Additionally, proximity to more natural habitat types such as forests, wetlands, or grasslands can provide additional resources for pollinators and thus increase pollinator use of nearby garden patches in an urban context. Below, we provide more specific evidence to guide landscape-scale planning for pollinators.

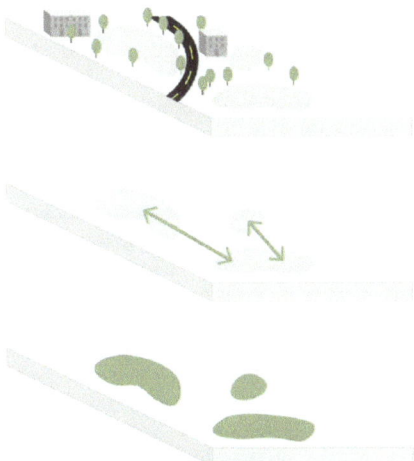

Figure 1. The inter-related levels to consider in planning pollinator habitat, from bottom to top: the size and quality of individual gardens or green spaces, the arrangement and connection among green spaces, and the matrix or context surrounding the green spaces.

5.1. Matrix Quality

Overall, landscape-level variables are frequently not as strong as garden-level characteristics at predicting pollinator population metrics; however, the effect of the surrounding landscape matrix type may differ by pollinator guild. In the urban habitats of Chicago, Tonietto et al. [59] found that bee species richness positively correlates with the proportion of natural area within a 500 m radius. Pardee and Philpott [56] found that cavity-nesting bee abundance was higher in urban gardens with more natural areas within 1 km, and they hypothesized that a mix of natural and man-made resources may assist with nest building. Ground-nesting bees were more abundant where wetlands were within 1 km, which suggests an association with wet habitats. Increased forest cover within 500 m and 2 km was associated with increased abundance of both cavity- and ground-nesting bees, respectively [56]. The authors speculate that the smaller-bodied cavity-nesting species they captured had smaller foraging distances and thus were more associated with nearby forest resources, whereas larger-bodied ground-nesting bees could travel farther and utilize more distant resources. These studies suggest that the effect of the landscape matrix is species-specific and influences life cycle needs and foraging distance.

In addition to the quantity of different land cover types within the matrix, the quality and permeability of the matrix may influence pollinator populations. As opposed to the rigid definitions of usable patches and linear corridors within an unusable matrix, matrix permeability describes a more fluid gradient of use that "spills over" into and within the matrix. A meta-analysis of wildlife habitat creation in agricultural areas suggests that increasing the permeability of a matrix by improving the quality of the matrix may be a more effective method to increase fragment connectivity and reduce the negative effects of patch isolation [104]. With this in mind, perhaps urban pollinator habitat improvement efforts should consider not just creating more patches or corridors of habitat but also improve the quality of the surrounding matrix.

5.2. Size, Arrangement, and Connectivity of Green Space within the Urban Matrix

The quality, size, and isolation of an urban green space can interact to determine its value for supporting insect pollinators. While the species–area relationship suggests that larger fragments have higher species richness, when it comes to urban pollinators, the relationship between size and quality is more complex. As noted above, high-quality small gardens can have an equivalent or higher pollinator habitat value than larger green space [52,92], even if isolated [59]. Reviews on the optimal size of green space to support insect pollinators suggests that it depends on the mobility of the pollinator [17,18,105]. Pollinator foraging distances are species-specific and influenced by body size, foraging specificity, and eusocial lifestyle [47]. While smaller-bodied species may persist within small patches of habitat more characteristic of urban areas, larger-bodied species are able to travel further and cover larger distances in search of resources, especially within a fragmented urban landscape [20,106]. For example, larger, colony-nesting bees such as bumblebees are considered more generalist pollinators and have larger foraging ranges than smaller, solitary bees. Bumblebees (10–23 mm body length) can forage up to one mile from the nest, while smaller mason bees (6–11 mm body length) forage within 300 feet from the nest [47]. Hinners et al. [49], studying suburban green space fragments in Colorado, found that species richness increased with area up to a point, but they also found that there was a shift in bee species composition from small to larger areas, some of which reflects mobility differences. Smaller areas harbored more bees that were eusocial, small-bodied, and cavity-nesters, whereas larger areas shifted to dominance by solitary, large-bodied, and ground-nesting species. Even if these same patterns do not apply elsewhere, the important lesson for planners is that a diversity of green space sizes across the landscape can increase the diversity of pollinators supported overall, due to differences in the response to green space characteristics.

Patch *isolation* is one measure that could negatively impact pollinator use of any one patch and the metapopulation of pollinators as a whole. Several studies have found that as a habitat patch becomes more isolated from a natural habitat, pollinator populations begin to decrease [107,108]. However, in the urban grassland areas of Berlin, Fischer et al. [109] found that only one pollinator species (*Bombus terrestris*) was affected by degree of habitat isolation (negatively), but the authors note that the goodness-of-fit for this model was low and the results should be used cautiously. One possible solution is the use of small "stepping stones" of pollinator habitat to connect isolated patches across a landscape [110]. While flight distances vary by body size and local site conditions, using estimates of flight distance may provide a starting point to determining how far away a functional stepping-stone habitat should be (Figure 2). Creating stepping stones between habitat patches could essentially increase the quality of the matrix and facilitate pollinator movement across the landscape. Especially for smaller pollinators with a limited foraging range, the proximity of abundant floral resources is critically important for survival and successful reproduction. Even across a small area, such as an urban or campus promenade, having frequent patches of dense foraging resources will ensure that pollinators with small foraging ranges will have accessible resources.

Although much of the research on pollinator corridor habitat has focused on agriculture, corridors of linear habitat may serve to connect habitat patches and facilitate pollinator movement across the urban landscape as well. In croplands, linear habitat in ditches can increase pollen dispersal between isolated habitat patches [111]. A similar study of linear hedgerows in cropland in southern England suggests that both hedgerow quality (absence of gaps, high species diversity, and an abundant, flowering understory layer) and landscape context (hedgerows were more valuable in intensively managed landscapes) influenced the value of hedgerows to pollinators and other insects [112]. Comparisons can be made between an inhospitable matrix of monoculture crops and a matrix of resource-poor urban environments such as traditional lawns or parking lots. Linear corridors of unbroken, floral-rich habitat may facilitate movement through intensively managed landscapes devoid of pollinator resources, however, the strategy of increasing matrix permeability, as discussed above, should also be considered.

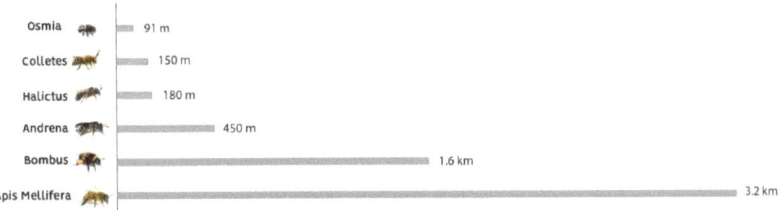

Figure 2. Estimated flight distances of bees [47] can provide insights into stepping stone spacing of pollinator habitat.

In conclusion, much research suggests that landscape-scale variables are less important than garden-scale variables in determining pollinator populations; however, pollinator foraging distances and proximity to natural habitats may be helpful factors to consider when prioritizing locations for new pollinator habitat. Where feasible, improving matrix quality and creating linear habitat corridors are two potential strategies that may improve overall urban landscape connectivity for pollinators, and these are areas needing more research. Interestingly, there is some evidence that the heterogeneous, dynamic, and cosmopolitan nature of urban landscapes has actually increased the number of species that can thrive there and perhaps is evolutionarily selecting for species that are more tolerant of these conditions [113].

6. Management—What Urban Management Practices Would Support Pollinator Ecosystem Services?

Individual decisions in urban yard and garden management are critical to consider in supporting pollinator habitat [114,115]. Currently, many common urban green space management practices threaten biodiversity in cities [32]. These include the continued maintenance of turf grass lawns, which leads to a lack of foraging resources, the application of pesticides and herbicides, and tree and shrub pruning/leaf removal (especially as "fall/spring clean ups," which can greatly diminish nesting habitat for pollinators).

Mowing management, as discussed above, has been studied in a way that allows for specific applications to improve practices for pollinator ecosystem services. Less intensively managed urban lawns host more plant species [116]. If coordinated, then even a small percentage of adoptees of a lower mowing frequency, delayed mowing, or no mowing can scale up and might have positive conservation implications for bee habitat [24,114]. A reduction in mowing frequency from every few weeks to only once or twice per season causes a species turnover and increases the plant species richness of urban lawns by 30% [85]. Moreover, the change in management also increases the spatial heterogeneity within and between lawns. Lerman et al. [87] suggest a 'lazymower' approach as a practical, economical, and time-saving alternative that also helps to promote bee conservation. They argue this approach might garner broad public support (compared with lawn reduction or replacement), because it more closely aligns with current single-family homeowner motivations for adopting lawn-dominated yardscapes. However, less is not always better. A very low mowing frequency might exceed the aesthetic tolerance of many homeowners and their neighbors.

7. Conclusions and Planning Recommendations

We have framed the current literature within actionable design questions that aim to guide urban design for pollinator ecosystem services. It is clear from this review that urban green spaces present ample opportunity for meaningful—and quantifiable—improvements in pollinator habitat in the face of increasing urbanization. Here, we summarize the recommendations based on the evidence reviewed above and provide evidence-based and feasible planting typologies.

7.1. Summary of Planting Recommendations at the Local Scale

Urbanized areas, especially less densely built areas more characteristic of "suburban" or "urban sprawl" that still have a high proportion of vegetated areas, represent a matrix for some bee species, especially large-bodied, generalist, and cavity-nesting species, while many specialist species may have habitat requirements that are incompatible with most levels of urbanization. Urban environments may be able to support a diverse assemblage of bee communities by improving habitat quality at a relatively small scale. Actions to diversify urban bee community assemblages and improve the genetic diversity of existing populations should emphasize provisioning habitat and resource requirements for smaller-bodied and ground-nesting species, which make up 75% of bee species [25].

For urban habitats, it is not necessarily the degree of "urban-ness" but instead the specific characteristics of each urban green space that ultimately influence pollinator abundance and diversity. Urban areas with a higher floral diversity and floral abundance tend to have a higher pollinator abundance and/or diversity. Native gardens typically feature greater floral abundance, taller vegetation, more cover, and more potential nesting sites that likely attract pollinators. Incorporating native plants into the urban landscape is critically important for supporting abundant, diverse pollinator populations. However, even though plant nativity has been found to positively affect bee abundance and richness, it need not be the only consideration in floral resource management. Many non-native, ornamental plants serve the dual purpose of providing pollinator resources (e.g., nectar, pollen, structure) and signaling important social cues, including legibility and familiarity, that influence the acceptance, care, and longevity of gardens. Public park managers, landscapers, and even homeowners should prioritize native flowering species, but they should not be afraid to fill in gaps in flowering periods with non-native species when needed [57], if the risks discussed above are considered.

Floral morphology is another important factor to consider while improving the quality of pollinator habitat. A garden with diverse flower shapes, colors, and other features will attract a more diverse assemblage of pollinators. Moreover, providing a diversity of flowers across the entire growing season will support a greater diversity of pollinators throughout their life cycles. The seasonality of flowering times and food availability throughout the growing season is essential to ensure that a diverse community of bees can persist even within a relatively small area [65]. An array of plant species that allow for consistent flowering, from spring ephemerals to late-season bloomers, has been shown to be of greater significance to bee populations than high plant nativity. Though plant selection is often focused on herbaceous species, trees represent significant sources of floral resources when in bloom due to the size and abundance of inflorescences. Including native flowering tree species into the landscape can provide a substantial pollinator resource within a very small footprint [73] as well as woody vegetation for cavity-nesters [56]. Leaf litter, plant stems, dead and decaying plant debris, bare soil, as well as other habitats that are beyond floral resources, provide valuable nesting and overwintering spots for pollinators.

Given the research findings reviewed, to support pollinators in urban green spaces through planting design, in summary, we recommend the following:

- Incorporate high densities of flowering plants, including flowering trees.
- Incorporate a high diversity of flowering species by varying the species and floral features in each bed across the landscape.

- Utilize species with flowering times that span the entire growing season, including beneficial spring bulbs and late-blooming fall flowers.
- In areas with high aesthetic and maintenance requirements, utilize non-native, ornamental, or "nativar" plants that provide pollinator resources but do not pose a risk of invasive spread.
- In areas with lower aesthetic and maintenance requirements, utilize as many native plants as possible.
- Seek out native plants that are well-adapted to the harsh conditions of urban environments. Some desirable characteristics include salt, heat, and drought tolerance and higher pH tolerance.
- Avoid the widespread use of wood mulch; instead, create a mix of bare ground and leaf litter cover to support bees with different nesting requirements.
- Where appropriate, use plants as "living mulch" to maximize plant resources and reduce weeding and watering requirements.
- Look for opportunities to "improve the matrix" between garden beds. Consider bee-friendly lawns, flowering trees, and functional planters that can provide additional support between garden beds.

Taking these recommendations and the research into consideration, we designed four replicable garden typologies that provide multifunctional benefits to people and pollinators in an urban or suburban setting. A typology describes a design solution for a particular set of conditions. We propose typologies that address four main design goals for urban pollinator gardens: pollinator needs, safety, ease of maintenance, and cultural aesthetics (Figure 3). The four proposed typologies (Figures 4–7) represent planting templates for four common urban site conditions with plants specific to the midwestern United States. In terms of pollinator needs, the typologies consider key features such as high floral diversity and abundance, high diversity and density of native plants, sufficient floral resources throughout the growing season, and diverse floral shapes. In terms of safety concerns, it may be an urban requirement that vegetation height be less than 3 feet or canopy height greater than 6 feet to provide clear visibility. The typologies also consider ease of maintenance by using seven species or less planted at a high density that will form a tightly growing mass to shade out weeds. Several cultivated native plants (i.e., "nativars") are used that were selected for their superior landscape performance characteristics, such as drought tolerance, tidy habit, or more attractive blooms. Finally, the designs feature plants that are visually attractive, including beautiful flowers and foliage, with an emphasis on plants that provide multiple seasons of interest. These ecological and social goals inform the proposed typologies, creating design solutions that provide numerous benefits when implemented in an urban setting (Figure 8).

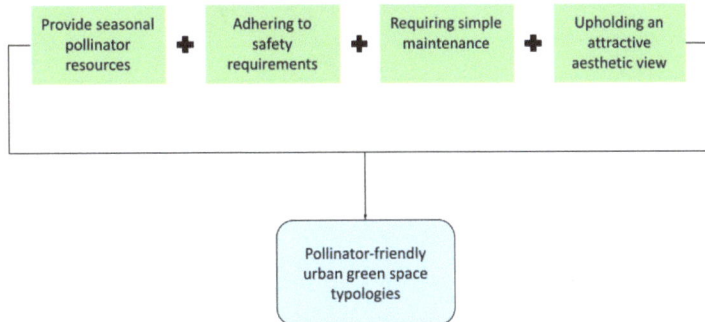

Figure 3. Guiding principles for creating informed pollinator garden typologies in an urbanized setting.

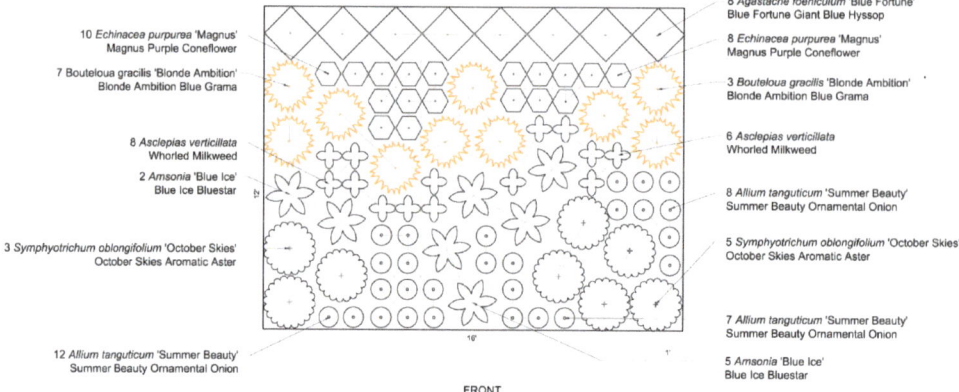

Figure 4. Typology 1: Full sun, medium moisture, high visibility (e.g., showy sidewalk garden).

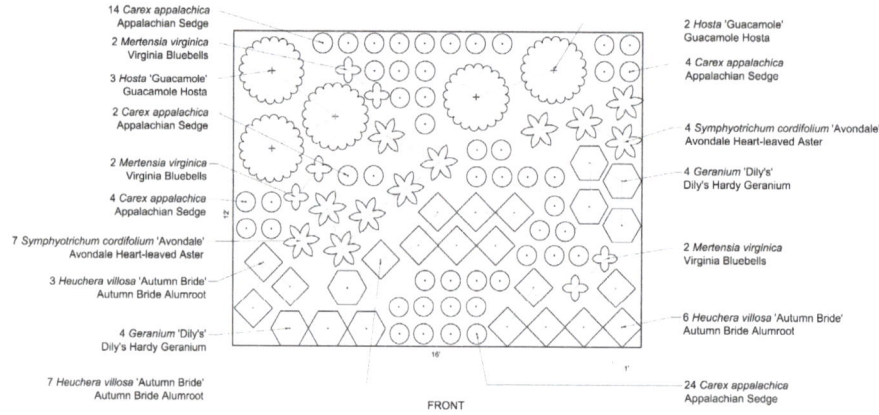

Figure 5. Typology 2: Shade, dry, high visibility (e.g., under tree canopy).

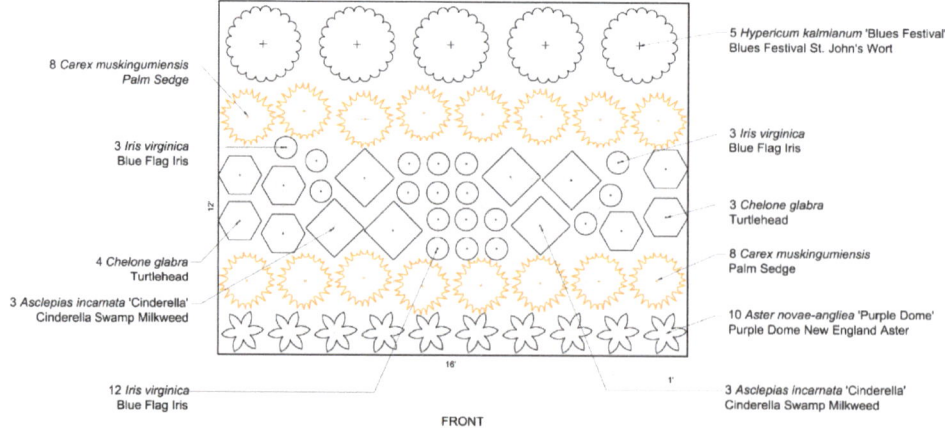

Figure 6. Typology 3: Partial sun, frequently wet (e.g., rain garden).

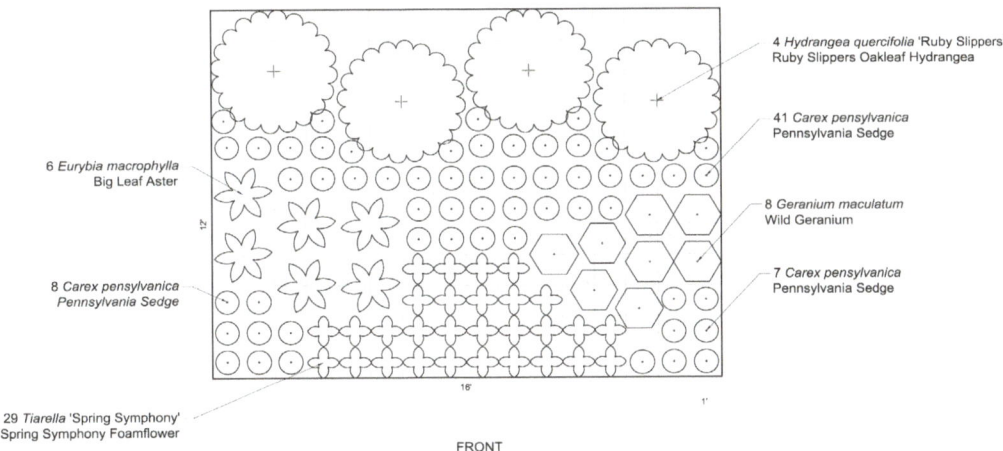

Figure 7. Typology 4: Shade, medium moisture, low visibility (e.g., foundation hedge).

Figure 8. Visual to illustrate application of Typology 3 and Typology 4 in an urban setting, where it provides multifunctional benefits of pollinator habitat, improved water infiltration (shown as arrows), and aesthetic enjoyment.

7.2. Summary of Landscape-Scale Recommendations for Urban Planning

While local garden-scale characteristics are critical for attracting and supporting pollinators, the consideration of a broader landscape-scale perspective can better inform urban planning to promote the ecosystem service value of pollinators. Pollinator foraging distances and proximity to natural habitats can be helpful factors to prioritize locations for new pollinator habitat and improve matrix quality. Linking quality, source habitats to the surrounding habitat is essential to support metapopulations of diverse bee species. Because the mean range for bee species is much smaller than for other animals, unpaved, unbuilt areas larger than 100 square meters can support abundant floral resources and may even represent viable habitat for small-bodied species [117], and even smaller spaces can be corridors. Small, unpaved surfaces in urban areas, such as along sidewalk edges or in-between buildings, should be planted with flowering herbs and trees to function as stepping-stones between habitat patches within an urban matrix [110]. By prioritizing existing urban green spaces for land-sparing and emphasizing land-sharing across the broader matrix through improvements in habitat quality, communities of more bee species may be able to establish large, connected, and persistent populations in urbanized environments. Beyond gardens designed for aesthetics, vegetable gardens can also provide the floral resources that pollinators require [56]. Property owners, especially those adjacent and near green spaces, should be encouraged to increase the overall structural and vegetative heterogeneity of their yards, but even for the most resistant of individuals, simply minimizing the presence and area of turf can be beneficial [118].

7.3. Conclusions

The field of urban pollinator landscape ecology is still developing, but there is no question that there is high potential for urban landscapes to support functional pollinator habitat. As the value of urban green spaces is increasingly being recognized, and we have accumulated a body of research on the specific features of effective pollinator habitat at the local and landscape scale, this knowledge can be directly applied to the design of individual gardens and planning across urban landscapes.

Author Contributions: All authors participated in the review and creation of the paper. B.W. and R.R. wrote the original manuscript, Z.B. and Z.L. contributed new portions and prepared the paper for publication. S.K.S. revised and edited the original and final versions and added to the literature. All authors have read and agreed to the published version of the manuscript.

Funding: This work received funding from internal sources of the School for Environment and Sustainability at the University of Michigan.

Data Availability Statement: No new data were created or analyzed in this study. Data sharing is not applicable to this article.

Conflicts of Interest: The authors declare no conflict of interest.

References

1. Potts, S.G.; Biesmeijer, J.C.; Kremen, C.; Neumann, P.; Schweiger, O.; Kunin, W.E. Global Pollinator Declines: Trends, Impacts and Drivers. *Trends Ecol. Evol.* **2010**, *25*, 345–353. [CrossRef]
2. Pleasants, J.M.; Oberhauser, K.S. Milkweed Loss in Agricultural Fields Because of Herbicide Use: Effect on the Monarch Butterfly Population. *Insect Conserv. Divers.* **2013**, *6*, 135–144. [CrossRef]
3. Goulson, D.; Nicholls, E.; Botías, C.; Rotheray, E.L. Bee Declines Driven by Combined Stress from Parasites, Pesticides, and Lack of Flowers. *Science* **2015**, *347*, 1255957. [CrossRef]
4. Nicolson, S.W.; Wright, G.A. Plant—Pollinator Interactions and Threats to Pollination. *Funct. Ecol.* **2017**, *31*, 22–25. [CrossRef]
5. IPBES. *The Assessment Report of the Intergovernmental Science-Policy Platform on Biodiversity and Ecosystem Services on Pollinators, Pollination and Food Production*; Potts, S.G., Imperatriz-Fonseca, V.L., Ngo, H.T., Eds.; Secretariat of the Intergovernmental Science-Policy Platform on Biodiversity and Ecosystem Services: Bonn, Germany, 2016; 552p.
6. Biesmeijer, J.C.; Roberts, S.P.M.; Reemer, M.; Ohlemüller, R.; Edwards, M.; Peeters, T.; Schaffers, A.P.; Potts, S.G.; Kleukers, R.; Thomas, C.D.; et al. Parallel Declines in Pollinators and Insect-Pollinated Plants in Britain and the Netherlands. *Science* **2006**, *313*, 351–354. [CrossRef] [PubMed]
7. Burkle, L.A.; Marlin, J.C.; Knight, T.M. Plant-Pollinator Interactions over 120 Years: Loss of Species, Co-Occurrence, and Function. *Science* **2013**, *339*, 1611–1615. [CrossRef] [PubMed]
8. Cameron, S.A.; Lozier, J.D.; Strange, J.P.; Koch, J.B.; Cordes, N.; Solter, L.F.; Griswold, T.L. Patterns of Widespread Decline in North American Bumble Bees. *Proc. Natl. Acad. Sci.* **2011**, *108*, 662–667. [CrossRef] [PubMed]
9. Martins, A.C.; Gonçalves, R.B.; Melo, G.A.R. Changes in Wild Bee Fauna of a Grassland in Brazil Reveal Negative Effects Associated with Growing Urbanization during the Last 40 Years. *Zoologia (Curitiba)* **2013**, *30*, 157–176. [CrossRef]
10. Balmford, A.; Bruner, A.; Cooper, P.; Costanza, R.; Farber, S.; Green, R.E.; Jenkins, M.; Jefferiss, P.; Jessamy, V.; Madden, J.; et al. Ecology: Economic Reasons for Conserving Wild Nature. *Science* **2002**, *297*, 950–953. [CrossRef]
11. Leemans, R.; Groot, R.S. *de Millennium Ecosystem Assessment: Ecosystems and Human Well-Being: A Framework for Assessment*; Island Press: Washington, DC, USA, 2003; ISBN 978-1-55963-402-1.
12. Ollerton, J.; Winfree, R.; Tarrant, S. How Many Flowering Plants Are Pollinated by Animals? *Oikos* **2011**, *120*, 321–326. [CrossRef]

13. Klein, A.-M.; Vaissière, B.E.; Cane, J.H.; Steffan-Dewenter, I.; Cunningham, S.A.; Kremen, C.; Tscharntke, T. Importance of Pollinators in Changing Landscapes for World Crops. *Proc. R. Soc. B Biol. Sci.* **2007**, *274*, 303–313. [CrossRef]
14. Gallai, N.; Salles, J.-M.; Settele, J.; Vaissière, B.E. Economic Valuation of the Vulnerability of World Agriculture Confronted with Pollinator Decline. *Ecol. Econ.* **2009**, *68*, 810–821. [CrossRef]
15. Reilly, J.R.; Artz, D.R.; Biddinger, D.; Bobiwash, K.; Boyle, N.K.; Brittain, C.; Brokaw, J.; Campbell, J.W.; Daniels, J.; Elle, E.; et al. Crop Production in the USA Is Frequently Limited by a Lack of Pollinators. *Proc. R. Soc. B Biol. Sci.* **2020**, *287*, 20200922. [CrossRef] [PubMed]
16. Fuller, R.A.; Irvine, K.N.; Devine-Wright, P.; Warren, P.H.; Gaston, K.J. Psychological Benefits of Greenspace Increase with Biodiversity. *Biol. Lett.* **2007**, *3*, 390–394. [CrossRef] [PubMed]
17. Harrison, T.; Winfree, R. Urban Drivers of Plant-Pollinator Interactions. *Funct. Ecol.* **2015**, *29*, 879–888. [CrossRef]
18. Ayers, A.C.; Rehan, S.M. Supporting Bees in Cities: How Bees Are Influenced by Local and Landscape Features. *Insects* **2021**, *12*, 128. [CrossRef]
19. Prendergast, K.S.; Dixon, K.W.; Bateman, P.W. A Global Review of Determinants of Native Bee Assemblages in Urbanised Landscapes. *Insect Conserv. Divers.* **2022**, *15*, 385–405. [CrossRef]
20. Ahrné, K.; Bengtsson, J.; Elmqvist, T. Bumble Bees (*Bombus* spp.) along a Gradient of Increasing Urbanization. *PLoS ONE* **2009**, *4*, e5574. [CrossRef]
21. Geslin, B.; Gauzens, B.; Thébault, E.; Dajoz, I. Plant Pollinator Networks along a Gradient of Urbanisation. *PLoS ONE* **2013**, *8*, e63421. [CrossRef]
22. Jones, E.L.; Leather, S.R. Invertebrates in Urban Areas: A Review. *EJE* **2013**, *109*, 463–478. [CrossRef]
23. Baldock, K.C.; Goddard, M.A.; Hicks, D.M.; Kunin, W.E.; Mitschunas, N.; Osgathorpe, L.M.; Potts, S.G.; Robertson, K.M.; Scott, A.V.; Stone, G.N. Where Is the UK's Pollinator Biodiversity? The Importance of Urban Areas for Flower-Visiting Insects. *Proc. R. Soc. B Biol. Sci.* **2015**, *282*, 20142849. [CrossRef]
24. Hall, D.M.; Camilo, G.R.; Tonietto, R.K.; Ollerton, J.; Ahrné, K.; Arduser, M.; Ascher, J.S.; Baldock, K.C.R.; Fowler, R.; Frankie, G.; et al. The city as a refuge for insect pollinators. *Conserv. Biol.* **2017**, *31*, 24–29. [CrossRef]
25. Wenzel, A.; Grass, I.; Belavadi, V.V.; Tscharntke, T. How Urbanization Is Driving Pollinator Diversity and Pollination—A Systematic Review. *Biol. Conserv.* **2020**, *241*, 108321. [CrossRef]
26. Potter, A.; LeBuhn, G. Pollination Service to Urban Agriculture in San Francisco, CA. *Urban Ecosyst.* **2015**, *18*, 885–893. [CrossRef]
27. Goulson, D.; Lepais, O.; O'connor, S.; Osborne, J.L.; Sanderson, R.A.; Cussans, J.; Goffe, L.; Darvill, B. Effects of Land Use at a Landscape Scale on Bumblebee Nest Density and Survival. *J. Appl. Ecol.* **2010**, *47*, 1207–1215. [CrossRef]
28. Langellotto, G.A.; Melathopoulos, A.; Messer, I.; Anderson, A.; McClintock, N.; Costner, L. Garden Pollinators and the Potential for Ecosystem Service Flow to Urban and Peri-Urban Agriculture. *Sustainability* **2018**, *10*, 2047. [CrossRef]
29. Silva, V.H.D.; Gomes, I.N.; Cardoso, J.C.F.; Bosenbecker, C.; Silva, J.L.S.; Cruz-Neto, O.; Oliveira, W.; Stewart, A.B.; Lopes, A.V.; Maruyama, P.K. Diverse Urban Pollinators and Where to Find Them. *Biol. Conserv.* **2023**, *281*, 110036. [CrossRef]
30. Baldock, K.C. Opportunities and Threats for Pollinator Conservation in Global Towns and Cities. *Curr. Opin. Insect Sci.* **2020**, *38*, 63–71. [CrossRef] [PubMed]
31. Xerces.org. Available online: https://xerces.org/publications/guidelines/pollinator-friendly-parks (accessed on 23 May 2023).
32. Aronson, M.F.; Lepczyk, C.A.; Evans, K.L.; Goddard, M.A.; Lerman, S.B.; MacIvor, J.S.; Nilon, C.H.; Vargo, T. Biodiversity in the City: Key Challenges for Urban Green Space Management. *Front. Ecol. Environ.* **2017**, *15*, 189–196. [CrossRef]
33. Llodrà-Llabrés, J.; Cariñanos, P. Enhancing Pollination Ecosystem Service in Urban Green Areas: An Opportunity for the Conservation of Pollinators. *Urban For. Urban Green.* **2022**, *74*, 127621. [CrossRef]
34. Davis, A.Y.; Lonsdorf, E.V.; Shierk, C.R.; Matteson, K.C.; Taylor, J.R.; Lovell, S.T.; Minor, E.S. Enhancing Pollination Supply in an Urban Ecosystem through Landscape Modifications. *Landsc. Urban Plan.* **2017**, *162*, 157–166. [CrossRef]
35. Bertuol-Garcia, D.; Morsello, C.; El-Hani, C.N.; Pardini, R. A Conceptual Framework for Understanding the Perspectives on the Causes of the Science–Practice Gap in Ecology and Conservation. *Biol. Rev.* **2018**, *93*, 1032–1055. [CrossRef]
36. Enquist, C.A.; Jackson, S.T.; Garfin, G.M.; Davis, F.W.; Gerber, L.R.; Littell, J.A.; Tank, J.L.; Terando, A.J.; Wall, T.U.; Halpern, B.; et al. Foundations of Translational Ecology. *Front. Ecol. Environ.* **2017**, *15*, 541–550. [CrossRef]
37. Toomey, A.H.; Knight, A.T.; Barlow, J. Navigating the Space between Research and Implementation in Conservation. *Conserv. Lett.* **2017**, *10*, 619–625. [CrossRef]
38. Nascimento, V.T.; Agostini, K.; Souza, C.S.; Maruyama, P.K. Tropical Urban Areas Support Highly Diverse Plant-Pollinator Interactions: An Assessment from Brazil. *Landsc. Urban. Plan.* **2020**, *198*, 103801. [CrossRef]
39. Maruyama, P.K.; Bonizário, C.; Marcon, A.P.; D'Angelo, G.; da Silva, M.M.; da Silva Neto, E.N.; Oliveira, P.E.; Sazima, I.; Sazima, M.; Vizentin-Bugoni, J.; et al. Plant-Hummingbird Interaction Networks in Urban Areas: Generalization and the Importance of Trees with Specialized Flowers as a Nectar Resource for Pollinator Conservation. *Biol. Conserv.* **2019**, *230*, 187–194. [CrossRef]
40. Diniz, U.M.; Lima, S.A.; Machado, I.C.S. Short-Distance Pollen Dispersal by Bats in an Urban Setting: Monitoring the Movement of a Vertebrate Pollinator through Fluorescent Dyes. *Urban Ecosyst.* **2019**, *22*, 281–291. [CrossRef]
41. Smith, T.J.; Saunders, M.E. Honey Bees: The Queens of Mass Media, despite Minority Rule among Insect Pollinators. *Insect Conserv. Divers.* **2016**, *9*, 384–390. [CrossRef]

42. Adamson, N.; Roulston, T.; Fell, R.; Mullins, D. From April to August—Wild Bees Pollinating Crops through the Growing Season in Virginia, USA. *Environ. Entomol.* **2012**, *41*, 813–821. [CrossRef]
43. Paini, D.R. Impact of the Introduced Honey Bee (*Apis mellifera*) (Hymenoptera: Apidae) on Native Bees: A Review. *Austral. Ecol.* **2004**, *29*, 399–407. [CrossRef]
44. MacInnis, G.; Normandin, E.; Ziter, C.D. Decline in Wild Bee Species Richness Associated with Honey Bee (*Apis mellifera* L.) Abundance in an Urban Ecosystem. *PeerJ* **2023**, *11*, e14699. [CrossRef]
45. Senapathi, D.; Biesmeijer, J.C.; Breeze, T.D.; Kleijn, D.; Potts, S.G.; Carvalheiro, L.G. Pollinator Conservation—The Difference between Managing for Pollination Services and Preserving Pollinator Diversity. *Curr. Opin. Insect Sci.* **2015**, *12*, 93–101. [CrossRef]
46. Senapathi, D.; Goddard, M.A.; Kunin, W.E.; Baldock, K.C.R. Landscape Impacts on Pollinator Communities in Temperate Systems: Evidence and Knowledge Gaps. *Funct. Ecol.* **2017**, *31*, 26–37. [CrossRef]
47. Holm, H. *Pollinators of Native Plants*; Pollination Press LLC: Minnetonka, MN, USA, 2014; ISBN 97809913563001.
48. Geslin, B.; Le Féon, V.; Folschweiller, M.; Flacher, F.; Carmignac, D.; Motard, E.; Perret, S.; Dajoz, I. The Proportion of Impervious Surfaces at the Landscape Scale Structures Wild Bee Assemblages in a Densely Populated Region. *Ecol. Evol.* **2016**, *6*, 6599–6615. [CrossRef] [PubMed]
49. Hinners, S.J.; Kearns, C.A.; Wessman, C.A. Roles of Scale, Matrix, and Native Habitat in Supporting a Diverse Suburban Pollinator Assemblage. *Ecol. Appl.* **2012**, *22*, 1923–1935. [CrossRef] [PubMed]
50. Vaughan, M.; Black, S.H. *Enhancing Nest Sites For Native Bee Crop Pollinators. Agroforestry note 34*; USDA National Agroforestry Center: Lincoln, NE, USA, 2007.
51. Bates, A.J.; Sadler, J.P.; Fairbrass, A.J.; Falk, S.J.; Hale, J.D.; Matthews, T.J. Changing Bee and Hoverfly Pollinator Assemblages along an Urban-Rural Gradient. *PLoS ONE* **2011**, *6*, e23459. [CrossRef]
52. Daniels, B.; Jedamski, J.; Ottermanns, R.; Ross-Nickoll, M. A "Plan Bee" for Cities: Pollinator Diversity and Plant-Pollinator Interactions in Urban Green Spaces. *PLoS ONE* **2020**, *15*, e0235492. [CrossRef]
53. Persson, A.S.; Ekroos, J.; Olsson, P.; Smith, H.G. Wild Bees and Hoverflies Respond Differently to Urbanisation, Human Population Density and Urban Form. *Landsc. Urban Plan.* **2020**, *204*, 103901. [CrossRef]
54. Chong, K.Y.; Teo, S.; Kurukulasuriya, B.; Chung, Y.F.; Rajathurai, S.; Tan, H.T.W. Not All Green Is as Good: Different Effects of the Natural and Cultivated Components of Urban Vegetation on Bird and Butterfly Diversity. *Biol. Conserv.* **2014**, *171*, 299–309. [CrossRef]
55. Dylewski, Ł.; Maćkowiak, Ł.; Banaszak-Cibicka, W. Are All Urban Green Spaces a Favourable Habitat for Pollinator Communities? Bees, Butterflies and Hoverflies in Different Urban Green Areas. *Ecol. Entomol.* **2019**, *44*, 678–689. [CrossRef]
56. Pardee, G.L.; Philpott, S.M. Native Plants Are the Bee's Knees: Local and Landscape Predictors of Bee Richness and Abundance in Backyard Gardens. *Urban Ecosyst.* **2014**, *17*, 641–659. [CrossRef]
57. Salisbury, A.; Armitage, J.; Bostock, H.; Perry, J.; Tatchell, M.; Thompson, K. Editor's Choice: Enhancing Gardens as Habitats for Flower-Visiting Aerial Insects (Pollinators): Should We Plant Native or Exotic Species? *J. Appl. Ecol.* **2015**, *52*, 1156–1164. [CrossRef]
58. Wojcik, V.A.; McBride, J.R. Common Factors Influence Bee Foraging in Urban and Wildland Landscapes. *Urban Ecosyst.* **2012**, *15*, 581–598. [CrossRef]
59. Tonietto, R.; Fant, J.; Ascher, J.; Ellis, K.; Larkin, D. A Comparison of Bee Communities of Chicago Green Roofs, Parks and Prairies. *Landsc. Urban Plan.* **2011**, *103*, 102–108. [CrossRef]
60. Tallamy, D.W.; Narango, D.L.; Mitchell, A.B. Do Non-Native Plants Contribute to Insect Declines? *Ecol. Entomol.* **2020**, *46*, 729–742. [CrossRef]
61. Erickson, E.; Adam, S.; Russo, L.; Wojcik, V.; Patch, H.M.; Grozinger, C.M. More Than Meets the Eye? The Role of Annual Ornamental Flowers in Supporting Pollinators. *Environ. Entomol.* **2020**, *49*, 178–188. [CrossRef]
62. Fukase, J. Increased Pollinator Activity in Urban Gardens with More Native Flora. *Appl. Ecol. Env. Res.* **2016**, *14*, 297–310. [CrossRef]
63. Rollings, R.; Goulson, D. Quantifying the Attractiveness of Garden Flowers for Pollinators. *J. Insect. Conserv.* **2019**, *23*, 803–817. [CrossRef]
64. Frankie, G.W.; Thorp, R.W.; Schindler, M.; Hernandez, J.; Ertter, B.; Rizzardi, M. Ecological Patterns of Bees and Their Host Ornamental Flowers in Two Northern California Cities. *J. Kans. Entomol. Soc.* **2005**, *78*, 227–246. [CrossRef]
65. Wojcik, V.A.; Frankie, G.W.; Thorp, R.W.; Hernandez, J.L. Seasonality in Bees and Their Floral Resource Plants at a Constructed Urban Bee Habitat in Berkeley, California. *J. Kans. Entomol.* **2008**, *81*, 15–28. [CrossRef]
66. Praz, C.J.; Müller, A.; Dorn, S. Specialized Bees Fail to Develop on Non-Host Pollen: Do Plants Chemically Protect Their Pollen. *Ecology* **2008**, *89*, 795–804. [CrossRef]
67. Winfree, R.; Fox, W.J.; Williams, N.M.; Reilly, J.R.; Cariveau, D.P. Abundance of Common Species, Not Species Richness, Drives Delivery of a Real-World Ecosystem Service. *Ecol. Lett.* **2015**, *18*, 626–635. [CrossRef] [PubMed]
68. Morales, C.L.; Traveset, A. A Meta-Analysis of Impacts of Alien vs. Native Plants on Pollinator Visitation and Reproductive Success of Co-Flowering Native Plants. *Ecol. Lett.* **2009**, *12*, 716–728. [CrossRef]
69. von der Lippe, M.; Kowarik, I. Do Cities Export Biodiversity? Traffic as Dispersal Vector across Urban–Rural Gradients. *Divers. Distrib.* **2008**, *14*, 18–25. [CrossRef]

70. McLean, P.; Gallien, L.; Wilson, J.R.U.; Gaertner, M.; Richardson, D.M. Small Urban Centres as Launching Sites for Plant Invasions in Natural Areas: Insights from South Africa. *Biol. Invasions* **2017**, *19*, 3541–3555. [CrossRef]
71. Garbuzov, M.; Ratnieks, F.L.W. The Strengths and Weaknesses of Lists of Garden Plants to Help Pollinators: The Strengths and Weaknesses of Lists of Garden Plants to Help Pollinators. *BioScience* **2014**, *64*, 1019–1026. [CrossRef]
72. Nassauer, J.I. Messy Ecosystems, Orderly Frames. *Landsc. J.* **1995**, *14*, 161–170. [CrossRef]
73. Somme, L.; Moquet, L.; Quinet, M.; Vanderplanck, M.; Michez, D.; Lognay, G.; Jacquemart, A.-L. Food in a Row: Urban Trees Offer Valuable Floral Resources to Pollinating Insects. *Urban Ecosyst.* **2016**, *19*, 1149–1161. [CrossRef]
74. Lowenstein, D.M.; Matteson, K.C.; Minor, E.S. Evaluating the Dependence of Urban Pollinators on Ornamental, Non-Native, and 'Weedy' Floral Resources. *Urban Ecosyst.* **2019**, *22*, 293–302. [CrossRef]
75. Staab, M.; Pereira-Peixoto, M.H.; Klein, A.-M. Exotic Garden Plants Partly Substitute for Native Plants as Resources for Pollinators When Native Plants Become Seasonally Scarce. *Oecologia* **2020**, *194*, 465–480. [CrossRef]
76. Davis, M.A.; Chew, M.K.; Hobbs, R.J.; Lugo, A.E.; Ewel, J.J.; Vermeij, G.J.; Brown, J.H.; Rosenzweig, M.L.; Gardener, M.R.; Carroll, S.P.; et al. Don't Judge Species on Their Origins. *Nature* **2011**, *474*, 153–154. [CrossRef] [PubMed]
77. Coombs, G. Monarda Research Report. Available online: https://issuu.com/mtcuba/docs/monarda-report-final/1 (accessed on 17 February 2021).
78. Baisden, E.C.; Tallamy, D.W.; Narango, D.L.; Boyle, E. Do Cultivars of Native Plants Support Insect Herbivores? *HortTechnology* **2018**, *28*, 596–606. [CrossRef]
79. Fenster, C.B.; Armbruster, W.S.; Wilson, P.; Dudash, M.R.; Thomson, J.D. Pollination Syndromes and Floral Specialization. *Annu. Rev. Ecol. Evol. Syst.* **2004**, *35*, 375–403. [CrossRef]
80. Robbins, P.; Birkenholtz, T. Turfgrass Revolution: Measuring the Expansion of the American Lawn. *Land Use Policy* **2003**, *20*, 181–194. [CrossRef]
81. Ignatieva, M.; Ahrné, K.; Wissman, J.; Eriksson, T.; Tidåker, P.; Hedblom, M.; Kätterer, T.; Marstorp, H.; Berg, P.; Eriksson, T. Lawn as a Cultural and Ecological Phenomenon: A Conceptual Framework for Transdisciplinary Research. *Urban For. Urban Green.* **2015**, *14*, 383–387. [CrossRef]
82. Robbins, P. *Lawn People: How Grasses, Weeds, and Chemicals Make Us Who We Are*; Temple University Press: Philadelphia, PA, USA, 2012; ISBN 9781592135806.
83. Jenkins, V. *The Lawn: A History of an American Obsession*; Smithsonian Institution: Washington, DC, USA, 1994; ISBN 9781560984061.
84. Nassauer, J.I.; Wang, Z.; Dayrell, E. What Will the Neighbors Think? Cultural Norms and Ecological Design. *Landsc. Urban Plan.* **2009**, *92*, 282–292. [CrossRef]
85. Sehrt, M.; Bossdorf, O.; Freitag, M.; Bucharova, A. Less Is More! Rapid Increase in Plant Species Richness after Reduced Mowing in Urban Grasslands. *Basic Appl. Ecol.* **2020**, *42*, 47–53. [CrossRef]
86. Fetridge, E.D.; Ascher, J.S.; Langellotto, G.A. The Bee Fauna of Residential Gardens in a Suburb of New York City (Hymenoptera: Apoidea). *Ann. Entomol. Soc. Am.* **2008**, *101*, 1067–1077. [CrossRef]
87. Lerman, S.B.; Contosta, A.R.; Milam, J.; Bang, C. To Mow or to Mow Less: Lawn Mowing Frequency Affects Bee Abundance and Diversity in Suburban Yards. *Biol. Conserv.* **2018**, *221*, 160–174. [CrossRef]
88. Knight, T.M.; Ashman, T.-L.; Bennett, J.M.; Burns, J.H.; Passonneau, S.; Steets, J.A. Reflections on, and Visions for, the Changing Field of Pollination Ecology. *Ecol. Lett.* **2018**, 1282–1295. [CrossRef]
89. Smith, R.M.; Gaston, K.J.; Warren, P.H.; Thompson, K. Urban Domestic Gardens (VIII): Environmental Correlates of Invertebrate Abundance. *Biodivers. Conserv.* **2006**, *15*, 2515–2545. [CrossRef]
90. Hülsmann, M.; von Wehrden, H.; Klein, A.-M.; Leonhardt, S.D. Plant Diversity and Composition Compensate for Negative Effects of Urbanization on Foraging Bumble Bees. *Apidologie* **2015**, *46*, 760–770. [CrossRef]
91. Keasar, T. The Spatial Distribution of Nonrewarding Artificial Flowers Affects Pollinator Attraction. *Anim. Behav.* **2000**, *60*, 639–646. [CrossRef] [PubMed]
92. Normandin, É.; Vereecken, N.J.; Buddle, C.M.; Fournier, V. Taxonomic and Functional Trait Diversity of Wild Bees in Different Urban Settings. *PeerJ* **2017**, *5*, e3051. [CrossRef]
93. Plascencia, M.; Philpott, S.M. Floral Abundance, Richness, and Spatial Distribution Drive Urban Garden Bee Communities. *Bull. Entomol. Res.* **2017**, *107*, 658–667. [CrossRef]
94. Quistberg, R.D.; Bichier, P.; Philpott, S.M. Landscape and Local Correlates of Bee Abundance and Species Richness in Urban Gardens. *Environ. Entomol.* **2016**, *45*, 592–601. [CrossRef]
95. O'Toole, C.; Raw, A. *Bees of the World*; Facts on File: New York, NY, USA, 2004; ISBN 9780816057122.
96. Michener, C.D. *The Bees of the World*, 2nd ed.; The Johns Hopkins University Press: Baltimore, MD, USA, 2007; ISBN 9780801885730.
97. MacIvor, J.S.; Packer, L. 'Bee hotels' as tools for native pollinator conservation: A premature verdict? *PLoS ONE* **2015**, *10*, e0122126. [CrossRef]
98. Geslin, B.; Gachet, S.; Deschamps-Cottin, M.; Flacher, F.; Ignace, B.; Knoploch, C.; Meineri, É.; Robles, C.; Ropars, L.; Schurr, L.; et al. Bee Hotels Host a High Abundance of Exotic Bees in an Urban Context. *Acta Oecolog.* **2020**, *105*, 103556. [CrossRef]
99. Neame, L.A.; Griswold, T.; Elle, E. Pollinator Nesting Guilds Respond Differently to Urban Habitat Fragmentation in an Oak-Savannah Ecosystem. *Insect Conserv. Divers.* **2013**, *6*, 57–66. [CrossRef]
100. Fortel, L.; Henry, M.; Guilbaud, L.; Mouret, H.; Vaissière, B.E. Use of human-made nesting structures by wild bees in an urban environment. *J. Insect Conserv.* **2016**, *20*, 239–253. [CrossRef]

101. Cane, J.H. Landscaping pebbles attract nesting by the native ground-nesting bee *Halictus rubicundus* (Hymenoptera: Halictidae). *Apidologie* **2015**, *46*, 728–734. [CrossRef]
102. Baldock, K.C.R.; Goddard, M.A.; Hicks, D.M.; Kunin, W.E.; Mitschunas, N.; Morse, H.; Osgathorpe, L.M.; Potts, S.G.; Robertson, K.M.; Scott, A.V.; et al. A Systems Approach Reveals Urban Pollinator Hotspots and Conservation Opportunities. *Nat. Ecol. Evol.* **2019**, *3*, 363–373. [CrossRef]
103. Majewska, A.A.; Altizer, S. Planting Gardens to Support Insect Pollinators. *Conserv. Biol.* **2020**, *34*, 15–25. [CrossRef] [PubMed]
104. Donald, P.F.; Evans, A.D. Habitat Connectivity and Matrix Restoration: The Wider Implications of Agri-Environment Schemes. *J. Appl. Ecol.* **2006**, *43*, 209–218. [CrossRef]
105. Egerer, M.H.; Arel, C.; Otoshi, M.D.; Quistberg, R.D.; Bichier, P.; Philpott, S.M. Urban Arthropods Respond Variably to Changes in Landscape Context and Spatial Scale. *J. Urban Ecol.* **2017**, *3*, jux001. [CrossRef]
106. Steffan-Dewenter, I.; Tscharntke, T. Effects of habitat isolation on pollinator communities and seed set. *Oecologia* **1999**, *121*, 432–440. [CrossRef] [PubMed]
107. Garibaldi, L.A.; Steffan-Dewenter, I.; Kremen, C.; Morales, J.M.; Bommarco, R.; Cunningham, S.A.; Carvalheiro, L.G.; Chacoff, N.P.; Dudenhöffer, J.H.; Greenleaf, S.S.; et al. Stability of Pollination Services Decreases with Isolation from Natural Areas despite Honey Bee Visits. *Ecol. Lett.* **2011**, *14*, 1062–1072. [CrossRef]
108. Kennedy, C.M.; Lonsdorf, E.; Neel, M.C.; Williams, N.M.; Ricketts, T.H.; Winfree, R.; Bommarco, R.; Brittain, C.; Burley, A.L.; Cariveau, D.; et al. A Global Quantitative Synthesis of Local and Landscape Effects on Wild Bee Pollinators in Agroecosystems. *Ecol. Lett.* **2013**, *16*, 584–599. [CrossRef]
109. Fischer, L.K.; Eichfeld, J.; Kowarik, I.; Buchholz, S. Disentangling Urban Habitat and Matrix Effects on Wild Bee Species. *PeerJ* **2016**, *4*, e2729. [CrossRef] [PubMed]
110. Van Rossum, F.; Triest, L. Stepping-Stone Populations in Linear Landscape Elements Increase Pollen Dispersal between Urban Forest Fragments. *Plant Ecol. Evol.* **2012**, *145*, 332–340. [CrossRef]
111. Van Geert, A.; Van Rossum, F.; Triest, L. Do linear landscape elements in farmland act as biological corridors for pollen dispersal? *J. Ecol.* **2010**, *98*, 178–187. [CrossRef]
112. Garratt, M.P.; Senapathi, D.; Coston, D.J.; Mortimer, S.R.; Potts, S.G. The benefits of hedgerows for pollinators and natural enemies depends on hedge quality and landscape context. *Agric. Ecosyst. Environ.* **2017**, *247*, 363–370. [CrossRef]
113. Sattler, T.; Duelli, P.; Obrist, M.K.; Arlettaz, R.; Moretti, M. Response of arthropod species richness and functional groups to urban habitat structure and management. *Landsc. Ecol.* **2010**, *25*, 941–954. [CrossRef]
114. Goddard, M.A.; Dougill, A.J.; Benton, T.G. Scaling up from Gardens: Biodiversity Conservation in Urban Environments. *Trends Ecol. Evol.* **2010**, *25*, 90–98. [CrossRef] [PubMed]
115. Shwartz, A.; Muratet, A.; Simon, L.; Julliard, R. Local and Management Variables Outweigh Landscape Effects in Enhancing the Diversity of Different Taxa in a Big Metropolis. *Biol. Conserv.* **2013**, *157*, 285–292. [CrossRef]
116. Rudolph, M.; Velbert, F.; Schwenzfeier, S.; Kleinebecker, T.; Klaus, V.H. Patterns and Potentials of Plant Species Richness in High-and Low-maintenance Urban Grasslands. *Appl. Veg. Sci.* **2017**, *20*, 18–27. [CrossRef]
117. Blaauw, B.R.; Isaacs, R. Larger patches of diverse floral resources increase insect pollinator density, diversity, and their pollination of native wildflowers. *Basic Appl. Ecol.* **2014**, *15*, 701–711. [CrossRef]
118. Turo, K.J.; Gardiner, M.M. Effects of urban greenspace configuration and native vegetation on bee and wasp reproduction. *Conserv. Biol.* **2021**, *35*, 1755–1765. [CrossRef]

Disclaimer/Publisher's Note: The statements, opinions and data contained in all publications are solely those of the individual author(s) and contributor(s) and not of MDPI and/or the editor(s). MDPI and/or the editor(s) disclaim responsibility for any injury to people or property resulting from any ideas, methods, instructions or products referred to in the content.

Article

Prediction of Urban Trees Planting Base on Guided Cellular Automata to Enhance the Connection of Green Infrastructure

Yi Le * and Sheng-Yang Huang

Bartlett School of Architecture, University College London, London WC1E 6BT, UK; ucfnhua@ucl.ac.uk
* Correspondence: zczlly7@ucl.ac.uk

Abstract: Urbanization and climate change pose significant challenges to urban ecosystems, underscoring the necessity for innovative strategies to enhance urban green infrastructure. Tree planting, a crucial aspect of green infrastructure, has been analyzed for optimized positioning using data metrics, priority scoring, and GIS. However, due to the dynamic nature of environmental information, the accuracy of current approaches is compromised. This study aims to present a novel approach integrating deep learning and cellular automata to prioritize urban tree planting locations to anticipate the optimal urban tree network. Initially, GIS data were collated and visualized to identify a suitable study site within London. CycleGAN models were trained using cellular automata outputs and forest mycorrhizal network samples. The comparison validated cellular automata's applicability, enabled observing spatial feature information in the outputs and guiding the parameter design of our 3D cellular automata system for predicting tree planting locations. The locations were optimized by simulating the network connectivity of urban trees after planting, following the spatial-behavioral pattern of the forest mycorrhizal network. The results highlight the role of robust tree networks in fostering ecological stability and cushioning climate change impacts in urban contexts. The proposed approach addresses existing methodological and practical limitations, providing innovative strategies for optimal tree planting and prioritization of urban green infrastructure, thereby informing sustainable urban planning and design. Our findings illustrate the symbiotic relationship between urban trees and future cities and offer insights into street tree density planning, optimizing the spatial distribution of trees within urban landscapes for sustainable urban development.

Keywords: carbon emissions; urban planting; ecological system; urban forestry; green infrastructure

Citation: Le, Y.; Huang, S.-Y. Prediction of Urban Trees Planting Base on Guided Cellular Automata to Enhance the Connection of Green Infrastructure. *Land* **2023**, *12*, 1479. https://doi.org/10.3390/land12081479

Academic Editors: Alessio Russo and Giuseppe T. Cirella

Received: 12 June 2023
Revised: 17 July 2023
Accepted: 22 July 2023
Published: 25 July 2023

Copyright: © 2023 by the authors. Licensee MDPI, Basel, Switzerland. This article is an open access article distributed under the terms and conditions of the Creative Commons Attribution (CC BY) license (https://creativecommons.org/licenses/by/4.0/).

1. Introduction

Carbon dioxide plays an important role in ecosystems [1,2]. Since pre-industrial times, seasonal mean temperatures have been anomalous over most land areas and atmospheric CO_2 has been steadily increasing, leading to global warming and more frequent natural disasters [3–5]. The Intergovernmental Panel on Climate Change (IPCC) concluded in its Climate Change 2001 report that "humans have a clear impact on the global climate" [6]. The increasing concentrations of carbon dioxide (CO_2), ozone (O_3), methane (CH_4) and nitrous oxide (NO) in the atmosphere make it difficult for the heat radiated by the sun to radiate into the air, resulting in higher temperatures near the surface and causing the greenhouse effect [7,8]. The rise in carbon emissions is driven by the burning of fossil fuels, the manufacture of commodities, deforestation, the use of transport, food production, the use of electricity in buildings, etc. [9,10]. The world's main sources of carbon emissions are concentrated in three main regions—the USA, China and Europe—and the highest emissions are spread around cities [11,12]. Excessive urban carbon emission makes the temperature in the city center significantly higher than that in the surrounding areas, which increases the temperature difference between day and night and leads to the urban heat island effect, and also aggravates the frequency of natural disasters [13–15]. The growing demand for transportation in urban life causes the imbalance of the urban ecosystem and damages the urban environment [7,16]. Climate change has become one of the greatest challenges facing humanity in the 21st century [17,18].

Forests are the largest plant communities on land and play an important role in the absorption of CO_2 [19,20]. Trees in forests, from growth to death, absorb carbon dioxide through photosynthesis and respiration and fix it in the vegetation and soil, and the capacity of different parts of the forest tree to absorb carbon varies [21]. There are different types of forests on the planet, such as tropical rainforests, temperate rainforests, temperate deciduous broadleaf forests, and temperate coniferous forests. The biodiversity and carbon storage capacity of forests at different latitudes also differ [22,23]. The Amazon basin is particularly rich and is the largest ecosystem carbon sink on Earth that could help mitigate carbon emissions [24,25]. The richness of the forest hierarchy helps to build solid forest ecosystems [26,27]. However, with human deforestation and forest degradation, the Amazon's carbon sink capacity has gradually diminished and the growth rate of above-ground biomass in the forest has fallen by a third, releasing large amounts of carbon emissions into the air that cannot be trapped, a shift that has turned the once carbon-dioxide-absorbing forest into a source of global warming [25,28]. Governments around the world are currently seeking solutions to reduce carbon emissions, with net-zero carbon emissions becoming the focus of global climate change research [29,30]. In the Paris Agreement, net-zero carbon emissions is described as a system that "balances anthropogenic emissions by sources and removals by sinks". Many European countries have started to develop policies to achieve this goal [31].

Some scholars have proposed the concept of urban forestry to further strengthen the urban ecological cycle system by optimizing urban green infrastructure in pursuit of sustainable development [32–34]. Although European countries have a long history in the design and management of urban green space, there is still controversy on the specific content of the concept of urban forestry [35]. In the broad sense, natural resource management activities such as forestry plantations are supposed to take place in suburban clearings but, in reality, such activities can take place in any tree-growing area of the city [36]. A more comprehensive definition of urban forests is networks or systems of all trees in a city, including green infrastructure or individual trees [37–40]. In China, research has addressed several issues related to the benefits of urban forests in relation to air quality, forest cover, and spatial pattern [41–43]. Meanwhile, in Europe and the USA, studies have explored the diversity of tree composition in urban forests and the relationship between forests and people [44,45]. Part of the urban forest focuses on the potential of urban economic benefits, biodiversity conservation, and urban climate regulation [36,46–48], which provides ample evidence of the role of urban forests in the human living environment.

Previous researchers have demonstrated the importance of optimizing the location of urban trees by analyzing data indicators related to urban trees, setting priority standard classification scores or using prioritized geographic information systems [49,50]. Although adequate use was made of existing urban tree data, the data variables changed in real time and the lack of correlation between the data meant that the final data-oriented results could be biased. Other studies have analyzed and counted urban natural resources to improve tree survival by constructing comprehensive indicators and providing a tree planting priority index [51,52]. Alternatively, a design model approach has been used to try to change the relationship between the location of urban trees and the roadway to increase the comfort of the habitat [53,54]. However, the fact that improper planting of urban trees will reduce the ecological value of trees and cause environmental problems and potential risks has been ignored [55].

It is worth noting that previous approaches to urban planning have used computer models to predict future urban change and to help justify urban planning from a holistic perspective [56,57]. Computational results from model simulations confirm that cellular automata perform better in computational urban simulation models [58,59]. Cellular automata have the ability to simulate dynamic processes, and are suitable for considering neighborhood relationships and the urban spatial dimension, and are widely used in predicting the urban expansion process and land use planning [60–62]. This approach was previously used in early urban studies, where tree roads were generated using cellular automata, and plots of land to be developed were placed on both sides of the road [63,64]. Cellular automata are capable of simple rule making based on the local urban environment, reflecting the spatial organization of the city in a dynamic process [63]. However, most of these studies have used formulae and urban tree data to calculate comparisons that can only be obtained over a wide range of tree planting areas, or data over a real period of time to explore index relationships. There are also limitations in the dynamic iterations of cellular automata, which lead to uncertainty and uncontrollability of the iterations [65]. Cellular automata have been used to test a large range of cities, and the predictions were altered at the urban texture level, but failed to optimize urban green infrastructure to improve urban climatic issues from an urban ecological sustainability perspective. Few studies have attempted to model the precise location of tree planting in a block, particularly in terms of the connectivity between urban green infrastructure at the regional scale, ignoring the location of existing and new trees in the city.

This study addresses the following questions: (1) How can potential urban tree planting sites be identified to face the current situation of fragmented tree planting in urban forestry? (2) How can the connectivity between green infrastructures be strengthened? (3) How can deep learning and cellular automata, representing connectionist and behaviorist AI, respectively, be combined to innovate urban forms? This paper aims to address these questions with the goals of elucidating the significance of urban trees in urban ecosystems, bolstering the design and management of green infrastructure to mitigate the impact of urban climate change, and laying a foundation for future sustainable urban development.

2. Materials and Methods

2.1. Introduction and Definition of the Main Methodological Components

With the rapid development of deep learning technology in recent years, implicit learning [66–68] has been widely used to handle the dynamism of information input, wherein dynamic cognitive models replace predefined ones. Consequently, we harness the implicit learning and generative capabilities of deep learning as an enhancement, guiding the computational simulation of discrete models to reason complex, dynamic behavior in response to environmental dynamics. The combination of CycleGAN and cellular automata techniques is adopted to establish the methodological framework of this study.

Cellular automata are a type of computational discrete model introduced by S. Ulan and J. von Neumann in the late 1940s [69]. The advantage of the cellular automata model is that it is able to model complex discrete dynamical systems [70], for instance, integrating the spatial and temporal dimensions of a city. Early scholars first proposed the application of cellular spatial models to geographical modeling [71]. In the 1980s and 1990s, cellular automata began to be used to simulate urban sprawl as the computational power and concepts of cellular automata models were updated [72,73]. The spatial patterns formed by the iterative process of cellular automata and the development of theories have facilitated the design of simulation models for urban evolution, allowing the cellular automata method to be used to test the assumptions of urban theories and to simulate the urban form [62,72]. In previous studies, researchers have adapted different transition rules to fit the study plots based on the model, for example, cellular automata model testing based on a strict rule-based transformation where the cellular grid space was set to 250 m and urban spatial changes were simulated by changing the rule setting [74]. Other researchers transformed rules based on urban morphology to use the model for visualizing future urban growth [75]. However, in response to the development of urban cellular automata models, some scholars have suggested that this may lead to problems in practice when a high number of influences are included within the model and questioned whether extensive rule adjustments can actually constitute cellular automata models [69]. Such experiments with large urban scales usually have a large range of individual cell space settings in the cellular automata model, which can easily lead to a lack of precision. In addition, cellular automata models lack a standard method for defining transformation rules, which can be aided by incorporating a CycleGAN model.

This paper's exploration of neurally guided cellular automata involves the development of a validation method based upon the hierarchical identification of urban data in learning zones. It also investigates the potential of combining cellular automata with deep learning, extending the research methods of cellular automata models beyond traditional urban theory, with an aim to bridge the gaps in past methods. The CycleGAN model can execute powerful image generation by completing image-to-image translation using cycle-consistent adversarial networks [76]. Specifically, this technical approach can generate a potential representation of an image X by identifying a corresponding representation and presenting this potential as a style Y. Other researchers have similarly employed adversarial loss for training to complete image-to-image translations [77]. The generated output can provide an initial design guide for model experiments, aiding in generating the desired target urban morphology. This can assist in defining the transition rules in the cellular automata model to find an approach better suited to improving the accuracy of tree planting in the study area.

2.2. Study Framework

As the first country in Europe to plan for urban forestry, England has a long-term plan for urban trees and a vision of zero greenhouse gas emissions by 2050 [36]. London, as the capital of the UK, is a good candidate for this study as it is vital to improve its urban forestry. In order to improve the ecological value of urban trees in urban forestry, a program was designed to predict the best planting position of trees in future urban forestry by learning the connection relation of the underground tree network in primary forest and the basic rules of cellular automata, including the following stages (Figure 1). The first stage focused on collecting carbon emission data and spatial coverage of green space in different London boroughs, visualizing the data using GIS spatial data analysis, comparing the data, selecting areas with high carbon emission and low green space based on the visualization results, selecting learning areas, and analyzing the data related to street trees. In the second stage, CycleGAN was used to propose a hypothesis of future urban morphological changes and to try to realize an ecological construction orientation of the urban forest, surrounded by trees and buildings. The third stage used the rules and principles of cellular automata to translate the programming language into the parametric software Grasshopper to simulate and more accurately predict the generated results by adding field site constraints. The fourth stage compared the predicted iterations of the cellular automata to analyze the results of the iterations under different variable settings, to see the network connections between urban trees after planting new trees, and to test the feasibility of previous assumptions.

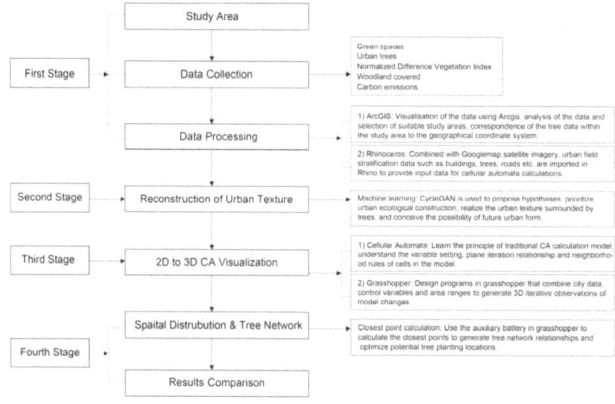

Figure 1. The workflow of data processing in our proposed framework.

2.3. Study Area

The study area excludes parcels with large urban ecological parks, focusing on parcels with a predominantly built-up distribution and a small and scattered distribution of green spaces and trees, which were then further selected according to the road hierarchy characteristics of the urban neighborhood. The study area is located in the northern part of the Camden district at a scale of 1:80 m (51°56′ N–51°55′ N, 0°12′ W–0°11′ W) and contains roads with predominantly service functions, and the urban distribution analysis includes water, green infrastructure, buildings, and roads (Figure 2). The distribution map shows that more traffic arteries are located in communities with little green infrastructure and scattered urban street trees, making it difficult to cope with the carbon emissions from daily traffic.

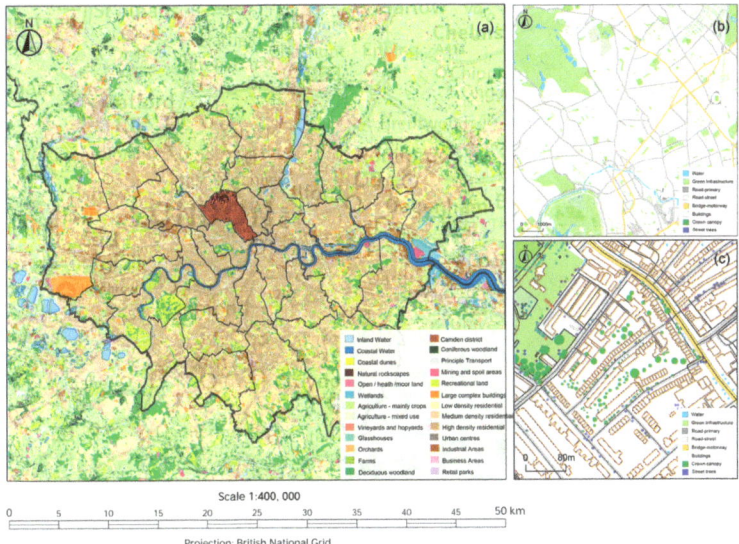

Figure 2. (**a**) Camden in Greater London; (**b**) 1 KM × 1 KM map of Camden; (**c**) study area.

2.4. Data Source and Processing

The GIS spatial data analysis method can help analyze the collected spatial data through the geographic information system. In this study, spatial data analysis can be used to more intuitively visualize the data and compare the differences between different regions. Spatial data analysis can be flexibly applied according to Excel tables to identify different categories, and the size of data in the classification can be matched according to geographical coordinates without being restricted by the regional area. The Office for National Statistics has collected and provided a table of annual average carbon dioxide emissions for London in 2019 and the woodland cover area for local authority areas in London in the same year (Figure 3). The information from the Excel spreadsheet was combined to match the UK regions' geographic coordinate system and then visualized using ArcGIS. The boroughs near downtown London tend to have the highest carbon emissions per square meter, according to the annual statistics on carbon emissions. Additionally, the distribution of green spaces in London, as well as the amount of tree cover in each London borough, were entered in the same coordinate system for comparison. The data visualization's findings indicate that, with a high concentration of real estate development, over half of all London boroughs currently have a proportion of tree cover of only 0–8%. The distribution of green space from the urban fringes of London towards the city center is characterized by an over-representation of large whole areas of green space to small pockets of fragmented green space. At the same time, through the comparison of the three groups of data, we find that there are administrative districts with a wide distribution of green space and low carbon emissions, but there are still some administrative districts with important roads, resulting in an imbalance between the green space and the average annual carbon emissions.

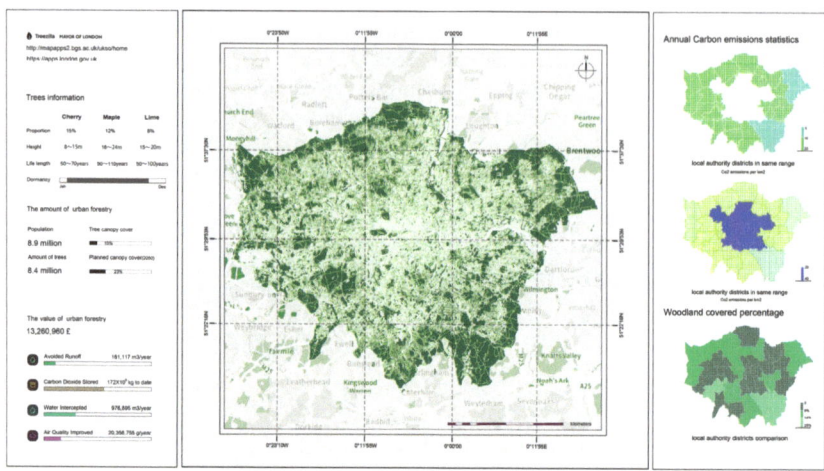

Figure 3. Comparison of London carbon emissions, woodland cover, and green spaces.

Camden has a higher average annual carbon footprint than the rest of London and less tree cover in the area. A small and medium-sized community with abundant data information and dense traffic network was selected from the 1 KM × 1 KM region in Camden district. Data analysis included water conservation (normalized difference moisture index, NDMI); plant health (normalized difference vegetation index, NDVI; sunlight (direct sun hours of the site); and existing tree canopy and tree species. To gain an in-depth understanding of the vegetation health status of the site from multiple perspectives, the street tree data in the learning area were imported into Rhino by tree species stratification and merged with building, road, and other data to restore the status quo of the site. The position of each tree was corresponded to the geographical coordinate system individually to prepare for the subsequent data calculation of cellular automata (Table 1).

Table 1. Street trees of study area, Camden.

Gla_id	Tree Species and Scientific Names	Longitude	Latitude
glaid_682290	Plane (Platanus hispanica)	−0.12786102449052	51.5675131207381
glaid_682291	Plane (Platanus hispanica)	−0.12788095908132	51.5675441929378
glaid_682292	Ash (Fraxinus excelsior)	−0.12772311698794	51.5675262975231
glaid_682293	Plane (Platanus hispanica)	−0.12762451064078	51.5675697763205
glaid_682294	Pear (Pyrus communis)	−0.12755616291557	51.5676037562588
glaid_682295	Plane (Platanus hispanica)	−0.12747623124352	51.5676387203248
glaid_682296	Plane (Platanus hispanica)	−0.12735170738228	51.5677124495487
glaid_682297	Plane (Platanus hispanica)	−0.12726708752885	51.5677561512526
glaid_682298	Plane (Platanus hispanica)	−0.12719287232343	51.5678064935142
glaid_682299	Cherry (Prunus genus)	−0.12715312606818	51.5678958735313
glaid_682300	Cherry (Prunus genus)	−0.12710738384612	51.5679380374256
glaid_682301	Cherry (Prunus genus)	−0.12706859431744	51.5679374183132
glaid_682394	Cherry (Prunus genus)	−0.12700819278152	51.5676059818775
glaid_682398	Whitebeam (Sorbus aria)	−0.12713078708266	51.5672386188778
glaid_682399	Whitebeam (Sorbus aria)	−0.12711551260981	51.5672240770995
glaid_682400	Whitebeam (Sorbus aria)	−0.12714728020639	51.5672235050338
glaid_682401	Whitebeam (Sorbus aria)	−0.12713378040792	51.5672079124889
glaid_682402	Ash (Fraxinus excelsior)	−0.12705398809426	51.5671341598225

Table 1. *Cont.*

Gla_id	Tree Species and Scientific Names	Longitude	Latitude
glaid_682618	Hawthorn (Crataegus)	−0.12798605750201	51.5679855102208
glaid_688705	Cherry (Prunus genus)	−0.12737171242051	51.5670113580804
glaid_697556	Lime (Tilia europaea)	−0.12774688769460	51.5673551009506
glaid_697557	Cherry (Prunus genus)	−0.12779069456203	51.5673635333951
glaid_697558	Cherry (Prunus genus)	−0.12777973154878	51.5674161441307
glaid_697559	Lime (Tilia europaea)	−0.12742623316460	51.5676018627646
glaid_697560	Cherry (Prunus genus)	−0.12748077097416	51.5674334053296
glaid_697561	Pear (Pyrus communis)	−0.12745816111797	51.5674253110274
glaid_697562	Pear (Pyrus communis)	−0.12744097518780	51.5674151450884
glaid_697563	Cherry (Prunus genus)	−0.12744327917633	51.5673590689984
glaid_697564	Pear (Pyrus communis)	−0.12737621039910	51.5673162737598
glaid_697565	Pear (Pyrus communis)	−0.12734530308466	51.5672959072194
glaid_697566	Cherry (Prunus genus)	−0.12751479831096	51.5673778354607
glaid_697567	Pear (Pyrus communis)	−0.12756027765596	51.5673455589041
glaid_697568	Pear (Pyrus communis)	−0.12758720845964	51.5673327697381

2.5. Methods

We investigated the possibility of using cellular automata (CA) to alter urban spaces by training CycleGAN with the outputs of CA and the samples from the forest mycorrhizal network. By comparing these two, we validated the applicability of CA and visually collected spatial feature information to guide the parameter setting of CA design. Subsequently, we utilized this information to set up a 3D cellular automata for predicting tree planting locations for the site.

2.5.1. CycleGAN Image-to-Image Translation

To enhance urban green ecology and combat urban climate change, CycleGAN reshapes existing urban surfaces through image style transformation, creating a diversity of future urban forms and providing guidance for urban design. The morphological design of urban neighborhoods greatly affects the outdoor environment. In this paper, satellite images containing learning areas of 1 KM × 1 KM and two different image samples were selected and tested separately to find ways to enhance the effective construction of urban forestry. This was used as a design guide to improve the management of urban forestry in the city. CycleGAN is commonly used to solve migration problems between images [78]. This method performs image transformation from reference image domain X to target image domain Y without relying on paired images. In this case, G and F are mapping functions between two image domains X and Y. The model includes two discriminators D_Y and D_x. D_Y promotes G to translate X into outputs that are identical to domain Y, and vice versa for D_x, F, and X. CycleGAN also uses the adversary loss [79] and cyclic consistency loss, which are two loss functions that are expressed respectively in the following formulas [76]:

$$\mathcal{L}_{GAN}(G, D_Y, X, Y) = \mathbb{E}_{y \sim p_{dat}(y)}[\log D_Y(y)] + \mathbb{E}_{x \sim p_{data}(x)}[\log(1 - D_Y(G(x)))], \quad (1)$$

$$\mathcal{L}_{cyc}(G, F) = \mathbb{E}_{x \sim p_{data}(x)}[\| F(G(x)) - x \|_1] + \mathbb{E}_{y \sim p_{dat}(y)}[\| G(F(y)) - y \|_1]. \quad (2)$$

CycleGAN requires the collection of hundreds of sets in each example folder. Due to the limitations of collecting samples of the same type of data, this paper cuts the samples into 1000 sheets each and selects 300 (180 × 180 pixels) of the field features that need to be preserved as the training dataset for transformation training.

2.5.2. Calculation Principles of Cellular Automata

Cellular automata are made up of a grid of cells, the size of which can be changed according to the requirements of the setup. Each cell's life and death relationship are controlled by setting rules in the grid [80,81]. The basic principle in cellular automata is that if the life and death relationship of a cell changes, other cells in the vicinity of that cell will also be affected [82]. Some of the cells in the lattice are given an initial state (usually time t = 0), while others are given a state (advance t by 1). The cells in the grid are housed in separate compartments but are closely related to each other, with a neighborhood effect, just as the trees in a forest exist as a whole. There are two common types of communities in which cellular automata identify neighbors, named after the theorists who invented them, Moore neighbors and Von Neumann neighbors (Figure 4). The Moore neighborhood consists of eight orthogonally adjacent unit cells, and the von Neumann neighborhood consists of four diagonally adjacent unit cells. The two differ in their results for visualizing changes in cellular automata [83]. Some scholars have examined the urban matrix layout of satellite images of residential areas in Australia to determine the applicability of the rules of the cellular automata model, and have proposed that Moore neighbors allow diagonal or vertical access to cellular space, whereas Von Neumann neighbors allow only vertical access to the space [84]. In terms of the more regular grid layout of modern cities, Moore neighbors possess stronger accessibility characteristics to help achieve the spatial distribution of urban trees. Therefore, we chose Moore neighbors to conduct the urban form experiment for deep learning.

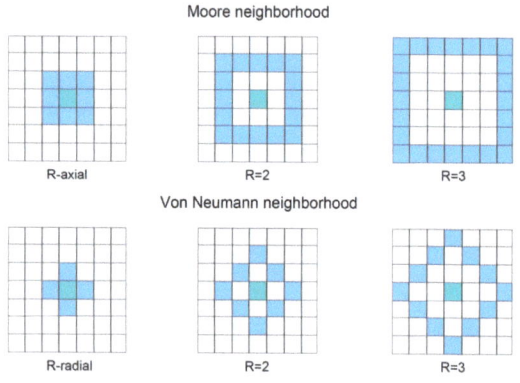

Cyan: central (considered) cell; Blue: adjacent unit (boundary) cells.

Figure 4. Rules of cellular automata.

2.5.3. Urban Reconstruction by CycleGAN

In this study, two sets of tests were conducted using CycleGAN. The first group (Figure 5) is an image transformation between satellite images of London city (sample A) and the forest mycorrhizal network intention map (sample B), in order to make the transformation results achieve the urban design orientation of green space as the main feature and the rest as secondary. In this training set, the images containing more green space features were selected. From the final training results obtained, although the green areas cover the largest area, the results cover the original building sites and roads, forming an urban surface with traffic networks cutting through the green areas and not realizing the urban green infrastructure construction with trees surrounding buildings.

The second group (Figure 6) uses the original sample A, but this time selecting feature points with 50 percent each of the images having a green space or building feature to ensure that the building footprint was not completely covered and lost in the final results. Sample B was chosen as the result of one iteration in the cellular automata. The samples used Moore neighbors, which modeled the canopy layer growth competition between trees in an urban forest, with common features between them and the problem of urban land competition. The results of the second image transformation test achieved a similar area of green space occupation to that of building occupation. Some of these areas formed a more coherent image of urban green space surrounding the building sites, but the main roads were completely covered in the results. The outcomes produced by the CycleGAN model demonstrate that the experimental results, derived from the Moore neighbor samples, can alter urban morphological features. Furthermore, they maintain several recognizable spatial features and deliver more reasonable spatial layouts compared to group 1 outputs.

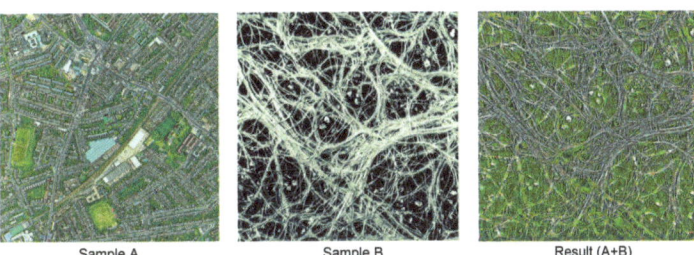

Figure 5. CycleGAN output group 1.

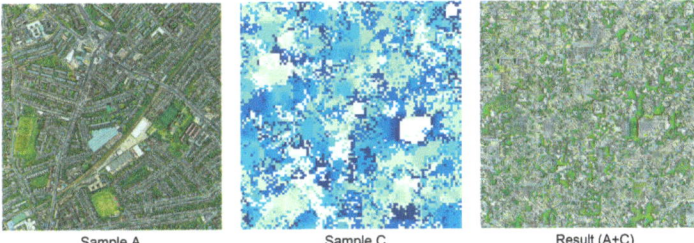

Figure 6. CycleGAN output group 2.

2.5.4. Grasshopper Transformation and Tree Networks

The cellular automata were implemented in the code as a two-dimensional spatial iteration, using previously prepared site data to set up the grid according to the size of the study area. In Grasshopper, the two-dimensional iterative change procedure is built according to the rules of cellular automata, setting up adjustable parameter entries and iteration rules, and the initial state of each grid cell is divided into two types: alive and dead. In the original code, the starting point for the calculation of cellular automata is randomly generated, whereas in Grasshopper the starting point can be set manually by combining the existing tree location data from the Rhino with the box selection to increase the controllability of the calculation. In this cellular automata model, we set the Moore neighborhood rules as follows: (1) a dead cell becomes alive when it is surrounded by exactly three living neighbors; (2) a cell becomes dead when it is surrounded by a single or four or more living neighbors; (3) a living cell continues to live until the next iteration when it is surrounded by two or three living neighbors. Cellular automata also have the ability to pause and restart, and the relationship between cell life and death changes continuously after multiple iterations. In Grasshopper, the grid is set up to look at the life and death status of cells centered on a single cell, with a set rule to find the neighboring cells in each row and column moving in the direction of the surrounding cells. The cellular automata will first look for cells in the grid around existing trees in the city and calculate whether the cellular cells are likely to be symbiotic. In the design program, the two-dimensional iterations are made three-dimensional to allow visual comparison of the planting changes between old and new trees. In addition, a new constraint was added to the conversion procedure to accurately calculate the location of new trees. Cells cannot be calculated in grids where buildings or main roads are distributed. This method enhances the urban ecology while preserving the existing buildings and traffic on the site.

After obtaining the iterative results of the cellular automata, it is still necessary to determine the validity of the urban tree planting locations. By combining the commonalities between tree dimensional networks in forests and cellular automata, urban tree networks are created using the program Closest point and Graft tree to analyze and optimize the reasonableness of the computational results. Scientists have experimentally demonstrated that trees in a forest interact with each other to form a large interconnected community. A team of researchers utilized DNA analysis to map a fungal network in a patch of Canadian forest [85] (Figure 7(a1,a2)). Model simulations revealed that more connections are lost when some trees are removed (Figure 7(b1,b2)). We categorized the different canopy sizes in the mycorrhizal network and viewed the tree connectivity relationships hierarchically (Figure 7(c1,c2,d1,d2)). These trees act as important hubs in the urban transport network, communicating with neighboring trees [85,86], supporting the energy transfer between the rest of the small trees, and enhancing the ecological stability of the urban green infrastructure area.

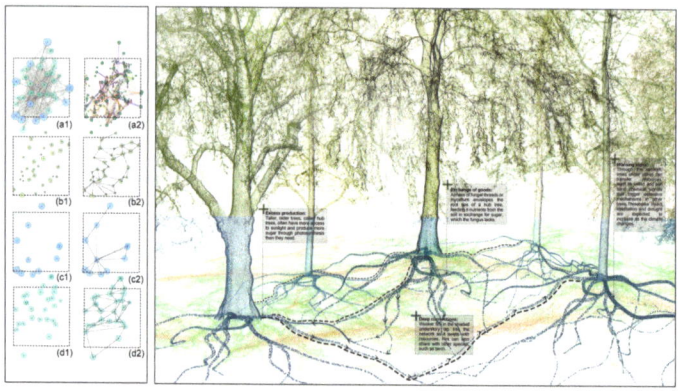

Figure 7. Tree networks and communication.

3. Experimental Results

3.1. Relationship between CycleGAN Results and Cellular Automata

According to its experimental output, CycleGAN can help to adjust the cellular automata model. From the first set of CycleGAN experimental outputs, it can be seen that restrictions should be set before the calculation of the urban cellular automata to exclude cellular grids that cannot be used for the calculation in order to keep the residential areas or major roads in the study area, so as to avoid the situation that the urban tree planting will cover all the cellular space in the cellular automata at the later stage of the calculation. The second set of CycleGAN outputs shows more intuitively that the grid size of the cellular automata model affects the distribution pattern of green space in the city when a larger city range is selected. With a larger spatial extent of a single cell, its green space distribution area may be prone to a scattered distribution, weakening the aggregation connection between green spaces. The current results suggest that this kind of urban surface remodeling orientation is more suitable for small and medium-sized urban ecological development, such as four-level roads in urban communities, and that such urban side roads are more preferable to urban traffic arteries for ecological development. The results of the test orientation were rationalized and applied to the cellular automata model. Therefore, the range of the setup grid was reduced in this model setup, and the individual cell space was set to 10 M × 10 M for calculation with reference to the canopy size of the most planted tree species in the study area. Furthermore, from examining the image generation pattern, the generation pattern of the Moore neighborhood extending from a single cell in all directions constitutes a more ideal tree planting condition, which can support part of the assumption that the tree surrounds the building. The experimental results show that the cellular automata model can be used for simulation when both CycleGAN output and urban satellite data maintain some common features.

3.2. Iteration of Cellular Automata

The results of the cellular automata iterations were transformed from 2D planes to 3D stereoscopic images using a design program to transfer the tree data within the learning area (Figure 8). The 3D iterative results provide a clearer view of the iterative life and death relationships between each generation of the cellular automata than the 2D flat images, making it easier to adjust the model parameters.

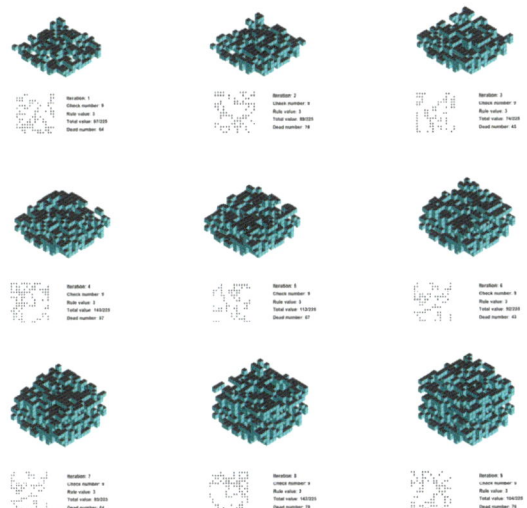

Figure 8. From 2D to 3D visualization.

Cellular automata can set different starting points for the test results obtained according to the tree classification. According to the statistics of the number of tree species in the learning area, this paper selected three most common street trees in London, namely cherry, maple, and whitebeam, for calculation. The final result of the first generation is the superposition of the three tree species in separate iterations, and the same is true for the third and sixth generations (Figure 9). The number of iterated trees increases gradually with the number of iterations compared to the 3D iterations without classification. The number of iterations is positively correlated without taking into account the life cycle of the tree and the precise location of the tree planting in the grid. Urban forestry construction is prioritized but access to feeder roads within urban communities needs to be reclassified.

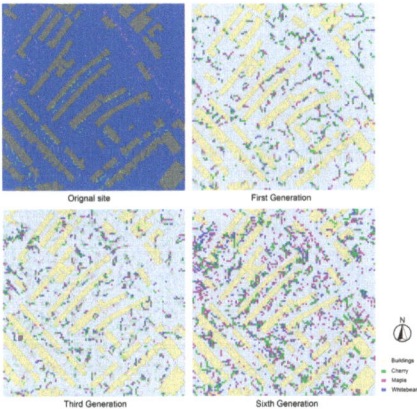

Figure 9. Calculation of tree species classification.

3.3. Trees Network Connection

Predicting tree network connections from the iterative results can further optimize urban tree locations and enhance urban green ecology. One of the tree location iterations is randomly selected from the results of the cellular automata iteration using the closest point assisted network calculation (Figure 10). According to the results, in the first single tree location connectivity network, some of the tree locations were scattered within the cellular automata grid at a distance from the densely populated areas of trees. In the end, the tree network was compiled over a long distance and there was an unreasonable tree network structure. In the second tree network test, the tree network connections were recalculated with the addition of scrub data from the urban green infrastructure. The results show that when the proportion of the iteration result data reaches a certain level, the tree network connection begins to rationalize and the density of the tree network becomes more concentrated. The distance from planting is judged based on the connectivity results, and the tree planting location is improved.

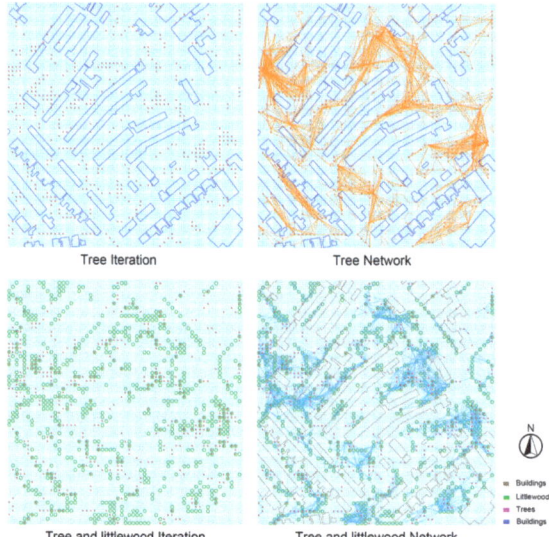

Figure 10. Network connection.

3.4. Results Comparison of Cellular Automata and CycleGAN

The cellular automata model calculates the weaving range of the network composition that can inform the construction of green infrastructure in cities, and the comparison with CycleGAN results also confirms the possibility of trees surrounding buildings. We compare the cellular automata output with the CycleGAN output at the same range, observing the changes in the reshaped urban morphology. We find that the model simulation results are oriented towards clustering and tightly connecting the otherwise fragmented distribution of green spaces in a piecemeal form compared to the traditional urban form, creating an ideal urban form that prioritizes urban green spaces more in accordance with the urban forest (Figure 11). Remarkably, we found that the network relationships simulated by the cellular automata constitute a morphological distribution of urban green space infrastructure with striking similarities to the CycleGAN output, an approach that alters the priority hierarchical order of the traditional urban planning distribution. In the original urban morphological distribution of the site, urban green spaces, buildings, and urban public spaces are divided by roads. With the rapid development of global urbanization, urban expansion and regeneration have led to an increase in urban areas and roads, and the development of urban road networks has had an impact on the urban form. The buildings in the study area of this experiment are residential, and with the original residential areas unchanged, we chose to prioritize the optimization of the green infrastructure in the site and calculate its connectivity, before reprogramming the auxiliary roads in the residential areas to ensure the connectivity between the community and the rest of the main roads.

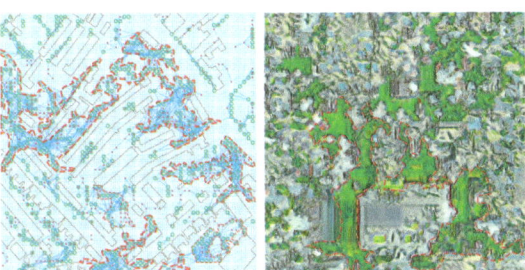

Figure 11. Comparison of green space composition.

4. Discussion

4.1. Urban Planning and Design Implication

Determining where to plant trees will be an important issue in the future sustainable development of cities. The findings of this paper highlight strategies to further optimize the spatial distribution of potential future urban tree planting locations based on existing trees on the site to help restore urban ecosystems. Other studies emphasizing the role of urban trees and ecosystems are also supported [87,88], complementing the approach to pinpointing specific locations for tree planting in space and improving biodiversity by creating green infrastructure patches, while providing a habitat for birds, insects, etc. Regulation of biodiversity is one of the influencing factors in the stability of urban ecosystems [88]. Furthermore, urban trees play an important role in urban ecosystems. Increased planting of trees can better conserve soil moisture and help reduce temperatures near the ground, thereby mitigating the effects of urban climate change, such as the urban heat island effect. Developing a strategy for well-planted trees can help protect the urban environment and enhance the eco-efficiency of urban ecosystems. It creates a healthy spatial environment for the daily activity space of city dwellers and reduces the interaction distance between people and the natural environment. It can also help alleviate people's daily work anxiety and other problems at a spiritual and psychological level, and connect humans with nature to provide more entertaining spatial environments that promote health and well-being.

This study proposes a new approach combining CycleGAN and cellular automata techniques to prioritize urban tree planting locations to predict the optimal urban tree network to help enhance connectivity between urban infrastructure developments. The results of the current work may be important for urban planners, designers, and researchers related to urban sustainability in urban planning and design, and advocate exploring the value of urban forest planning methods in future green space development. Firstly, the results of the cycle-consistent adversarial network training in CycleGAN show the diversity and possibilities of urban morphological change. Instead of setting standards for urban morphology or using inertial thinking in planning and design, we should constantly optimize design principles to suit the current urban situation. The trained CycleGAN model differs from traditional urban planning and design perspectives by creating urban forms where trees surround community buildings, changing the status quo where green spaces are fragmented by buildings. Green infrastructure zones are considered on a regional scale to reshape the distribution of urban residents, green spaces, and roads. Secondly, trees are an important part of the green infrastructure in the urban form. It is essential to enhance urban forestry by improving the location of potential tree planting in future urban planning and design management. Botanists emphasize that we need to plant trees in the right places [89]. Cellular automata calculation results elucidate the urban ecological construct relationships between different levels of data by distinguishing between urban tree species and urban forest hierarchies. The study provided a reference for improving the way street tree planting density is planned and managed in urban forestry by pinpointing the potential planting locations of different tree species. In addition, sustainable urban ecological construction is the current goal of urban planning and design. A well-connected network of urban trees will contribute to the stability of urban ecosystems in the face of increasing urban CO_2 emissions. Previous studies have shown that street trees in urban centers are scattered and fragmented, lacking a holistic approach to tree connectivity [90,91]. The small size of the tree planting may affect the root growth and the smoothness of the road surface, resulting in bending and deformation [92]. It also affects the life cycle of trees, reducing survival rates and ecological benefits [93]. In contrast to traditional forestry, tree networks do not focus on individual trees, but form a large, closely-knit community. The transfer of nutrient energy through underground rhizomes increases tree survival and builds a strong and stable ecosystem [85,86]. Study results on tree network connectivity encourage the construction of green infrastructure areas with a high density of tree networks, which will act as a link to maintain the ecological health of their surroundings.

4.2. Limitations and Further Research

This article has identified potential locations for tree planting to support the design and management of green infrastructure and enhance its connectivity, but there are still limitations to the ecological value of tree species planted in this study. For example, the determination of urban tree species requires an analysis of urban soil conditions, canopy size, and tree life cycle to maximize the ecological benefits of urban forestry in response to urban climate change, which is not considered in the current work. It was also found during field research and street tree data collection in the study area that the urban tree database was not up to date with information on newly planted trees, and the limited data available may lead to some uncertainty in the calculation results. In addition, it was found during the cellular automata simulation that this method may not be applicable to areas with a large building occupation area, and the large number of buildings occupying the grid space may lead to unsatisfactory results in the final iterative calculations. In future studies, it might be necessary to test the classification of different site occupation areas, for instance, blocks with mainly public facilities land or industrial land. At the same time, there is a need for further experimentation and validation to determine the feasibility of urban green infrastructure in terms of its eco-efficiency. It is worth noting that, in conjunction with the comprehensive analysis of the current urban situation, the connectivity of the urban tree dimensional network will pose a challenge to the distribution of minor roads and branch roads within urban communities in the future. The branch roads within urban communities should be further explored in order to guarantee community access and connectivity to surrounding roads while prioritizing green infrastructure, and to rethink the symbiotic relationship between urban trees and the future city. The environment in which trees grow in cities is different from that in forests. The environment in which urban trees grow is influenced by human activities, so continuous awareness-raising on urban forestry is required to increase the ecological resilience of cities.

5. Conclusions

This study, by employing a multitude of methodologies, scrutinizes selected learning zones and lays the groundwork for ideas aimed at the future of urban ecology. One significant outcome is the design orientation that fuses urban buildings with trees, effectively reimagining the current urban planning priorities from an urban design perspective. Our findings strongly emphasize the transformative potential of machine learning tools in reshaping the urban landscape. By employing CycleGAN and cellular automata models, we demonstrated a novel method to prioritize and optimize urban tree planting locations, hence paving the way towards more sustainable and eco-friendly urban environments.

This research challenges the existing paradigm where urban land use is primarily dominated by traffic and buildings, proposing instead a shift towards prioritizing urban ecology. Such a shift has the potential to transform future urban transport and road planning, offering more sustainable and eco-friendly alternatives. Building on field data from the study area, we utilized a design program to simulate and calculate potential tree planting locations for future urban forestry. These computations, combined with features of the forest ecology network, aided in determining the network connections between urban trees, thus enhancing urban green infrastructure.

The results suggest that the integration of these tools can reshape urban landscapes, fostering green infrastructure and prioritizing urban forests. This approach creates an urban form where green spaces surround and interact with buildings, challenging the traditional urban planning methods where green spaces are often fragmented by infrastructure. Furthermore, the dynamic iterations of the cellular automata principle offered a unique lens to simulate urban tree health and mortality. The specificity of the results to the site area of each tree bolsters the confidence and adaptability of the simulation outcomes. The insights gleaned from the three-dimensional network connections facilitated the further optimization of potential tree-planting locations. When juxtaposed with machine learning results, the network connections not only confirm the feasibility of the proposed design orientation, but also enrich our understanding of urban green infrastructure network connections.

This research demonstrated a novel approach to computationally guide urban planning in enhancing urban forestry, leading to a reduction in carbon emissions and mitigating the impacts of urban climate change. Notably, in this study, the application of CycleGAN to the rule-making process of cellular automata occurs at the level of qualitative reference based on visual observations of the CycleGAN outputs. Theoretically, the rules can be customized according to the visual feature representation of the training image data, i.e., the feature maps, extracted within the CycleGAN model. Such customized rules may enable the cellular automata to output more desired states of urban landscapes. This concept warrants further exploration in future research.

Author Contributions: Conceptualization, Y.L.; methodology, Y.L. and S.-Y.H.; software, Y.L.; validation, Y.L.; formal analysis, Y.L.; investigation, Y.L.; resources, Y.L.; data curation, Y.L.; writing—original draft preparation, Y.L.; writing—review and editing, Y.L. and S.-Y.H.; visualization, Y.L.; supervision, S.-Y.H. There is no project management or access to funds here. All authors have read and agreed to the published version of the manuscript.

Funding: This research received no external funding.

Data Availability Statement: Not applicable.

Acknowledgments: The authors would like to acknowledge all the reviewers and editors.

Conflicts of Interest: The authors declare no conflict of interest.

References

1. Di Vita, G.; Pilato, M.; Pecorino, B.; Brun, F.; D'Amico, M. A review of the role of vegetal ecosystems in CO_2 capture. *Sustainability* **2017**, *9*, 1840. [CrossRef]
2. Wang, N.; Zhao, Y.; Song, T.; Zou, X.; Wang, E.; Du, S. Accounting for China's Net Carbon Emissions and Research on the Realization Path of Carbon Neutralization Based on Ecosystem Carbon Sinks. *Sustainability* **2022**, *14*, 14750. [CrossRef]
3. Sima, S.; Crişciu, A.V.; Secuianu, C. Phase Behavior of Carbon Dioxide+ Isobutanol and Carbon Dioxide+ tert-Butanol Binary Systems. *Energies* **2022**, *15*, 2625. [CrossRef]
4. Reichle, D.E. *The Global Carbon Cycle and Climate Change: Scaling Ecological Energetics from Organism to the Biosphere*; Elsevier: Amsterdam, The Netherlands, 2023.
5. Walther, G.-R.; Post, E.; Convey, P.; Menzel, A.; Parmesan, C.; Beebee, T.J.; Fromentin, J.-M.; Hoegh-Guldberg, O.; Bairlein, F. Ecological responses to recent climate change. *Nature* **2002**, *416*, 389–395. [CrossRef] [PubMed]
6. Metz, B.; Davidson, O.; Swart, R.; Pan, J. *Climate Change 2001: Mitigation: Contribution of Working Group III to the Third Assessment Report of the Intergovernmental Panel on Climate Change*; Cambridge University Press: Cambridge, UK, 2001.
7. Lianhe, J. Global carbon cycle: From fundamental scientific problem to green responsibility. *Science* **2021**, *73*, 39–43.

8. Badiou, P.; McDougal, R.; Pennock, D.; Clark, B. Greenhouse gas emissions and carbon sequestration potential in restored wetlands of the Canadian prairie pothole region. *Wetl. Ecol. Manag.* **2011**, *19*, 237–256. [CrossRef]
9. Pan, Y.; Weng, G.; Li, C.; Li, J. Coupling coordination and influencing factors among tourism carbon emission, tourism economic and tourism innovation. *Int. J. Environ. Res. Public Health* **2021**, *18*, 1601. [CrossRef]
10. Lv, Z.; Shi, Y.; Zang, S.; Sun, L. Spatial and temporal variations of atmospheric CO_2 concentration in China and its influencing factors. *Atmosphere* **2020**, *11*, 231. [CrossRef]
11. Udara Willhelm Abeydeera, L.H.; Wadu Mesthrige, J.; Samarasinghalage, T.I. Global research on carbon emissions: A scientometric review. *Sustainability* **2019**, *11*, 3972. [CrossRef]
12. Wang, C.; Wang, F.; Zhang, H.; Ye, Y.; Wu, Q.; Su, Y. Carbon emissions decomposition and environmental mitigation policy recommendations for sustainable development in Shandong province. *Sustainability* **2014**, *6*, 8164–8179. [CrossRef]
13. Manisalidis, I.; Stavropoulou, E.; Stavropoulos, A.; Bezirtzoglou, E. Environmental and health impacts of air pollution: A review. *Front. Public Health* **2020**, *8*, 14. [CrossRef] [PubMed]
14. Bowler, D.E.; Buyung-Ali, L.; Knight, T.M.; Pullin, A.S. Urban greening to cool towns and cities: A systematic review of the empirical evidence. *Landsc. Urban Plan.* **2010**, *97*, 147–155. [CrossRef]
15. Imhoff, M.L.; Zhang, P.; Wolfe, R.E.; Bounoua, L. Remote sensing of the urban heat island effect across biomes in the continental USA. *Remote Sens. Environ.* **2010**, *114*, 504–513. [CrossRef]
16. Salam, M.A.; Noguchi, T. Impact of human activities on carbon dioxide (CO_2) emissions: A statistical analysis. *Environmentalist* **2005**, *25*, 19–30. [CrossRef]
17. Allen, C.D.; Breshears, D.D.; McDowell, N.G. On underestimation of global vulnerability to tree mortality and forest die-off from hotter drought in the Anthropocene. *Ecosphere* **2015**, *6*, 1–55. [CrossRef]
18. Ahmed Ali, K.; Ahmad, M.I.; Yusup, Y. Issues, impacts, and mitigations of carbon dioxide emissions in the building sector. *Sustainability* **2020**, *12*, 7427. [CrossRef]
19. Lee, S.J.; Yim, J.S.; Son, Y.M.; Son, Y.; Kim, R. Estimation of forest carbon stocks for national greenhouse gas inventory reporting in South Korea. *Forests* **2018**, *9*, 625. [CrossRef]
20. Francini, S.; D'Amico, G.; Vangi, E.; Borghi, C.; Chirici, G. Integrating GEDI and Landsat: Spaceborne lidar and four decades of optical imagery for the analysis of forest disturbances and biomass changes in Italy. *Sensors* **2022**, *22*, 2015. [CrossRef]
21. Zhao, H.; Yan, Y.; Zhang, C.; Zhang, D. Three modes involved in forest carbon cycle: Mechanism and selection. *Sci. Silvae Sin.* **2014**, *50*, 134–139.
22. Dang, H.N.; Ba, D.D.; Trung, D.N.; Viet, H.N.H. A Novel Method for Estimating Biomass and Carbon Sequestration in Tropical Rainforest Areas Based on Remote Sensing Imagery: A Case Study in the Kon Ha Nung Plateau, Vietnam. *Sustainability* **2022**, *14*, 16857. [CrossRef]
23. Zekeng, J.C.; van der Sande, M.T.; Fobane, J.L.; Mphinyane, W.N.; Sebego, R.; Mbolo, M.M.A. Partitioning main carbon pools in a semi-deciduous rainforest in eastern Cameroon. *For. Ecol. Manag.* **2020**, *457*, 117686. [CrossRef]
24. Hoorn, C.; Wesselingh, F.P.; Ter Steege, H.; Bermudez, M.; Mora, A.; Sevink, J.; Sanmartín, I.; Sanchez-Meseguer, A.; Anderson, C.; Figueiredo, J. Amazonia through time: Andean uplift, climate change, landscape evolution, and biodiversity. *Science* **2010**, *330*, 927–931. [CrossRef] [PubMed]
25. Brienen, R.J.; Phillips, O.L.; Feldpausch, T.R.; Gloor, E.; Baker, T.R.; Lloyd, J.; Lopez-Gonzalez, G.; Monteagudo-Mendoza, A.; Malhi, Y.; Lewis, S.L. Long-term decline of the Amazon carbon sink. *Nature* **2015**, *519*, 344–348. [CrossRef] [PubMed]
26. Mensah, S.; du Toit, B.; Seifert, T. Diversity–biomass relationship across forest layers: Implications for niche complementarity and selection effects. *Oecologia* **2018**, *187*, 783–795. [CrossRef] [PubMed]
27. Peña-Claros, M. Changes in forest structure and species composition during secondary forest succession in the Bolivian Amazon1. *Biotropica* **2003**, *35*, 450–461. [CrossRef]
28. Gatti, L.V.; Basso, L.S.; Miller, J.B.; Gloor, M.; Gatti Domingues, L.; Cassol, H.L.; Tejada, G.; Aragão, L.E.; Nobre, C.; Peters, W. Amazonia as a carbon source linked to deforestation and climate change. *Nature* **2021**, *595*, 388–393. [CrossRef]
29. Davis, S.J.; Lewis, N.S.; Shaner, M.; Aggarwal, S.; Arent, D.; Azevedo, I.L.; Benson, S.M.; Bradley, T.; Brouwer, J.; Chiang, Y.-M. Net-zero emissions energy systems. *Science* **2018**, *360*, eaas9793. [CrossRef]
30. Khan, R.; Awan, U.; Zaman, K.; Nassani, A.A.; Haffar, M.; Abro, M.M.Q. Assessing hybrid solar-wind potential for industrial decarbonization strategies: Global shift to green development. *Energies* **2021**, *14*, 7620. [CrossRef]
31. Pye, S.; Li, F.G.; Price, J.; Fais, B. Achieving net-zero emissions through the reframing of UK national targets in the post-Paris Agreement era. *Nat. Energy* **2017**, *2*, 17024. [CrossRef]
32. Ostoić, S.K.; van den Bosch, C.C.K. Exploring global scientific discourses on urban forestry. *Urban For. Urban Green.* **2015**, *14*, 129–138. [CrossRef]
33. Jorgensen, E. *Urban Forestry: Some Problems and Proposals*; Faculty of Forestry, University of Toronto: Toronto, ON, Canada, 1967.
34. French, J. The concept of urban forestry. *Aust. For.* **1975**, *38*, 177–182. [CrossRef]
35. Konijnendijk, C.C. A decade of urban forestry in Europe. *For. Policy Econ.* **2003**, *5*, 173–186. [CrossRef]
36. Defra. *The England Trees Action Plan 2021–2024*; Department for Environment, Food & Rural Affairs, UK Government Office: UK, 2021. Available online: https://www.gov.uk/government/publications/england-trees-action-plan-2021-to-2024 (accessed on 12 May 2023).

37. Barona, C.O.; Devisscher, T.; Dobbs, C.; Aguilar, L.O.; Baptista, M.D.; Navarro, N.M.; da Silva Filho, D.F.; Escobedo, F.J. Trends in urban forestry research in Latin America & the Caribbean: A systematic literature review and synthesis. *Urban For. Urban Green.* **2020**, *47*, 126544.
38. Nowak, D.J. Historical vegetation change in Oakland and its implications for urban forest management. *J. Arboric.* **1993**, *19*, 313–319. [CrossRef]
39. Dobbs, C.; Escobedo, F.J.; Zipperer, W.C. A framework for developing urban forest ecosystem services and goods indicators. *Landsc. Urban Plan.* **2011**, *99*, 196–206. [CrossRef]
40. Threlfall, C.G.; Kendal, D. The distinct ecological and social roles that wild spaces play in urban ecosystems. *Urban For. Urban Green.* **2018**, *29*, 348–356. [CrossRef]
41. Duan, W.; Wang, C.; Pei, N.; Zhang, C.; Gu, L.; Jiang, S.; Hao, Z.; Xu, X. Spatiotemporal ozone level variation in urban forests in Shenzhen, China. *Forests* **2019**, *10*, 247. [CrossRef]
42. Duan, Q.; Tan, M.; Guo, Y.; Wang, X.; Xin, L. Understanding the spatial distribution of urban forests in China using Sentinel-2 images with Google Earth Engine. *Forests* **2019**, *10*, 729. [CrossRef]
43. Zhou, W.; Zhang, S.; Yu, W.; Wang, J.; Wang, W. Effects of urban expansion on forest loss and fragmentation in six megaregions, China. *Remote Sens.* **2017**, *9*, 991. [CrossRef]
44. Blood, A.; Starr, G.; Escobedo, F.; Chappelka, A.; Staudhammer, C. How do urban forests compare? Tree diversity in urban and periurban forests of the southeastern US. *Forests* **2016**, *7*, 120. [CrossRef]
45. Referowska-Chodak, E. Pressures and threats to nature related to human activities in European urban and suburban forests. *Forests* **2019**, *10*, 765. [CrossRef]
46. Livesley, S.J.; Escobedo, F.J.; Morgenroth, J. The biodiversity of urban and peri-urban forests and the diverse ecosystem services they provide as socio-ecological systems. *Forests* **2016**, *7*, 291. [CrossRef]
47. Song, J.; Feng, Q.; Wang, X.; Fu, H.; Jiang, W.; Chen, B. Spatial association and effect evaluation of CO_2 emission in the Chengdu-Chongqing urban agglomeration: Quantitative evidence from social network analysis. *Sustainability* **2018**, *11*, 1. [CrossRef]
48. Wolf, K.L.; Lam, S.T.; McKeen, J.K.; Richardson, G.R.; van den Bosch, M.; Bardekjian, A.C. Urban trees and human health: A scoping review. *Int. J. Environ. Res. Public Health* **2020**, *17*, 4371. [CrossRef]
49. Strohbach, M.W.; Arnold, E.; Haase, D. The carbon footprint of urban green space—A life cycle approach. *Landsc. Urban Plan.* **2012**, *104*, 220–229. [CrossRef]
50. Locke, D.H.; Grove, J.M.; Lu, J.W.; Troy, A.; O'Neil-Dunne, J.P.; Beck, B.D. Prioritizing preferable locations for increasing urban tree canopy in New York City. *Cities Environ.* **2011**, *3*, 4.
51. Lin, J. Developing a composite indicator to prioritize tree planting and protection locations. *Sci. Total Environ.* **2020**, *717*, 137269. [CrossRef] [PubMed]
52. Morani, A.; Nowak, D.J.; Hirabayashi, S.; Calfapietra, C. How to select the best tree planting locations to enhance air pollution removal in the MillionTreesNYC initiative. *Environ. Pollut.* **2011**, *159*, 1040–1047. [CrossRef] [PubMed]
53. Lusk, A.C.; da Silva Filho, D.F.; Dobbert, L. Pedestrian and cyclist preferences for tree locations by sidewalks and cycle tracks and associated benefits: Worldwide implications from a study in Boston, MA. *Cities* **2020**, *106*, 102111. [CrossRef]
54. Milošević, D.D.; Bajšanski, I.V.; Savić, S.M. Influence of changing trees locations on thermal comfort on street parking lot and footways. *Urban For. Urban Green.* **2017**, *23*, 113–124. [CrossRef]
55. Lawrence, A.; De Vreese, R.; Johnston, M.; Van Den Bosch, C.C.K.; Sanesi, G. Urban forest governance: Towards a framework for comparing approaches. *Urban For. Urban Green.* **2013**, *12*, 464–473. [CrossRef]
56. Kamusoko, C.; Gamba, J. Simulating urban growth using a Random Forest-Cellular Automata (RF-CA) model. *ISPRS Int. J. Geo-Inf.* **2015**, *4*, 447–470. [CrossRef]
57. Zheng, Q.; Yang, X.; Wang, K.; Huang, L.; Shahtahmassebi, A.R.; Gan, M.; Weston, M.V. Delimiting urban growth boundary through combining land suitability evaluation and cellular automata. *Sustainability* **2017**, *9*, 2213. [CrossRef]
58. Li, X.; Chen, Y.; Liu, X.; Xu, X.; Chen, G. Experiences and issues of using cellular automata for assisting urban and regional planning in China. *Int. J. Geogr. Inf. Sci.* **2017**, *31*, 1606–1629. [CrossRef]
59. Liu, X.; Li, X.; Shi, X.; Zhang, X.; Chen, Y. Simulating land-use dynamics under planning policies by integrating artificial immune systems with cellular automata. *Int. J. Geogr. Inf. Sci.* **2010**, *24*, 783–802. [CrossRef]
60. Shafizadeh-Moghadam, H.; Asghari, A.; Tayyebi, A.; Taleai, M. Coupling machine learning, tree-based and statistical models with cellular automata to simulate urban growth. *Comput. Environ. Urban Syst.* **2017**, *64*, 297–308. [CrossRef]
61. Gonzalez, P.B.; Aguilera-Benavente, F.; Gomez-Delgado, M. Partial validation of cellular automata based model simulations of urban growth: An approach to assessing factor influence using spatial methods. *Environ. Model. Softw.* **2015**, *69*, 77–89. [CrossRef]
62. Batty, M. *Understanding Cities with Cellular Automata, Agent Based Models, and Fractals*; MIT Press: Cambridge, MA, USA, 2005.
63. Batty, M. Cellular automata and urban form: A primer. *J. Am. Plan. Assoc.* **1997**, *63*, 266–274. [CrossRef]
64. Reps, J.W. *The Making of Urban America: A History of City Planning in the United States*; Princeton University Press: Princeton, NY, USA, 1965.
65. Li, X.; Yeh, A.G.-O. Neural-network-based cellular automata for simulating multiple land use changes using GIS. *Int. J. Geogr. Inf. Sci.* **2002**, *16*, 323–343. [CrossRef]
66. Reber, A.S. Implicit learning and tacit knowledge. *J. Exp. Psychol. Gen.* **1989**, *118*, 219. [CrossRef]

67. Perruchet, P.; Pacton, S. Implicit learning and statistical learning: One phenomenon, two approaches. *Trends Cogn. Sci.* **2006**, *10*, 233–238. [CrossRef]
68. Daikoku, T.; Yatomi, Y.; Yumoto, M. Implicit and explicit statistical learning of tone sequences across spectral shifts. *Neuropsychologia* **2014**, *63*, 194–204. [CrossRef]
69. Santé, I.; García, A.M.; Miranda, D.; Crecente, R. Cellular automata models for the simulation of real-world urban processes: A review and analysis. *Landsc. Urban Plan.* **2010**, *96*, 108–122. [CrossRef]
70. Wolfram, S. Cellular automata as models of complexity. *Nature* **1984**, *311*, 419–424. [CrossRef]
71. Tobler, W.R. Cellular geography. In *Philosophy in Geography*; Springer: Dordrecht, The Netherlands, 1979; pp. 379–386.
72. Batty, M.; Xie, Y. From cells to cities. *Environ. Plan. B Plan. Des.* **1994**, *21*, S31–S48. [CrossRef]
73. Couclelis, H. Cellular worlds: A framework for modeling micro—Macro dynamics. *Environ. Plan. A* **1985**, *17*, 585–596. [CrossRef]
74. Jenerette, G.D.; Wu, J. Analysis and simulation of land-use change in the central Arizona–Phoenix region, USA. *Landsc. Ecol.* **2001**, *16*, 611–626. [CrossRef]
75. Clarke, K.C.; Hoppen, S.; Gaydos, L. A self-modifying cellular automaton model of historical urbanization in the San Francisco Bay area. *Environ. Plan. B Plan. Des.* **1997**, *24*, 247–261. [CrossRef]
76. Zhu, J.-Y.; Park, T.; Isola, P.; Efros, A.A. Unpaired image-to-image translation using cycle-consistent adversarial networks. In Proceedings of the IEEE International Conference on Computer Vision, Venice, Italy, 22–29 October 2017; pp. 2223–2232.
77. Shrivastava, A.; Pfister, T.; Tuzel, O.; Susskind, J.; Wang, W.; Webb, R. Learning from simulated and unsupervised images through adversarial training. In Proceedings of the IEEE Conference on Computer Vision and Pattern Recognition, Honolulu, HI, USA, 21–26 July 2017; pp. 2107–2116.
78. Park, M.; Tran, D.Q.; Jung, D.; Park, S. Wildfire-detection method using DenseNet and CycleGAN data augmentation-based remote camera imagery. *Remote Sens.* **2020**, *12*, 3715. [CrossRef]
79. Zheng, K.; Wei, M.; Sun, G.; Anas, B.; Li, Y. Using vehicle synthesis generative adversarial networks to improve vehicle detection in remote sensing images. *ISPRS Int. J. Geo-Inf.* **2019**, *8*, 390. [CrossRef]
80. Ozturk, D. Urban growth simulation of Atakum (Samsun, Turkey) using cellular automata-Markov chain and multi-layer perceptron-Markov chain models. *Remote Sens.* **2015**, *7*, 5918–5950. [CrossRef]
81. Yüzer, M.A.; Yüzer, Ş. Cellular Automata Tabanlı LUCAM Modeli Ile Istanbul'un Gelişim ve Dönüşümüne Ilişkin Makro Form Simülasyonları. *J. Istanb. Kültür Univ.* **2006**, *4*, 231–244.
82. Liu, L.; Wang, X.; Eck, J.; Liang, J. Simulating crime events and crime patterns in a RA/CA model. In *Geographic Information Systems and Crime Analysis*; IGI Global: Hershey, PA, USA, 2005; pp. 197–213.
83. Kier, L.B.; Seybold, P.G.; Cheng, C.-K. *Modeling Chemical Systems Using Cellular Automata*; Springer Science & Business Media: Berlin/Heidelberg, Germany, 2005.
84. Ward, D.P.; Murray, A.T.; Phinn, S.R. A stochastically constrained cellular model of urban growth. *Comput. Environ. Urban Syst.* **2000**, *24*, 539–558. [CrossRef]
85. Simard, S. *Finding the Mother Tree: Uncovering the Wisdom and Intelligence of the Forest*; Penguin UK: London, UK, 2021.
86. Whitfield, J. Fungal roles in soil ecology: Underground networking. *Nature* **2007**, *449*, 136–139. [CrossRef]
87. Andersson, E.; McPhearson, T.; Kremer, P.; Gomez-Baggethun, E.; Haase, D.; Tuvendal, M.; Wurster, D. Scale and context dependence of ecosystem service providing units. *Ecosyst. Serv.* **2015**, *12*, 157–164. [CrossRef]
88. Cimburova, Z.; Pont, M.B. Location matters. A systematic review of spatial contextual factors mediating ecosystem services of urban trees. *Ecosyst. Serv.* **2021**, *50*, 101296. [CrossRef]
89. Vogt, J.; Hauer, R.J.; Fischer, B.C. The costs of maintaining and not maintaining the urban forest: A review of the urban forestry and arboriculture literature. *Arboric. Urban For.* **2015**, *41*, 293–323. [CrossRef]
90. Miller, R.W.; Hauer, R.J.; Werner, L.P. *Urban Forestry: Planning and Managing Urban Greenspaces*; Waveland Press: Long Grove, IL, USA, 2015.
91. Dan, H. It Goes on like a Forest. In *The Word for World Is Still Forest, Red*; Anna-Sophie Springer & Etienne Turpin: Berlin, Germany, 2017; Volume 4.
92. Roman, L.A.; Scatena, F.N. Street tree survival rates: Meta-analysis of previous studies and application to a field survey in Philadelphia, PA, USA. *Urban For. Urban Green.* **2011**, *10*, 269–274. [CrossRef]
93. Bartens, J.; Wiseman, P.E.; Smiley, E.T. Stability of landscape trees in engineered and conventional urban soil mixes. *Urban For. Urban Green.* **2010**, *9*, 333–338. [CrossRef]

Disclaimer/Publisher's Note: The statements, opinions and data contained in all publications are solely those of the individual author(s) and contributor(s) and not of MDPI and/or the editor(s). MDPI and/or the editor(s) disclaim responsibility for any injury to people or property resulting from any ideas, methods, instructions or products referred to in the content.

Article

Soil Dynamics in an Urban Forest and Its Contribution as an Ecosystem Service

María de la Luz Espinosa Fuentes [1,*], Oscar Peralta [1], Rocío García [1], Eugenia González del Castillo [1], Rosa María Cerón Bretón [2], Julia Griselda Cerón Bretón [2], Eric Tun Camal [2] and Faustino Zavala García [3]

[1] Atmospheric Sciences and Climate Change Institute, Autonomous National University of Mexico, Mexico City 04510, Mexico; oscar@atmosfera.unam.mx (O.P.); gmrocio@atmosfera.unam.mx (R.G.); eu.gonzaranda@atmosfera.unam.mx (E.G.d.C.)
[2] Chemistry Faculty, Autonomous University of Carmen, Ciudad del Carmen 24180, Mexico; rceron@pampano.unacar.mx (R.M.C.B.); jceron@pampano.unacar.mx (J.G.C.B.); 090004@mail.unacar.mx (E.T.C.)
[3] Marine Sciences and Limnology Institute, Autonomous National University of Mexico, Mexico City 04510, Mexico; zavgar@cmarl.unam.mx
* Correspondence: marilu@atmosfera.unam.mx

Abstract: Forests embedded in an urban matrix are an important site to investigate the effects of multiple anthropogenic influences that can lead to the modification of biogeochemical cycles and, consequently, of the ecosystem services they provide. In this study, the main soil properties, exchangeable cations, and heavy metal concentrations were measured to assess soil quality and fertility, as well as soil carbon stock (SCS) and CO_2 effluxes (Rs) at the Natural Protected Area Bosque de Tlalpan (BT). Four study zones were considered: strict protection zone (Z1), restricted use protection zone (Z2), extensive public use zone (Z3), and intensive public use zone (Z4) during three climatic seasons (rainy, dry-cold, and dry-warm seasons). The concentration of heavy metals in the BT soil showed that these elements are within the reference limits accepted by Mexican standards and are not considered toxic to the environment, except for mercury, which exceeded the standard with double the concentration. The results revealed significant variations in the SCS and soil organic matter (SOM) among the different sites. The highest mean values of SCS (3.01 ± 0.63 and 4.96 ± 0.19 kg m^{-2}) and SOM ($7.5 \pm 1.01\%$ and $8.7 \pm 0.93\%$) were observed in areas of high protection and extensive public use. CO_2 effluxes showed significant differences between sampling seasons, with fluxes being highest during the rainy season (3.14 ± 1.01 μmol·m^{-2}·s^{-1}). The results suggest that the level of conservation and effective management of the sites played an important role in the carbon storage capacity and in the physicochemical properties of the soil. This not only provided insights into the current state of an urban forest within a large urban area but also emphasized the significance of conserving such ecosystems.

Keywords: CO_2 effluxes; soil carbon stock; soil fertility; land management; urban soils

1. Introduction

Urban forests play a critical role in carbon cycling, with their ability to store carbon in both the short and long term and reduce greenhouse gas emissions. Moreover, the soils of urban forests offer essential environmental services such as flood mitigation, reduction in urban heat island effect, and provision of green spaces vital for physical and mental health. They also significantly contribute to nutrient cycling on a global scale [1–5], assisting the study of biogeochemical cycles [6–9] and the development of crucial ecosystem services [10].

Urban soil provides regulating services such as climate change mitigation through carbon sequestration, pollutant reduction, nutrient supply and retention, organic matter stabilization, greenhouse gas regulation, and biodiversity preservation [11]. These services help to maintain the health and sustainability of urban environments. Ecosystem services

Citation: Espinosa Fuentes, M.d.l.L.; Peralta, O.; García, R.; González del Castillo, E.; Cerón Bretón, R.M.; Cerón Bretón, J.G.; Tun Camal, E.; Zavala García, F. Soil Dynamics in an Urban Forest and Its Contribution as an Ecosystem Service. *Land* **2023**, *12*, 2098. https://doi.org/10.3390/land12122098

Academic Editors: Alessio Russo and Giuseppe T. Cirella

Received: 21 October 2023
Revised: 15 November 2023
Accepted: 19 November 2023
Published: 23 November 2023

Copyright: © 2023 by the authors. Licensee MDPI, Basel, Switzerland. This article is an open access article distributed under the terms and conditions of the Creative Commons Attribution (CC BY) license (https://creativecommons.org/licenses/by/4.0/).

are evaluated using soil quality indicators, such as bulk density (BD), electrical conductivity (EC), organic carbon, soil texture (clay and silt), cation exchange capacity (CEC), inorganic nitrogen concentration, pH, and concentrations of potentially toxic elements [12–15].

Carbon sequestration is a crucial soil function that supports vital ecosystem services such as soil fertility maintenance and climate change mitigation [16,17]. Soil fertility, one of the soil properties that determine productivity, is regarded as one of the ecosystem services that soil can provide for the benefit of humans [18] as well as for nutrient maintenance in natural ecosystems [19]. Soil fertility is determined by various factors, including texture, water retention capability, profile depth, nutrient availability, organic carbon content [20], and soil organic matter (SOM) [18,21–23].

Urban forests act as carbon reservoirs, sequestering CO_2 and integrating it as biomass into SOM [24]. Additionally, urban forest soils are subject to significant amounts of heavy metal, organic compounds, and acid compound deposition resulting from atmospheric pollution [2]. The retention of contaminants by soils is largely determined by SOM and pH. Heavy metal adsorption to soil components decreases when organic matter is decreased [25]. The mobility and bioavailability of heavy metals also reduce with an increase in pH as they are removed and taken up by colloids [26]. In addition, pH influences metal transport processes [27].

Urban forest soils with impermeable surfaces can accumulate nutrients, such as metals [28]. Previous studies have investigated urban forests in Mexico City. Fenn et al. [29] determined the concentration of heavy metals in Desierto de los Leones National Park. Santiago-Romero et al. [30] assessed the carbon content stored in above-ground biomass within plant communities in the Bosque de Tlalpan. Similarly, Hernández-Guillen et al. [31] estimated the carbon sequestration within trees in a section of the Chapultepec Forest.

Although urban soils are vital in providing ecosystem services, there has been no integrated evaluation of their quality in Mexico.

The goal of this study was to examine the role of soil as a regulation ecosystem service by comprehensively considering the soil functions that sustain it. To achieve this aim, (1) the soil quality was evaluated through physicochemical parameters, (2) the effect of physical and chemical parameters on the concentration and distribution of heavy metals, cations, and anions was determined, (3) carbon stock and CO_2 efflux were measured in the BT, an urban forest in Mexico City under different management regimes.

2. Materials and Methods

2.1. Study Area

BT is a protected natural area with 253 ha located south of Mexico City (19°17′36″, 99°11′46″). It has an average altitude of 2389 m above sea level and a temperate subhumid climate with an average annual temperature of 15 °C and annual rainfall ranging from 850 to 911 mm. BT is in a volcanic terrain, where lithosols are the primary edaphic unit.

In BT, there are three main types of vegetation: xeric scrub, oak forest, and cultivated forest [32]. Xerophytic scrub is associated with basaltic substrates, shallow soils, and heterogeneous microhabitats that depend on the soil depth, the shading and humidity conditions, the amount of exposed basalt, and the cracks on the rocks [33]. Shrubs and herbs succulents predominate with introduced trees such as *Schinus molle*, *Eucalyptus* spp., *Pinus* spp., and *Cupressus* spp. Oak forests are patches dominated by oak species such as *Quercus rugosa*, *Q. laurina*, *Q. mexicana*, and *Q. crassipes*. Cultivated trees are in reforested areas with *Eucalyptus* spp., *Pinus* spp., *Cupressus* spp., and *Fraxinus uhdei*.

2.2. Sampling Campaign and Classification of Sampling Areas

The sampling campaigns took place in August 2021 (rainy season), January 2022 (dry-cold season), and April 2022 (dry-warm season). The BT study area was separated into four sites (Figure 1A,B) in accordance with the zoning proposed by the 2011 Forest Management Plan [32]. The various zones were determined based on environmental quality, current and potential use, and the impact of human intervention.

Figure 1. Location of the study area (**A**) and sampling sites (**B**) in the Natural Protected Area Bosque de Tlalpan.

The strict protection zone (Z1) has limited human intervention and consists of habitats with delicate flora and fauna resources that necessitate absolute protection owing to their fragility and high value for maintaining the aquifer's recharge capacity. Z1 spans an area of 97.5 ha and is characterized by scrub and oak trees present in different regions of the forest. On the other hand, the restricted use protection zone (Z2) encompasses an area of 53.66 ha and is composed of oak trees, cultivated trees, and xeric scrub species. Between 1970 and 1975, the area underwent clearance for a football pitch construction and was later utilized as a heliport from 1988 to 1994. Currently, the area is utilized for sporting activities and as a viewing point since it has been reforested. The extensive public use zone (Z3) spans 70.27 ha, introducing various tree species and wildlife. The zone has dense vegetation and grassy terrains with accessible paths and roads. The intensive public use zone (Z4) spans 16.28 ha. The recreational area, with jogging tracks, paths, and a playground, experiences significant anthropogenic influence. The vegetation cover primarily comprises cultivated trees [32].

2.3. Soil Sampling and Analyses

At each site, a 4 × 12 m plot was selected with three sampling points of 1 m^2, at each point, four soil core samples were taken from the shallowest layer (0–10 cm deep) using a core sampler of 193.3 cm^3 to obtain 3 composite samples per site, this procedure was repeated in each campaign giving a total of 36 samples.

After soil extraction, the roots and leaves visible on the label were removed and sealed, and the samples were sent to the laboratory. Each composite sample was air-dried and sieved in the laboratory with a 2 mm stainless steel mesh sieve. Soil moisture (SM) was measured according to Etchevers et al. [34], and soil temperature (Ts) was measured in situ with a Decagon Echo 5-TM sensor. The pH was determined with a HANNA (Instruments electronic) model pH211, with an Extech Instrument electrode. The pH meter was calibrated with two buffer solutions 7 and 4 Merck Certipur (DEU). The EC was measured using a TRANS instrument model HC3010, TDS-Conductivity-Salinity (USA), with an electrode TIPB10-0400 with K = 0.9660 and calibrated with a standard solution of 1.4 mS.

The pH and EC were measured from aqueous extracts and verified in duplicate after 1 min, followed by stirring the samples for 30 s. The aqueous extracts were prepared with deionized water (>18.2 MΩcm) and $CaCl_2$ 0.01 M ACS certified (Japan) ($CaCl_2 \cdot 2H_2O$, Fisher Scientific, Hampton, NH, USA) solution. The determination of soil texture was made following Bouyouco's method [35]. For BD (0–10 cm deep), an undisturbed sample of soil was taken using a steel cylinder (95.4 cm^3). BD was determined from the oven-dried (105 °C) mass of the core and the core volume.

Total organic carbon (TOC) was determined using a Shimadzu TOC-V CSN analyzer with an SSM-5000A solid sample module. The TOC analyzer is an analytical instrument that measures the total amount of organic carbon contained in many solid samples in addition to aqueous samples, including soil, sludge, and sediments. For the TOC analyzer, 20 mg of each soil sample was weighed into sample cups. To calibrate the analyzer, D-glucose Anhidra powder reagent (carbon concentration: 40%) was collected in a sample boat, and its TC was measured [36]. SOM was estimated from the TOC determination using a conversion factor of 1.724 based on the assumption that SOM contains 58% organic carbon [37].

The SCS (kg m^{-2}) was calculated using Equation (1)

$$SCS = [BD\ (g\ cm^{-3}) \times TOC\ (\%) \times SDI\ (cm)]/10 \quad (1)$$

where SDI is the soil depth interval (0–10 cm deep).

In each of the sampling plots and once per campaign, the soil CO_2 efflux, here called soil respiration (Rs), was monitored. At each sampling point, at least one week before the Rs measurement, 5 PVC collars (9.1 cm high) were partially inserted into the soil at 5 m intervals. The static closed chamber method was used. The chamber consists of an acrylic cylinder (14.6 cm inner diameter, 24.3 cm height) sealed at one end; a diffusion-based, air CO_2 mole fraction sensor (CARBOCAP GMP 343, Vaisala), and an air temperature and relative humidity sensor (CS215 Campbell Scientific) are attached to the sealed top of the chamber. At the start of the measurement, the open end of the chamber was placed on the collar, enclosing an air volume V of 0.0037 m^3 (the total volume of the chamber minus the volume occupied by the sensors). Due to microbial activity, plant root activity, and possible dissolution of calcium carbonate in the water present in the soil, CO_2 is released from the soil and accumulated in the chamber at a rate $\Delta c/\Delta t$. Water vapor evaporated from the soil also accumulates in the chamber after it is closed, diluting the CO_2 mole fraction c [38]. To account for this dilution effect, a corrected CO_2 mole fraction c' is calculated:

$$c' = \frac{c}{1-w} \quad (2)$$

where w is the water vapor mole fraction (mol/mol) in the air, computed from the relative humidity reported by the temperature and relative humidity sensor inside the chamber.

The efflux rate Rs from the soil is then estimated as the corrected rate of change using the following equation:

$$Rs = \frac{V}{A} \cdot \frac{\Delta c'}{\Delta t} \cdot \frac{P_0}{\mathcal{R} T_0} \quad (3)$$

where V is the chamber volume, A is the circular area of the chamber, P_0 is the initial atmospheric pressure, T_0 is the initial temperature, and $\mathcal{R}$ is the universal gas constant. The University Network of Atmospheric Observatories (RUOA) meteorological station, 4 km northeast of BT, measures the atmospheric pressure every minute with a Vaisala PTB110 sensor; these readings are considered P_0. The other variables were measured every 20 s for

10 min. The accumulation of CO_2 increased after 40 s. The Rs was calculated with 5 min data to reduce the bias of saturation of the chamber [39,40].

For the preparation of samples, the Official Mexican Standard [41] was followed. The digestion procedures and analytical determination of heavy metals in the soil were carried out according to the standard method of Digestion Procedure for Microwave Extraction for Ambient Filter Samples Method IO-3.1 [42], with control of the pressure and temperature. For the analysis of heavy metals, samples are previously processed for their acid digestion, weighing 0.5 g of soil per sample. A microwave digestion system (CEM MARS-5, Matthews, NC, USA) with Teflon-coated digestion containers was used. Samples were digested using a concentrated acid mixture: 15 mL HNO_3 65%, 2 mL H_2O_2 35% [43]. H_2O_2 was used to enhance the decomposition of organic matter in samples. Samples were digested at 180 °C for 30 min, with a heating rate of 10 °C min^{-1} [43]. After the microwave digestion process, the final solution was diluted to 25 mL with deionized water (>18.2 MΩcm). For the determination of Hg-Total (HgT), the atomic absorption spectrophotometry technique with a cold steam hydride generator (AAS-GH) was applied; for the determination of Cd, Cu, Ni, Pb, and Zn, the inductively coupled plasma optical emission spectrometry (ICP-OES) technique was applied [44]. The ionic species analyzed were Cl^-, NO_3^-, PO_4^{3-}, and SO_4^{2-} by anion exchange chromatography, and for the determination of cationic species: Na^+, K^+, Ca^{2+}, Mg^{2+}, by cation exchange chromatography, applying the high-resolution liquids chromatography technique (HPLC).

A solution of 45 mmol L^{-1} Na_2CO_3/14 mmol L^{-1} $NaHCO_3$ was used to elute anions and 20 mmol L^{-1} H_2SO_4 to elute cations at a flow rate of 0.25 mL min^{-1}. A 20 µL volume of soil solution was injected into the chromatograph. The chromatographic standard curve was prepared using certified Dionex solutions, whose concentrations ranged from 0.1 to 40 mg L^{-1} for anions and from 0.25 to 100 mg L^{-1} for cations. To verify the validity of the results, we used NIST standard reference materials SRM 1646a (Gaithersburg, MD, USA) and CRM-029 (Sigma-Aldrich, Burlington, MD, USA). The cation exchange capacity (CEC) was calculated by adding the exchangeable Na^+, K^+, Ca^{2+}, Mg^{2+}, and H charge equivalents.

2.4. Statistical Analysis

Statistical analysis was conducted using Statgraphics Centurion 19. The assumption of normality and homogeneity of variances was tested using the Shapiro–Wilk and Levene tests prior to the ANOVA analysis. The purpose of the ANOVA analysis was to determine the variations in soil parameters (such as physicochemical properties, heavy metals, SOM, SCS, and Rs) between sampling sites and monitoring seasons. Tukey's post hoc test was used to separate means where differences were significant ($p < 0.05$).

Pearson's correlation was utilized to determine the relationship between the soil variables. Additionally, a collinearity analysis was undertaken in SPSS to establish the associations between the dependent and independent variables in the soil parameters. This analysis discriminated the highly weighted variables in accordance with the Sarstedt criterion [45]. Furthermore, a principal component analysis (PCA) was conducted on the selected variables after the collinearity analysis.

3. Results

3.1. Environmental Parameters and Physicochemical Properties of the Soil

In 2021, the rainy season extended from May to October (Figure 2A); at the time of the measurement in August 2021, the BT soils had received 92 mm of rain in the previous month and a total of 546 mm since May. Consequently, the SM was highest during the rainy season (36.7 ± 5.7%), particularly at Z1 (Figure 2B), and lowest during the dry-cold season (6.5 ± 2.3%). Despite the long seasonal drought that preceded the dry-warm season measurements, several early rainfall events in the previous week wetted the soil to an intermediate moisture level (20.5 ± 4.2%). Ts varied less between seasons, being highest during the dry-warm season of April 2022 and, on average, only 4 °C less during the cold-dry season of February 2022, when the lowest temperatures were recorded (Figure 2C).

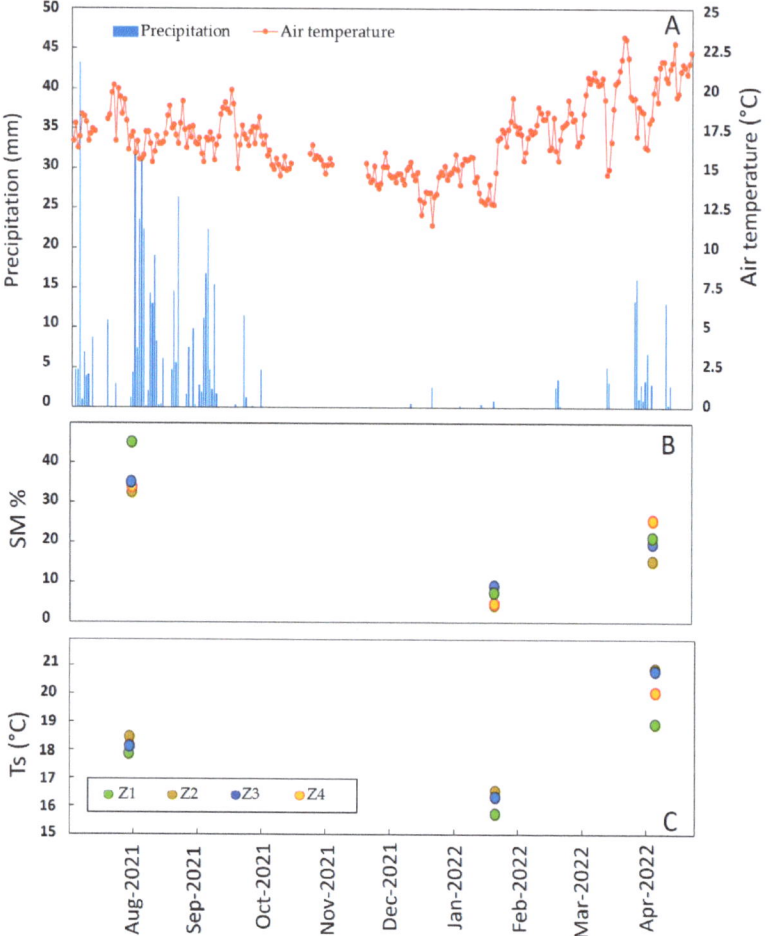

Figure 2. Precipitation and air temperature conditions (**A**), volumetric soil moisture SM% (**B**), and soil temperature Ts °C (**C**) prevalent at the sampling sites (Z1, Z2, Z3, Z4) in different months of the year.

In the study area, sand, clay, and silt ranged from 46 to 67%, 12 to 16%, and 21 to 40% respectively (Figure 3). Based on this, Z1, Z3, and Z4 were categorized as loam soil, and Z2 was categorized as sandy loam.

pH, EC, and BD remained without significant changes in the sampling sites and seasons (Table 1). pH ranged from 5.20 to 6.50, EC values ranged from 0.51 to 0.79 dS cm^{-1}, the BD values ranged from 0.83 to 1.26 g cm^{-3}, the sites with the highest BD values were Z3 (0.96 to 1.01 g cm^{-3}) and Z4 (1.24 to 1.26 g cm^{-3}) indicating more compact soils. The SOM showed differences between sites (F = 42.33; p = 0.0001); the mean values of Z1 and Z3 were 6.27 ± 1.43% and 8.55 ± 0.33%, respectively, Z2 and Z4 showed lower values of 2% in both sites. The CEC in the 0–10 cm horizon ranged from 6.1 to 21.86 Cmol (+) kg^{-1} (Table 1). Z2 had the lowest average values, and Z3 had the highest.

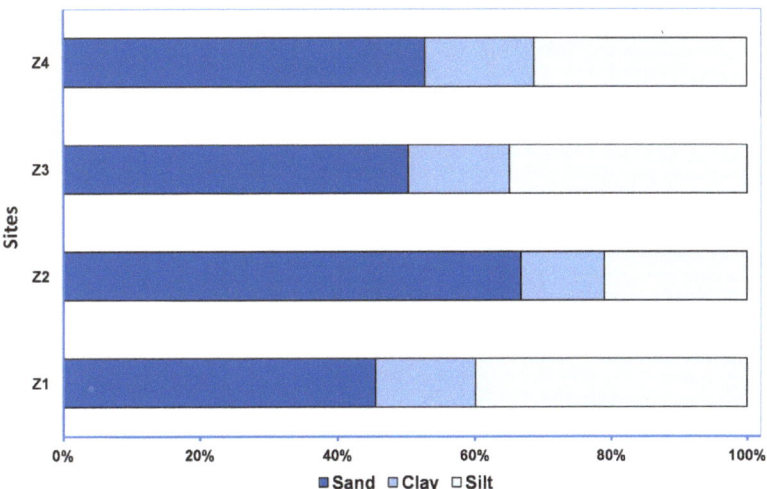

Figure 3. Soil particle size measured as a percentage contribution of sand, clay, and silt at the different BT sites (Z1, Z2, Z3, and Z4).

Table 1. Soil summary values of the soil profile (0 to 10 cm) at different sampling sites and seasons at BT. pH, electrical conductivity (EC), bulk density (BD), soil organic matter (SOM), and cation exchange capacity (CEC). Mean and standard deviation are given. Letters indicate a significant difference between sites ($p < 0.05$) according to Tukey's test.

Season	Sites	pH	EC (dS cm^{-1})	BD (g cm^{-3})	SOM (%)	CEC (C mol (+) kg^{-1})
Rainy	Z1	5.91 ± 0.36	0.54 ± 0.02	0.85 ± 0.02	7.49 ± 1.01 (a)	13.29 ± 0.28
	Z2	6.50 ± 0.21	0.67 ± 0.03	0.92 ± 0.01	2.88 ± 0.75 (b)	6.10 ± 0.12
	Z3	5.70 ± 0.21	0.69 ± 0.03	1.00 ± 0.06	8.66 ± 0.93 (c)	15.73 ± 0.36
	Z4	5.22 ± 0.28	0.56 ± 0.03	1.26 ± 0.01	2.91 ± 0.11 (b)	12.74 ± 0.31
Dry-cold	Z1	5.72 ± 0.35	0.75 ± 0.04	0.84 ± 0.01	6.64 ± 1.08 (a)	15.78 ± 0.36
	Z2	6.10 ± 0.68	0.51 ± 0.02	0.95 ± 0.01	2.56 ± 0.46 (b)	10.95 ± 0.36
	Z3	5.50 ± 0.21	0.53 ± 0.04	1.01 ± 0.08	8.18 ± 1.53 (c)	18.29 ± 0.24
	Z4	5.70 ± 0.24	0.60 ± 0.04	1.24 ± 0.01	2.80 ± 0.22 (b)	15.77 ± 0.39
Dry-warm	Z1	5.64 ± 0.58	0.61 ± 0.02	0.83 ± 0.02	4.68 ± 1.47 (a)	17.23 ± 0.58
	Z2	6.10 ± 0.36	0.60 ± 0.04	0.91 ± 0.03	2.78 ± 0.18 (b)	11.18 ± 0.14
	Z3	5.20 ± 0.28	0.79 ± 0.04	0.96 ± 0.02	8.82 ± 1.01 (c)	21.86 ± 0.27
	Z4	5.72 ± 0.48	0.66 ± 0.02	1.25 ± 0.01	2.84 ± 0.49 (b)	15.71 ± 0.25

3.2. Heavy Metals, Cations, and Anions in Soils

Significant differences in the metallic species, Cu (F = 3.12, p = 0.04) and Zn (F = 8.58, p = 0.007), were detected between the sampling sites. Cu and Zn were the metals with the highest concentrations in Z3 and Z4. Cd, Pb, and Ni showed similar concentrations at all sites. According to MON [41], the observed heavy metal concentrations are within acceptable limits and do not pose a threat to the environment. The Hg concentrations were high at all sites (46.5 to 62 mg kg^{-1}) and above the MON (23 mg kg^{-1}) [41] and international standards (6.6 mg kg^{-1}) [46] (Figure 4A).

Mg^{2+} concentrations were similar at all sites; K^+, Na^+, and Ca^{2+} showed significant variation between sites ($p < 0.05$). K^+ and Na^+ showed the highest values in Z4 and Ca^{2+} in Z1 (Figure 4B). On the anion side, PO_4^{3-} also showed differences between sites ($p < 0.05$), with the Z3 reporting phosphate concentrations of 184 mg kg^{-1}. SO_4^{2-}, NO_3^-, and Cl^- concentrations were similar at all sites (Figure 4C).

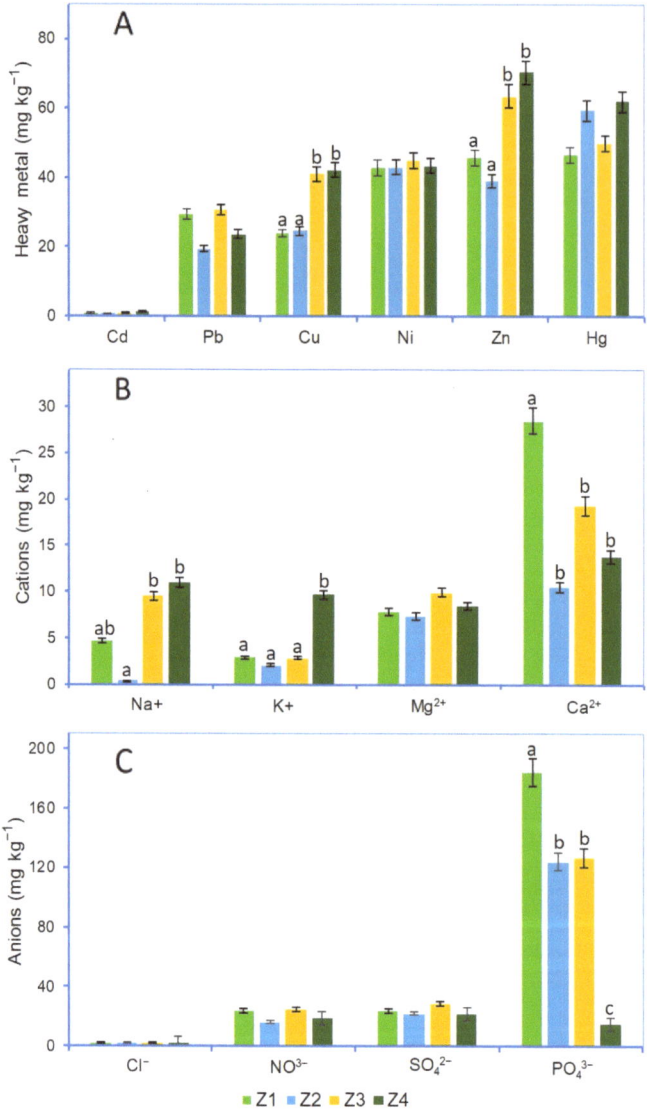

Figure 4. Heavy metals (**A**), cations (**B**), and anions (**C**) concentrations at the different BT sites (Z1, Z2, Z3, and Z4). Different letters above bars indicate statistical significance at $p < 0.05$ between sites according to Tukey's test.

3.3. Soil CO_2 Efflux (Rs) and Soil Carbon Stock

The average CO_2 efflux was 1.22 µmol·m^{-2}·s^{-1}. Collars measurements ranged from 0.18 to 5.72 µmol·m^{-2}·s^{-1}. Measurements between sampling seasons showed significant differences (F = 91.57, $p = 0.000$), with the highest mean effluxes recorded during the rainy season (3.14 ± 1.01 µmol·m^{-2}·s^{-1}) and the lowest during the dry-cold season (0.53 ± 0.34 µmol·m^{-2}·s^{-1}) (Figure 5). Among sites, the highest mean value was observed in Z1 (2.07 ± 0.57 µmol·m^{-2}·s^{-1}) and the lowest in Z4 (1.58 ± 0.93 µmol·m^{-2}·s^{-1}), while sites Z2 and Z3 had mean Rs of 1.88 µmol·m^{-2}·s^{-1} and 1.80 µmol·m^{-2}·s^{-1}, respectively (Figure 5).

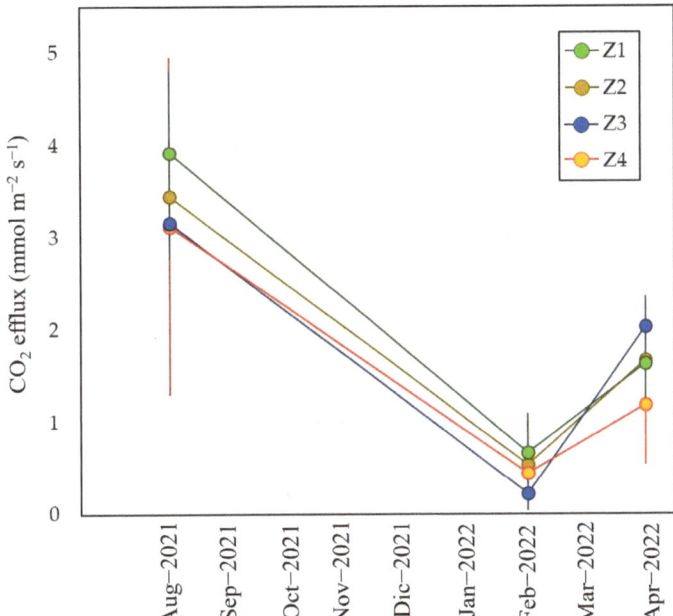

Figure 5. CO_2 efflux at the different periods and sampling sites in BT.

The SCS showed significant differences between sites (F = 42.33, p = 0.000), with values ranging from 1.4 to 5.1 kg m^{-2} (0–10 cm depth), with the highest values recorded in Z3, and the lowest values were observed in Z2 and Z4 (1.4 to 2.1 kg m^{-2}, respectively) (Figure 6). Pearson's correlation showed negative but not significant correlations between SCS and sand, pH and BD (Figure 7A).

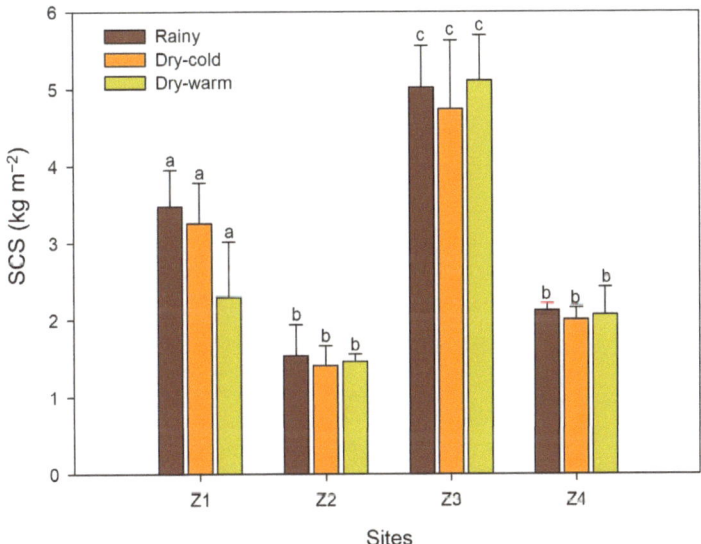

Figure 6. Soil carbon stock (SCS) at the different BT sites (Z1, Z2, Z3, and Z4) and seasons at 0–10 cm depth. Different letters above the bars indicate statistical significance of $p < 0.05$ between sites according to Tukey's test.

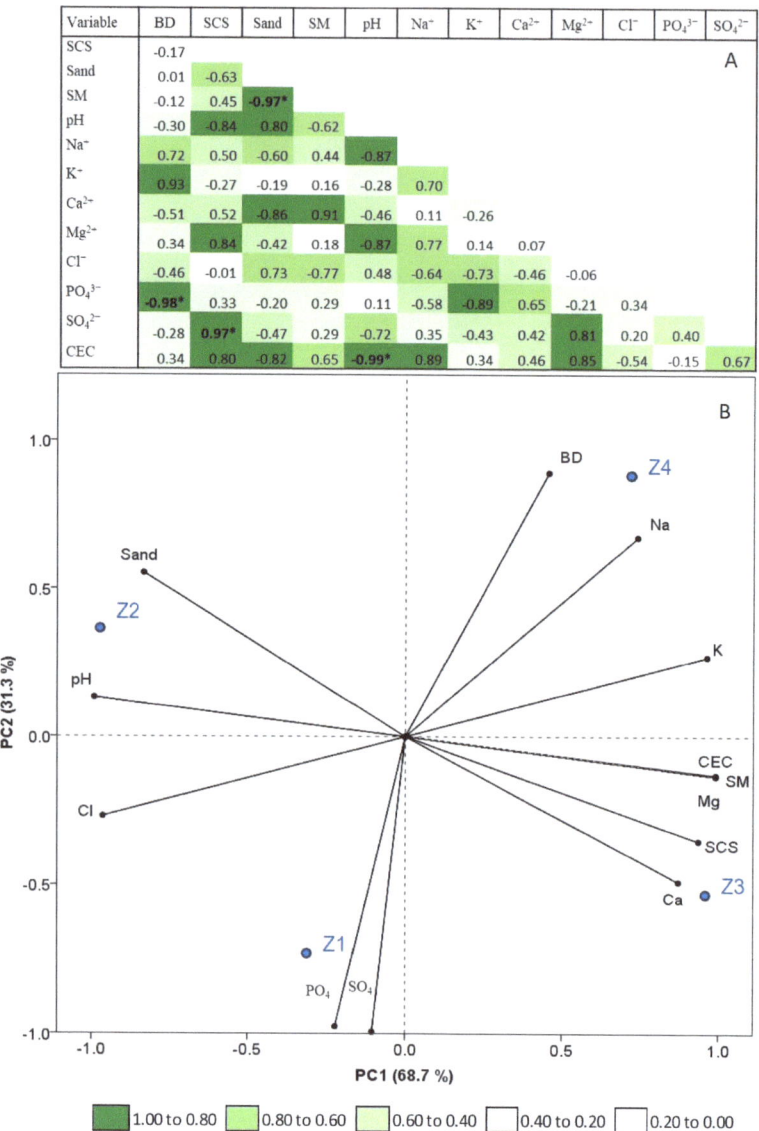

Figure 7. Pearson's correlation coefficients between soil variables (**A**) and principal component analysis of the most highly correlated variables according to the different sampling sites (**B**). * = Correlation significant at $p < 0.05$ (2-tailed). Z1, Z2, Z3 and Z4, are the different sampling sites.

3.4. Principal Component Analysis

The first two components of the PCA explained 100% of the variance, 68.7% for PC1 and 31.3% for PC2. The PCA showed that the soils of BT represent ensembles related to the soil characteristics of each sampling site (Figure 7B). In PC1, the most important factors with contributions between 10.38% and 11.1% were K^+, Mg^{2+}, and Cl^- ions, including CEC, SM, and pH (Table 2).

Table 2. Loadings and contribution of variables to the principal components.

Variable	PC1		PC2	
	Loadings	Contribution (%)	Loadings	Contribution (%)
Sand	−0.833	7.77	0.553	7.52
SM	0.991	11.00	−0.133	0.43
pH	−0.991	11.00	0.133	0.43
BD	0.457	2.33	0.890	19.46
SCS	0.936	9.81	−0.352	3.04
K^+	0.963	10.38	0.270	1.79
Ca^{2+}	0.872	8.51	−0.489	5.89
Mg^{2+}	0.991	11.00	−0.133	0.43
Cl^-	−0.964	10.39	−0.268	1.76
CEC	0.992	11.01	−0.129	0.41
PO_4^{3-}	−0.220	0.54	−0.975	23.39
SO_4^{2-}	−0.105	0.12	−0.994	24.30

Several factors contributed to the separation of the plots along PC2. The factors with the largest contribution (19.46% to 24.30%) were BD, SO_4^{2-}, and PO_4^{3-} (Table 2). The PCA showed that the nutrient status of the soils was the most important factor for the separation of the plots according to the different sampling sites.

4. Discussion

4.1. Soil Physicochemical Properties and Its Quality and Fertility

The soils had a loamy and sandy loam texture of sand > silt > clay. Soil texture influences the CEC for the retention and exchange of the cations Na^+, K^+, Ca^{2+}, and Mg^{2+} [47,48]. The highest CEC values were observed at sites where silt and clay predominated (Z1 and Z3); this condition has been observed in other studies [49–52]. According to MON [53], the CEC found in BT are soils with an acceptable exchange capacity, which is reflected in the concentrations of cations that ensure the availability of nutrients for plants. In this study, the presence of macronutrients such as K^+, Ca^{2+}, and Mg^{2+} and micronutrients like Cu and Zn can reflect the natural fertility of the soil by defining the potential of soil to provide these mineral elements [54].

According to Legout et al. [19], soil fertility in forest ecosystems is defined as the capacity of the soil to retain nutrients associated with organic matter and clay content to ensure the proper functioning of the soil-plant system, providing an ecosystem service of regulation. Ca^{2+}, Mg^{2+}, and Na^+ were dominant at all sites in the BT. Ca^{2+} is a function of soil properties; Mg^{2+} is attributed to a hydration ratio that is retained less than Ca^{2+} [55]. The presence of NO_3^-, SO_4^{2-}, and PO_4^{3-} also indicates a high organic load.

Concentrations of heavy metals in the soil were lower than at MON and were not considered toxic to the environment [53], except for Hg, which was twice the standard concentration. In the BT, Hg probably originates from anthropogenic activities and is trapped by organic matter enriching the surface layers of the soil [56]. The results show a strong relationship between heavy metals and SOM because organic matter can retain metals [57]. Chen et al. [58] suggest that the increase in trace elements is related to urbanization, which may be the case for BT.

BD is a dynamic soil property associated with texture and SOM [59]. The observed BD values are similar to those found by Saavedra-Romero et al. [60] in urban soils from the San Juan de Aragón forest, Mexico City (0.87 to 1.14 g cm^{-3}) and by Chávez-Aguilar et al. [61] (0.52 to 0.77 g cm^{-3}) in the temperate forest of Nevado de Toluca, Mexico. In urban parks, the BD is higher, reflecting soil compaction due to the influence of regular mowing and human trampling [62]. Studies in urban parks in other locations report mean values of 1.39 and 1.73 g cm^{-3} [63], 1.29 g cm^{-3} [64], and 0.97 g cm^{-3} [62], similar values to Z4. The compaction observed in Z2 and Z4 is due to the continuous human activity and the passage of people.

Sites with anthropogenic disturbance have low SOM values (Z2 and Z4). According to the MON [53], these sites are classified as very low class (<4%). Protected sites (Z1 and Z3) have SOM in the medium class (6.1 to 10.9%), indicating good soil quality.

4.2. Variability of Soil CO_2 Efflux

Soil CO_2 efflux, or Rs, is an important pathway in the global carbon cycle [65]. A number of physical (soil and air temperature, soil moisture) and biological (plant cover, plant phenology, and carbohydrate substrate supply from photosynthesis) factors have been recognized as regulators of Rs [66,67]. However, most process-based and analytical models only include the influence of Ts and SM as significant controls; typically, the effect of Ts on CO_2 efflux is modeled as an exponential function, with SM as an additive or multiplicative term. In this study, Rs did not differ between sampling sites, but there were differences between sampling periods, with higher values during the rainy season and lower values during the dry cold season. The sensitivity of Rs to SM, which modulates—and sometimes overrides—the effect of Ts, has been documented for a variety of water-limited environments such as deserts, Mediterranean ecosystems, and tropical dry forests [68–70]. A rapid increase in water availability leads to microbial reactivation, resulting in a pulse of root respiration and organic matter decomposition [71], carbon mineralization, and nutrient availability [72].

4.3. Influence of Management Type of BT on Soil Carbon Stock

The soils of BT are of volcanic origin, with lithosol and leptosol. They have an average depth of 25 cm, with stones of different sizes, so their potential to store carbon is limited.

The SCS in the study area shows a high variability between sites, ranging from 1.4 to 5.1 kg m^{-2} (0–10 cm). High spatial variability in carbon storage in urban forests and parks (0–10 cm) has also been found in other studies from New Zealand (2.7 to 4.8 kg m^{-2}) [73], Boston, MA, USA (2. 29 to 5.67 kg m^{-2}) [74], Baltimore, MD, USA (6.0 to 8.0 kg m^{-2}) [75], Republic of Korea (2.8 kg m^{-2}) [76], and Milan, Italy (0.75 to 6.48 kg m^{-2}) [64]. All these studies support the hypothesis that differences in management and land use within cities and land use types can explain the observed large intra-urban variability. SCS can regulate climate and influence other soil properties [77,78]. SOC is correlated with soil texture, SM, and CEC. The SCS was higher in soils with a clay texture, such as Z1 and Z3, so these sites are better adapted to sequester carbon [79,80].

Z1 showed the highest SCS, probably because it is the most protected site with very little anthropogenic influence, which is positive for soil and carbon dynamics. Z2 and Z4 had the lowest SCS, both sites having been historically degraded by anthropogenic activities such as recreational use and lawn mowing [32], which is reflected in the high sand content (53–67%) used to fill sites prior to reforestation and consequently higher soil compaction. These results are consistent with other studies that mention that soil compaction by human activities can reduce SCS reserves in urban forests [81,82].

Xu et al. [83] demonstrated that historical land use can have long-term effects on critical ecosystem processes, such as SCS accumulation. Therefore, the land use pattern is one of the most important determinants of SCS at the city or urban forest scale [64,76].

5. Conclusions

The physicochemical properties showed that the BT soil was of high quality, with fertile soils suitable for biomass production, based on SOM, pH, and soil texture as the most critical determinants of soil fertility since these parameters promote the solubility of Cd, Cu, Zn, Pb, Ni, and Hg, providing cations, anions, and nutrients.

The Rs values were higher during the rainy season, associated with higher soil moisture. Sites with high SCS were the least disturbed, so they have important long-term effects on forest carbon accumulation and storage processes.

The results of this study indicate that urban forests can act as carbon sinks if their soils are kept in good condition and under good conservation management. This can lead to the maintenance of the ecosystem services provided by the forest.

Author Contributions: Conceptualization, M.d.l.L.E.F. and E.G.d.C.; methodology, M.d.l.L.E.F., R.G., F.Z.G. and E.T.C.; software, F.Z.G. and E.T.C.; validation M.d.l.L.E.F., O.P. and E.G.d.C.; formal analysis, M.d.l.L.E.F., R.G., E.G.d.C., R.M.C.B. and J.G.C.B.; investigation, M.d.l.L.E.F., R.G., E.G.d.C. and O.P.; resources, M.d.l.L.E.F.; data curation, M.d.l.L.E.F., O.P. and F.Z.G.; writing—original draft preparation, M.d.l.L.E.F., O.P., R.G., E.G.d.C., R.M.C.B., J.G.C.B., F.Z.G. and E.T.C.; writing—review and editing, M.d.l.L.E.F., O.P., R.G., E.G.d.C., R.M.C.B., J.G.C.B., F.Z.G. and E.T.C.; visualization, M.d.l.L.E.F. and O.P.; supervision, M.d.l.L.E.F., O.P., R.G., E.G.d.C., R.M.C.B. and J.G.C.B.; project administration, M.d.l.L.E.F.; funding acquisition, M.d.l.L.E.F. All authors have read and agreed to the published version of the manuscript.

Funding: This research was funded by Project DGAPA-UNAM, PAPIIT IA102321 "Dinámica de los ciclos biogeoquímicos derivada de los impactos antropogénicos en un bosque urbano".

Data Availability Statement: Data are contained within the article.

Acknowledgments: The authors are grateful for the support of the administrators of the Tlalpan Forest for the facilities provided. The authors also thank María Isabel Saavedra, José Manuel Hernández Solis, and Moises López Carrasco for their assistance during laboratory analysis.

Conflicts of Interest: The authors declare no conflict of interest.

References

1. Pouyat, R.V.; Page-Dumroese, D.S.; Patel-Weynand, T.; Geiser, L.H. *Forest and Rangeland Soils of the United States under Changing Conditions: A Comprehensive Science Synthesis*; Springer Nature: Cham, Switzerland, 2020; p. 289.
2. Dovletyarova, E.A.; Mosina, L.V.; Vasenev, V.I.; Ananyeva, N.D.; Patlseva, A.; Ivashchenko, K.V. Monitoring and assessing anthropogenic influence on soil's health in urban forests: The case from Moscow City. In *Adaptive Soil Management: From Theory to Practices*; Springer: Berlin/Heidelberg, Germany, 2017; pp. 531–557.
3. O'Riordan, R.; Davies, J.; Stevens, C.; Quinton, J.N.; Boyko, C. The ecosystem services of urban soils: A review. *Geoderma* **2021**, *395*, 115076. [CrossRef]
4. Cram, S.; Cotler, H.; Morales, L.M.; Sommer, I.; Carmona, E. Identificación de los servicios ambientales potenciales de los suelos en el paisaje urbano del Distrito Federal. *Investig. Geográficas* **2008**, *66*, 81–104.
5. Denegri, G.; Rodríguez Vagaria, A.; Mijailoff, J.; Mársico, J.; Acciaresi, G. Bosques urbanos: Su aporte al turismo en la costa atlántica norte de Argentina. *Estud. Perspect. Tur.* **2018**, *27*, 316–335.
6. Galicia, L.; Gamboa Cáceres, A.M.; Cram, S.; Chávez Vergara, B.; Peña Ramírez, V.; Saynes, V.; Siebe, C. Almacén y dinámica del carbono orgánico del suelo en bosques templados de México. *Terra Latinoam.* **2016**, *34*, 1–29.
7. López-López, S.F.; Martínez-Trinidad, T.; Benavides-Meza, H.M.; García-Nieto, M.; Ángeles-Pérez, G. Reservorios de biomasa y carbono en el arbolado de la primera sección del Bosque de Chapultepec, Ciudad de México. *Madera y Bosques* **2018**, *24*, 1–1.4. [CrossRef]
8. Mañon de la Cruz, R.; Orozco Hernández, M.E.; Mireles Lezama, P. *Evaluación de los Servicios Ambientales del Parque Metropolitano Bicentenario, Toluca, México*; Revista Iberoamericana de Ciencias: Brownsville, TX, USA, 2018; pp. 6–21.
9. Bautista, R.J.D.; Baeza, A.T.; Acosta, S.D.C.R.; Morales, P.S.; Alcántara, A.G.; Rivera, A.A.; Hernández, R.S. Almacenamiento de carbono y agua en un área periurbana de Tabasco. *Rev. Terra Latinoam.* **2019**, *37*, 197. [CrossRef]
10. Hyun, J.; Kim, Y.J.; Kim, A.; Plante, A.F.; Yoo, G. Ecosystem services-based soil quality index tailored to the metropolitan environment for soil assessment and management. *Sci. Total. Environ.* **2022**, *820*, 153301. [CrossRef]
11. Morel, J.L.; Chenu, C.; Lorenz, K. Ecosystem services provided by soils of urban, industrial, traffic, mining, and military areas (SUITMAs). *J. Soils Sediments* **2014**, *15*, 1659–1666. [CrossRef]
12. Hyun, J.; Kim, Y.J.; Yoo, G. A method for soil quality assessment in the metropolitan greenery: A comprehensive view of ecosystem services and soil functions. *MethodsX* **2023**, *10*, 102102. [CrossRef]
13. Calzolari, C.; Tarocco, P.; Lombardo, N.; Marchi, N.; Ungaro, F. Assessing soil ecosystem services in urban and peri-urban areas: From urban soils survey to providing support tool for urban planning. *Land Use Policy* **2020**, *99*, 105037. [CrossRef]
14. Sefati, Z.; Khalilimoghadam, B.; Nadian, H. Assessing urban soil quality by improving the method for soil environmental quality evaluation in a saline groundwater area of Iran. *CATENA* **2018**, *173*, 471–480. [CrossRef]
15. Ziter, C.; Turner, M.G. Current and historical land use influence soil-based ecosystem services in an urban landscape. *Ecol. Appl.* **2018**, *28*, 643–654. [CrossRef]
16. Millennium Ecosystem Assessment. *Ecosystems and Human Well-Being*; Island Press: Washington, DC, USA, 2005; Volume 5, p. 563.

17. Blum, W.E. Functions of soil for society and the environment. *Rev. Environ. Sci. Bio/Technol.* **2005**, *4*, 75–79. [CrossRef]
18. Barroso-Tagua, R.; Alvarez, D.; Huera, T.; Changoluisa, D.; Bravo, C. La fertilidad del suelo como un servicio eco sistémico en cultivo de cacao (*Theobroma cacao* L.), en la provincia de Napo. In *Libro de Memorias: Simposio Internacional Sobre Manejo Sostenible de Tierras y Seguridad Alimentaria*; Alemán, R., Reyes, H., Bravo, C., Eds.; Universidad Estatal Amazónica: Puyo, Ecuador, 2017; pp. 99–106.
19. Legout, A.; Hansson, K.; van der Heijden, G.; Laclau, J.-P.; Mareschal, L.; Nys, C.; Nicolas, M.; Saint-André, L.; Ranger, J. Chemical fertility of forest ecosystems. Part 2: Towards redefining the concept by untangling the role of the different components of biogeochemical cycling. *For. Ecol. Manag.* **2020**, *461*, 117844. [CrossRef]
20. Bautista-Cruz, A.; del Castillo, R.F.; Etchevers-Barra, J.D.; Gutiérrez-Castorena, M.d.C.; Baez, A. Selection and interpretation of soil quality indicators for forest recovery after clearing of a tropical montane cloud forest in Mexico. *For. Ecol. Manag.* **2012**, *277*, 74–80. [CrossRef]
21. Estrada-Herrera, I.R.; Hidalgo-Moreno, C.; Guzmán-Plazola, R.; Almaraz Suárez, J.J.; Navarro-Garza, H.; Etchevers-Barra, J.D. Soil quality indicators to evaluate soil fertility. *Agrociencia* **2017**, *51*, 813–831.
22. Hansson, K.; Laclau, J.-P.; Saint-André, L.; Mareschal, L.; van der Heijden, G.; Nys, C.; Nicolas, M.; Ranger, J.; Legout, A. Chemical fertility of forest ecosystems. Part 1: Common soil chemical analyses were poor predictors of stand productivity across a wide range of acidic forest soils. *For. Ecol. Manag.* **2020**, *461*, 117843. [CrossRef]
23. Bikindou, F.D.A.; Gomat, H.Y.; Deleporte, P.; Bouillet, J.-P.; Moukini, R.; Mbedi, Y.; Ngouaka, E.; Brunet, D.; Sita, S.; Diazenza, J.-B.; et al. Are NIR spectra useful for predicting site indices in sandy soils under Eucalyptus stands in Republic of Congo? *For. Ecol. Manag.* **2012**, *266*, 126–137. [CrossRef]
24. Burbano Orjuela, H. El carbono orgánico del suelo y su papel frente al cambio climático. *Rev. Cienc. Agrícolas* **2018**, *35*, 82–96. [CrossRef]
25. Antoniadis, V.; Robinson, J.; Alloway, B. Effects of short-term pH fluctuations on cadmium, nickel, lead, and zinc availability to ryegrass in a sewage sludge-amended field. *Chemosphere* **2008**, *71*, 759–764. [CrossRef]
26. Huaraca-Fernandez, J.N.; Pérez-Sosa, L.; Bustinza-Cabala, L.S.; Pampa-Quispe, N.B. Organic amendments in the immobilization of cadmium in contaminated agricultural soils: A review. *Inf. Technol.* **2020**, *31*, 139–152. [CrossRef]
27. Kaninga, B.K.; Chishala, B.H.; Maseka, K.K.; Sakala, G.M.; Lark, M.R.; Tye, A.; Watts, M.J. Review: Mine tailings in an African tropical environment—Mechanisms for the bioavailability of heavy metals in soils. *Environ. Geochem. Health* **2019**, *42*, 1069–1094. [CrossRef]
28. Setälä, H.; Francini, G.; Allen, J.; Jumpponen, A.; Hui, N.; Kotze, D. Urban parks provide ecosystem services by retaining metals and nutrients in soils. *Environ. Pollut.* **2017**, *231*, 451–461. [CrossRef]
29. Fenn, M.E.; Castro-Servín, J.M.; Hernández-Tejeda, T.; Krage, N.; Goodson, C.; Meixner, T. Heavy metals in forest soils, vegetation, and drainage waters in the Basin of Mexico. In *Urban Air Pollution and Forests: Resources at Risk in the Mexico City Air Basin*; Springer: New York, NY, USA, 2002; pp. 194–221.
30. Romero, A.; García, F. Estimación del contenido de carbono en la zona ecológica y cultura Bosque de Tlalpan, Distrito Federal. In *Estado Actual del Conocimiento del Ciclo del Carbono y sus Interacciones en México: Síntesis a 2013*; Paz Pellat, F., Wong González, J., Bazan, M., Saynes, V., Eds.; Programa Mexicano del Carbono Colegio de Postgraduados Universidad Autónoma de Chapingo, Instituto Tecnológico y de Estudios Superiores de Monterrey: Texcoco, Mexico, 2013; pp. 149–154, ISBN 978-607-96490-1-2.
31. Hernández-Guillén, A.; Rojas-García, F.; Benavides-Meza, H. Estimación del contenido y captura de carbono en la segunda sección del Bosque de Chapultepec, Distrito Federal. In *Estado Actual del Conocimiento del Ciclo del Carbono y sus Interacciones en México: Síntesis a 2013*; Pellat, F.P., González, J.W., Bazán, M., Saynes, V., Eds.; Ecosistemas Terrestres; Instituto Tecnológico y de Estudios Superiores de Monterrey: Monterrey, Mexico, 2013.
32. GODF (Gaceta Oficial del Distrito Federal). Acuerdo por el que se Expide el Programa de Manejo del Área Natural Protegida "Bosque de Tlalpan". 2011; pp. 10–111. Available online: http://centro.paot.org.mx/centro/leyes/df/pdf/GODF/GODF_20_06_2011.pdf?b=ce (accessed on 20 October 2023).
33. Díaz-Limón, M.P.; Cano-Santana, Z.; Queijeiro-Bolaños, M.E. Mistletoe infection in an urban forest in Mexico City. *Urban For. Urban Green.* **2016**, *17*, 126–134. [CrossRef]
34. Etchevers, B.; Jorge, D. *Manual Para la Determinación de Carbono en la Parte Aérea y Subterránea de Sistemas de Producción en Laderas*; No. CP FE, FOLLETO 678; Colegio de Postgraduados: Montecillo, Mexico, 2005.
35. Gee, G.W.; Bauder, J.W. *Particle Size Analysis. Methods of Soil Analysis, Part 1*; ASA and SSSA: Madison, WI, USA, 1986; pp. 383–411.
36. TOC Aplication Handbook, Shimadzu. 2001. Available online: https://www.ssi.shimadzu.com/sites/ssi.shimadzu.com/files/pim/pim_document_file/ssi/applications/application_note/16415/TOC%20Application%20-%20TOC%20-%20%20TOC%20Application%20Handbook%20Version%202-PT.pdf (accessed on 20 October 2023).
37. Kerven, G.L.; Menzies, N.W.; Geyer, M.D. Soil carbon determination by high temperature combustion-a comparison with dichromate oxidation procedures and the influence of charcoal and carbonate carbon on the measured value. *Commun. Soil Sci. Plant Anal.* **2000**, *31*, 1935–1939. [CrossRef]
38. Welles, J.; Demetriades-Shah, T.; McDermitt, D. Considerations for measuring ground CO_2 effluxes with chambers. *Chem. Geol.* **2001**, *177*, 3–13. [CrossRef]
39. Davidson, E.; Savage, K.; Verchot, L.; Navarro, R. Minimizing artifacts and biases in chamber-based measurements of soil respiration. *Agric. For. Meteorol.* **2002**, *113*, 21–37. [CrossRef]

40. Kandel, T.P.; Lærke, P.E.; Elsgaard, L. Effect of chamber enclosure time on soil respiration flux: A comparison of linear and non-linear flux calculation methods. *Atmos. Environ.* **2016**, *141*, 245–254. [CrossRef]
41. NOM-147-SEMARNAT/SSA1-2004. (Norma Oficial Mexicana 2004). Que Establece Criterios para Determinar las Concentraciones de Remediación de Suelos Contaminados por Arsénico, Bario, Berilio, Cadmio, Cromo Hexavalente, Mercurio, níquel, Plata, Plomo, Selenio, Talio y/o Vanadio. Available online: https://www.gob.mx/cms/uploads/attachment/file/135331/48.-_NORMA_OFICIAL_MEXICANA_NOM-147-SEMARNAT-SSA1-2004.pdf (accessed on 20 October 2023).
42. U.S. EPA. "IO Compendium Method IO-3.1: Compendium of Methods for the Determination of Inorganic Compounds in Ambient Air: Selection, Preparation and Extraction of Filter Material." EPA/625/R-96/010a. 1999. Available online: https://www.epa.gov/sites/default/files/2015-07/documents/epa-io-3.1.pdf (accessed on 20 October 2023).
43. Bech, J.; Roca, N.; Tume, P.; Ramos-Miras, J.; Gil, C.; Boluda, R. Screening for new accumulator plants in potential hazards elements polluted soil surrounding Peruvian mine tailings. *CATENA* **2016**, *136*, 66–73. [CrossRef]
44. Benipal, G.; Harris, A.; Srirajayatsayai, C.; Tate, A.; Topalidis, V.; Eswani, Z.; Qureshi, M.; Hardaway, C.J.; Galiotos, J.; Douvris, C. Examination of Al, As, Cd, Cr, Cu, Fe, Mg, Mn, Ni, Pb, Sb, Se, V, and Zn in sediments collected around the downtown Houston, Texas area, using inductively coupled plasma-optical emission spectroscopy. *Microchem. J.* **2017**, *130*, 255–262. [CrossRef]
45. Sarstedt, M.; Mooi, E.; Sarstedt, M.; Mooi, E. Regression analysis. In *A Concise Guide to Market Research: The Process, Data, and Methods Using IBM SPSS Statistics*; Springer: Berlin/Heidelberg, Germany, 2019; pp. 209–256.
46. CEPA. *Canadian Soil Quality Guidelines for the Protection of Environmental and Human Health*; National Guidelines and Standards Office: Gatineau, QC, Canada, 2007.
47. Khaledian, Y.; Brevik, E.C.; Pereira, P.; Cerdà, A.; Fattah, M.A.; Tazikeh, H. Modeling soil cation exchange capacity in multiple countries. *CATENA* **2017**, *158*, 194–200. [CrossRef]
48. Mishra, G.; Sulieman, M.M.; Kaya, F.; Francaviglia, R.; Keshavarzi, A.; Bakhshandeh, E.; Loum, M.; Jangir, A.; Ahmed, I.; Elmobarak, A.; et al. Machine learning for cation exchange capacity prediction in different land uses. *CATENA* **2022**, *216*, 106404. [CrossRef]
49. Liao, K.; Xu, S.; Wu, J.; Zhu, Q.; An, L. Using support vector machines to predict cation exchange capacity of different soil horizons in Qingdao City, China. *J. Plant Nutr. Soil Sci.* **2014**, *177*, 775–782. [CrossRef]
50. Liao, K.; Xu, S.; Zhu, Q. Development of ensemble pedotransfer functions for cation exchange capacity of soils of Qingdao in China. *Soil Use Manag.* **2015**, *31*, 483–490. [CrossRef]
51. Ulusoy, Y.; Tekin, Y.; Tümsavaş, Z.; Mouazen, A.M. Prediction of soil cation exchange capacity using visible and near infrared spectroscopy. *Biosyst. Eng.* **2016**, *152*, 79–93. [CrossRef]
52. Sulieman, M.; Saeed, I.; Hassaballa, A.; Rodrigo-Comino, J. Modeling cation exchange capacity in multi geochronological-derived alluvium soils: An approach based on soil depth intervals. *CATENA* **2018**, *167*, 327–339. [CrossRef]
53. NOM-021-RECNAT-2000. Norma Oficial Mexicana que Establece las Especificaciones de Fertilidad, Salinidad y Clasificación de Suelos. Estudios, Muestreo y Análisis. Diario Oficial de la Federación. México, D.F. Available online: https://faolex.fao.org/docs/pdf/mex50674.pdf (accessed on 20 October 2023).
54. Legout, A.; Hansson, K.; Heijden, G.; Laclau, J.P.; Augusto, L.; Ranger, J. Chemical fertility of forest soils: Basic concepts. *Rev. For. Française* **2014**, *66*, 413–424. [CrossRef]
55. Yin, X.; Wang, X.; Wu, H.; Takahashi, H.; Inaba, Y.; Ohnuki, T.; Takeshita, K. Effects of NH_4^+, K^+, Mg^{2+}, and Ca^{2+} on the cesium adsorption/desorption in binding sites of vermiculitized biotite. *Environ. Sci. Technol.* **2017**, *51*, 13886–13894. [CrossRef]
56. Wang, J.; Feng, X.; Anderson, C.W.; Xing, Y.; Shang, L. Remediation of mercury contaminated sites—A review. *J. Hazard. Mater.* **2012**, *221*, 1–18. [CrossRef]
57. He, S.; He, Z.; Yang, X.; Stoffella, P.J.; Baligar, V.C. Soil biogeochemistry, plant physiology, and phytoremediation of cadmium-contaminated soils. *Adv. Agron.* **2015**, *134*, 135–225.
58. Chen, M.; Boyle, E.A.; Switzer, A.D.; Gouramanis, C. A century long sedimentary record of anthropogenic lead (Pb), Pb isotopes and other trace metals in Singapore. *Environ. Pollut.* **2016**, *213*, 446–459. [CrossRef]
59. Athira, M.; Jagadeeswaran, R.; Kumaraperumal, R. Influence of soil organic matter on bulk density in Coimbatore soils. *Int. J. Chem. Stud.* **2019**, *7*, 3520–3523.
60. Saavedra-Romero LD, L.; Alvarado-Rosales, D.; Martínez-Trinidad, T.; Hernández-de la Rosa, P. Propiedades físicas y químicas del suelo urbano del Bosque San Juan de Aragón, Ciudad de México. *Terra Latinoam.* **2020**, *38*, 529–540. [CrossRef]
61. Chávez-Aguilar, G.; Burrola-Aguilar, C.; González-Morales, A.; Pérez-Suárez, M. Almacén de carbono orgánico del suelo y abundancia ectomicorrízica bajo dos especies de coníferas en el Nevado de Toluca, México. *Agro Product.* **2020**, *13*. [CrossRef]
62. Edmondson, J.L.; Davies, Z.G.; McCormack, S.A.; Gaston, K.J.; Leake, J.R. Are soils in urban ecosystems compacted? A citywide analysis. *Biol. Lett.* **2011**, *7*, 771–774. [CrossRef]
63. Scharenbroch, B.C.; Lloyd, J.E.; Johnson-Maynard, J.L. Distinguishing urban soils with physical, chemical, and biological properties. *Pedobiologia* **2005**, *49*, 283–296. [CrossRef]
64. Canedoli, C.; Ferrè, C.; Abu El Khair, D.; Padoa-Schioppa, E.; Comolli, R. Soil organic carbon stock in different urban land uses: High stock evidence in urban parks. *Urban Ecosyst.* **2019**, *23*, 159–171. [CrossRef]
65. Huang, N.; Song, X.-P.; Black, T.A.; Jassal, R.S.; Myneni, R.B.; Wu, C.; Wang, L.; Song, W.; Ji, D.; Yu, S.; et al. Spatial and temporal variations in global soil respiration and their relationships with climate and land cover. *Sci. Adv.* **2020**, *6*, eabb8508. [CrossRef]
66. Fang, C.; Moncrieff, J. The dependence of soil CO_2 efflux on temperature. *Soil Biol. Biochem.* **2001**, *33*, 155–165. [CrossRef]

67. Baldocchi, D.; Tang, J.; Xu, L. How switches and lags in biophysical regulators affect spatial-temporal variation of soil respiration in an oak-grass savanna. *J. Geophys. Res. Biogeosci.* **2006**, *111*. [CrossRef]
68. Vargas, R.; Allen, M.F. Diel patterns of soil respiration in a tropical forest after Hurricane Wilma. *J. Geophys. Res. Biogeosci.* **2008**, *113*. [CrossRef]
69. Cable, J.M.; Ogle, K.; Lucas, R.W.; Huxman, T.E.; Loik, M.E.; Smith, S.D.; Tissue, D.T.; Ewers, B.E.; Pendall, E.; Welker, J.M.; et al. The temperature responses of soil respiration in deserts: A seven desert synthesis. *Biogeochemistry* **2010**, *103*, 71–90. [CrossRef]
70. Leon, E.; Vargas, R.; Bullock, S.; Lopez, E.; Panosso, A.R.; La Scala, N. Hot spots, hot moments, and spatio-temporal controls on soil CO_2 efflux in a water-limited ecosystem. *Soil Biol. Biochem.* **2014**, *77*, 12–21. [CrossRef]
71. Vargas-Terminel, M.L.; Flores-Rentería, D.; Sánchez-Mejía, Z.M.; Rojas-Robles, N.E.; Sandoval-Aguilar, M.; Chávez-Vergara, B.; Robles-Morua, A.; Garatuza-Payan, J.; Yépez, E.A. Soil Respiration is influenced by seasonality, forest succession and contrasting biophysical controls in a Tropical Dry Forest in Northwestern Mexico. *Soil Syst.* **2022**, *6*, 75. [CrossRef]
72. Grand, S.; Rubin, A.; Verrecchia, E.P.; Vittoz, P. Variation in soil respiration across soil and vegetation types in an alpine valley. *PLoS ONE* **2016**, *11*, e0163968. [CrossRef]
73. Weissert, L.; Salmond, J.; Schwendenmann, L. Variability of soil organic carbon stocks and soil CO_2 efflux across urban land use and soil cover types. *Geoderma* **2016**, *271*, 80–90. [CrossRef]
74. Raciti, S.M.; Hutyra, L.R.; Finzi, A.C. Depleted soil carbon and nitrogen pools beneath impervious surfaces. *Environ. Pollut.* **2012**, *164*, 248–251. [CrossRef]
75. Pouyat, R.V.; Yesilonis, I.; Golubiewski, N.E. A comparison of soil organic carbon stocks between residential turf grass and native soil. *Urban Ecosyst.* **2009**, *12*, 45–62. [CrossRef]
76. Bae, J.; Ryu, Y. Land use and land cover changes explain spatial and temporal variations of the soil organic carbon stocks in a constructed urban park. *Landsc. Urban Plan.* **2015**, *136*, 57–67. [CrossRef]
77. Vogel, H.-J.; Bartke, S.; Daedlow, K.; Helming, K.; Kögel-Knabner, I.; Lang, B.; Rabot, E.; Russell, D.; Stößel, B.; Weller, U.; et al. A systemic approach for modeling soil functions. *SOIL* **2018**, *4*, 83–92. [CrossRef]
78. Wiesmeier, M.; Urbanski, L.; Hobley, E.; Lang, B.; von Lützow, M.; Marin-Spiotta, E.; van Wesemael, B.; Rabot, E.; Ließ, M.; Garcia-Franco, N.; et al. Soil organic carbon storage as a key function of soils—A review of drivers and indicators at various scales. *Geoderma* **2018**, *333*, 149–162. [CrossRef]
79. Zacháry, D.; Filep, T.; Jakab, G.; Varga, G.; Ringer, M.; Szalai, Z. Kinetic parameters of soil organic matter decomposition in soils under forest in Hungary. *Geoderma Reg.* **2018**, *14*, e00187. [CrossRef]
80. Juhos, K.; Madarász, B.; Kotroczó, Z.; Béni, Á.; Makádi, M.; Fekete, I. Carbon sequestration of forest soils is reflected by changes in physicochemical soil indicators—A comprehensive discussion of a long-term experiment on a detritus manipulation. *Geoderma* **2021**, *385*, 114918. [CrossRef]
81. Wang, E.H.; Zhao, Y.S.; Chen, X.W.; Zhou, Y.Y.; Chai, Y.F.; Wang, Q.B.; Chen, M. Effect of heavy machinery operation on soil aggregates character in Phaeozem region. *Chin. Chin. J. Soil Sci.* **2009**, *40*, 756–760.
82. Zhang, P.; Cui, Y.; Zhang, Y.; Jia, J.; Wang, X.; Zhang, X. Changes in soil physical and chemical properties following surface mining and reclamation. *Soil Sci. Soc. Am. J.* **2016**, *80*, 1476–1485. [CrossRef]
83. Xu, X.; Wang, C.; Sun, Z.; Hao, Z.; Day, S. How do urban forests with different land use histories influence soil organic carbon? *Urban For. Urban Green.* **2023**, *83*, 127918. [CrossRef]

Disclaimer/Publisher's Note: The statements, opinions and data contained in all publications are solely those of the individual author(s) and contributor(s) and not of MDPI and/or the editor(s). MDPI and/or the editor(s) disclaim responsibility for any injury to people or property resulting from any ideas, methods, instructions or products referred to in the content.

Article

Analysis of Spatial Characteristics Contributing to Urban Cold Air Flow

Hyunsu Kim [1], Kyushik Oh [1,*] and Ilsun Yoo [2]

1. Department of Urban Planning and Engineering, Hanyang University, Seoul 04763, Republic of Korea; ho1ho1ho2@hanyang.ac.kr
2. Cheil Engineering Co., Ltd., Seoul 06779, Republic of Korea; ehrtm110@naver.com
* Correspondence: ksoh@hanyang.ac.kr

Abstract: To mitigate the urban heat island phenomenon at night, cool, fresh air can be introduced into the city to circulate and dissipate the heat absorbed during the day, thereby reducing high urban air temperatures. In other words, cold air flow (CAF) generated by mountainous and green areas should be introduced to as wide an area as possible within the city. To this end, it is necessary to first understand the characteristics of urban spatial factors that impact CAF, and to conduct concrete and quantitative analyses of how these urban spatial characteristics are contributing to air temperature reduction. In this study, the following are conducted: (1) an analysis of the relationship between cold air volume flux (CAVF) and the amount of air temperature reduction; (2) urban spatial categorization; (3) an analysis of the relationship between CAVF and the amount of air temperature reduction by urban spatial type; (4) a regression analysis between the amount of air temperature reduction and urban spatial characteristic factors that affect CAF; and finally, (5) the use of CAF to reduce urban air temperatures in urban planning and a design is proposed. Urban space was categorized into nine types using the results of the tertile analysis of CAVF and urban temperature reduction. It was determined that building height (BH) has a positive (+) influence on all urban spatial types, while building area ratio (BA) has a negative (−) effect. However, in the case of wall area index (WAI), the direction of influence varied depending on the development density; relatively low BA areas should focus on development that increases height to increase WAI, while relatively high BA areas should focus on development that reduces BA to reduce WAI by targeting development types closer to the tower type. And even in areas with similar development density, influence varies depending on the terrain elevation. Moreover, it is necessary to prepare improvement measures to increase the factors with CAF that positively influence air temperature reduction and decrease those with negative influence according to the characteristics of urban spatial types. Such results quantitatively and specifically confirmed the effects of spatial factors that affect CAF by urban spatial type on air temperature reduction. The results of this study can be used as useful information for the efficient use of CAF, a major element of urban ecosystem services.

Keywords: urban spatial characteristics; cold air flow; air temperature reduction; urban heat island; urban planning

Citation: Kim, H.; Oh, K.; Yoo, I. Analysis of Spatial Characteristics Contributing to Urban Cold Air Flow. *Land* **2023**, *12*, 2165. https://doi.org/10.3390/land12122165

Academic Editors: Alessio Russo and Giuseppe T. Cirella

Received: 13 October 2023
Revised: 7 December 2023
Accepted: 11 December 2023
Published: 14 December 2023

Copyright: © 2023 by the authors. Licensee MDPI, Basel, Switzerland. This article is an open access article distributed under the terms and conditions of the Creative Commons Attribution (CC BY) license (https:// creativecommons.org/licenses/by/ 4.0/).

1. Introduction

Urban heat islands are a representative environmental problem caused by rapid urbanization, where the air temperature of areas within a city is approximately 2K–3K higher than that of suburban areas [1]. The rapid urbanization substantially reduces the volume of the natural space of the urban ecosystem services within them, leading to stronger heat island intensities. For example, increased sensible and storage heat due to increased buildings, artificial heat emissions from human activities, and reduced evapotranspiration potential due to decreased vegetation and increased impervious surfaces have been identified as the primary causes of the urban heat island phenomenon [1–5]. In particular, closed urban spaces developed in high-rise and high-density areas block the circulation and inflow of air in the city, which worsens natural ventilation [6,7]. Furthermore, the heat stored in the urban fabric inside the city during the day cannot escape, causing the urban heat island phenomenon to continue even after sunset.

To mitigate the aforementioned urban heat island phenomenon at night, cool and fresh air can be transported into the city to circulate and dissipate the heat absorbed during the day, thereby reducing high urban air temperatures [8–14]. Cold air is generated by radiative cooling of surfaces, which initiates the transport of sensible heat from the air to the surface. Cold air flows (CAF) are driven by differences in air temperature and pressure [15–18]. This flow of cool, fresh air is defined as CAF [19], and it can be used to transport cool, fresh air generated in mountainous and green areas into the city while dissipating heat within.

CAF is generated between sunset and sunrise when the atmospheric temperature decreases due to decreased solar radiant energy [20], mainly in mountainous and green areas, and flows into the city with atmospheric currents at night. Several studies have been conducted to materialize the concept of CAF, focusing on the path and cold air volume flux (CAVF) generated by mountainous and green areas around cities and using them in urban planning and design [21–26]. Such studies emphasized that CAF can be effectively used to improve the thermal environment of urban spaces. In addition, there have been many studies that have simulated the impact of CAF on mitigating the urban thermal environment. In other words, it is necessary to devise specific measures to enable CAF generated by mountainous and green areas to reach the widest areas possible within the city [17,27,28].

However, from an urban planning and design perspective, there is still a lack of research on how the physical characteristics of urban spaces, such as the path of CAF from its entry into the city to its dissipation and the amount of CAF, can be used to reduce urban air temperatures. Additionally, efforts are needed to apply the flowing principles of CAF to urban planning and design based on scientific and quantitative analysis. To this end, it is necessary to first understand the characteristics of urban spatial factors that impact CAF, and to conduct concrete and quantitative analyses of how these urban spatial factors are contributing to air temperature reduction by urban spatial type. It is possible to establish a space-specific customized cold air supply plan based on characteristic analysis and analysis of cold air hindrances for each urban space. Ultimately, urban planning and design for urban air temperature reduction using cold air will be possible [25,27,29]. Thus, in this study, the relevance of CAVF that can be advected into urban spaces to the amount of air temperature reduction is analyzed. Second, the urban space is categorized according to a combination of CAVF and the amount of urban temperature reduction. Third, the relationship between the CAVF and the amount of temperature reduction for each urban spatial type is analyzed, along with the contribution of urban spatial characteristics that help CAF lead to urban air temperature reduction. Finally, based on the analysis results, a plan to use CAF to reduce urban air temperature in urban planning and a design are proposed.

2. Materials and Methods

The study has been conducted in the following steps (Figure 1): (1-1) Analysis of CAVF; (1-2) Identification of urban air temperature and amount of air temperature reduction; (2) Analysis of the relationship between CAVF and amount of air temperature reduction; (3) Categorization of urban spatial type considering CAVF and amount of air temperature reduction; (4) Analysis of the relationship between CAVF and the amount of air temperature reduction by urban spatial type; (5) Regression analysis between the amount of air temperature reduction and urban spatial characteristic factors that affect CAF by urban spatial type; and (6) Derivation of urban spatial characteristic factors as well as urban planning and design implications by urban spatial type.

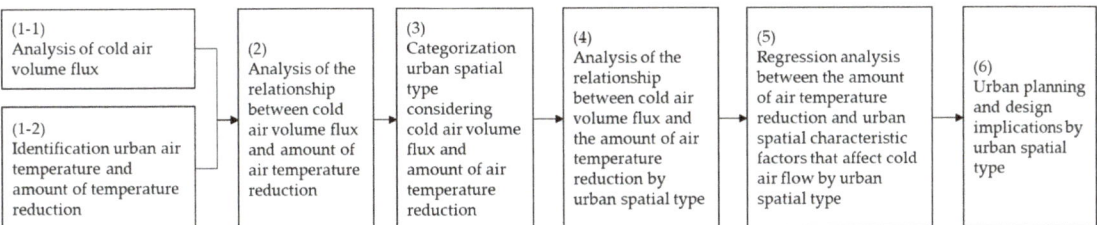

Figure 1. Study workflow.

A case study was carried out for Seoul, the capital and largest city in South Korea (Figure 2). Seoul is located in a temperate climate, is one of the most densely populated cities in the world, and is home to about 18.5% of the total population (approximately 9.5 million people). Seoul is a heat island city characterized by very high-density development.

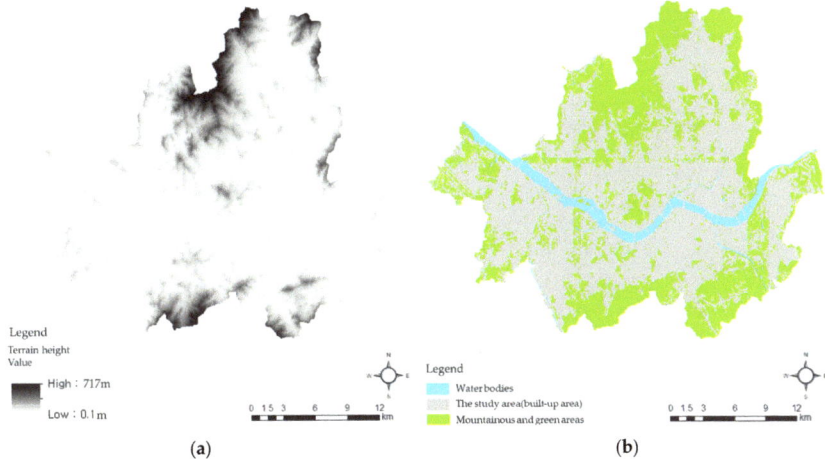

Figure 2. Terrain height (**a**) and land cover (**b**) in the study area (Seoul, Republic of Korea).

In this study, an analysis on the CAVF was performed with a 50 m × 50 m resolution, considering KLAM_21 modeling resolution. As the purpose of this study is to analyze the relationship between the CAVF and air temperature reduction in urban spaces and to identify CAF-influenced urban spatial characteristics that can reduce urban air temperature, a total of 50,950 grids (127.4 km^2) were used for analysis after excluding spaces that generate CAF such as water bodies (e.g., rivers and streams), mountainous areas, and green areas.

2.1. CAVF and the Amount of Urban Air Temperature Reduction

2.1.1. CAVF

There are four main methods for analyzing the CAF, namely, physical measurements, numerical modeling analysis, wind tunnel testing, and using theoretical analysis (Table 1). When using theoretical analysis for CAF analysis, the empirical constants required for the analysis may be inappropriate for the circumstances of the case study area [19]. There are several limiting factors while analyzing the CAF through physical measurements, such as limited observation points (point measurements), time, equipment, cost, and unpredictability. Furthermore, for the wind tunnel tests, scale models of real urban terrain were built, and artificial winds were applied to observe wind flow over terrain and urban structures. However, wind tunnel tests are not suitable for large cities such as Seoul. Alternatively, the analysis of the CAF using numerical models can reflect the urban spatial characteristics of the case study area.

Table 1. Analysis method of CAVF.

		Advantages	Disadvantages
Physical measurements		• Actual measured values • Complex terrain analysis possible	• Incurs installation and maintenance costs • Limited analysis scope • Unable to identify cold air source points
Wind tunnel testing		• Creating a scaled-down model of the actual terrain for wind environment analysis	• Unable to determine the presence of cold air currents • Mainly analyzes changes in wind environment such as building wind due to changes in air flow characteristics, and macro-scale analysis is not possible
Theoretical analysis		• Very straightforward analysis • It is possible to estimate the amount of cold air generation	• Unable to consider actual regional conditions • Macro-scale analysis not feasible
Numerical modeling analysis	KLAM_21	• Specialized simulation analysis for cold air flow, generation, and prediction • Analyzes large-scale areas up to 37.5×37.5 km^2 • Reflects actual urban information based on land use	• Unable to analyze temperature
	Envi-met	• Considers surface, vegetation, atmospheric environment, etc., comprehensively • Micro-scale analysis possible • Considers various environmental factors such as temperature, humidity, airflow, plant heat emission, and reflection	• Macro-scale analysis not feasible • Cold air flow analysis is not possible (focused on wind environment changes)
	MUKLIMO_3	• Analyzes wind flow considering terrain, buildings, vegetation, etc. • Micro-scale wind flow simulation possible	• Macro-scale analysis not feasible • Cold air flow analysis is not possible (focused on wind environment changes)

Representative numerical modeling programs to analyze CAF include KLAM_21, Envi-met, MUKLIOMO_3, REWIMET, and FITNAH. Among various numerical modeling programs, for this study, KLAM_21 Version 2.012 was applied to analyze CAF by considering various conditions such as the possibility of analyzing at the scale of a large city like Seoul, the case study area, the possibility of considering various land uses, and the possibility of high-resolution analysis. KLAM_21 is a modeling analysis tool for the CAVF developed by the German Meteorological Service (DWD, Deutscher Wetterdienst). KLAM_21 can be operated with up to 3000 × 3000 grid cells, and each grid can be analyzed with a resolution of 10–50 m [23].

For KLAM_21 modeling, elevation and land use-specific physical parameters are required. Physical parameters include roughness, building coverage area ratio (BA), building height (BH), wall area index (WAI), tree cover fraction, tree height, leaf area index (LAI), and local heat loss rate. It is important to accurately determine the parameter values depending on the properties of the case study area. However, because the appropriate parameters have not yet been established for Seoul, it was necessary to construct physical parameters that fit the characteristics of the case study area. For the modeling analysis, a spatial scope of analysis with a 50 × 50 m grid resolution was considered and physical parameters for the analysis were prepared using GIS spatial analysis (Table 2 and Figure A2). The KLAM_21 modeling run time was set from 9 p.m. (sunset) to 6 a.m. (sunrise), which is the time when cold air is generated in mountainous and green areas after sunset. A map of the CAVF from 9 p.m. to 6 a.m. was prepared.

Table 2. KLAM_21 modeling Physical Parameters.

Class	Z0	BA	BH	WAI	TA	TH	LAI	α
Forest	0.4	-	-	-	0.4	13.8	3.5	0.56
Semi-sealed	0.02	-	-	-	-	-	-	0.64
Industrial	0.08	0.3	6.3	1.34	-	-	-	0
Park	0.1	-	-	-	0.3	11	3.0	1.0
Open Space	0.05	-	-	-	-	-	-	1.0
Sealed	0.01	-	-	-	-	-	-	0.28
Water	0.001	-	-	-	-	-	-	0
Low rise (1–3 floors) 3–9 m	0.1	0.21	5.71	1.98	0.0	0.0	0.0	0.28
Low–mid rise (4–8 floors) 10–26 m	0.1	0.34	10.42	3.36	0.0	0.0	0.0	0.28
Mid rise (9–16 floors) 27–48 m	0.1	0.2	39.54	8.16	0.0	0.0	0.0	0.0
Mid–high rise (17–26 floors) 49–78 m	0.3	0.2	62.91	12.88	0.0	0.0	0.0	0.0
High rise (27–123 floors) 79–555 m	0.3	0.23	103.2	15.6	0.0	0.0	0.0	0.0

Z0: roughness length (m), BA: building coverage area ratio, BH: building height (m), WAI: wall area index, TA: tree cover fraction, TH: mean tree height (m), LAI: leaf area index, and α: relative local heat loss.

2.1.2. Urban air Temperature and Amount of Air Temperature Reduction

The urban heat island phenomenon is typically most pronounced two to three hours after sunset on cloudless and calm days [1]. The analysis time in this study considered these characteristics of urban heat islands. Eight days in the summer of 2021 were selected considering the weather conditions (cloud cover of at most 5/8, low wind speed ($\leq$2 m/s)) and KLAM_21 modeling analysis time (9 p.m. to 6 a.m.) (Table 3).

An air temperature map was prepared by using weather data collected from 26 Automatic Weather Stations (AWS) operated by the Korea Meteorological Administration (KMA) and about 1100 Smart City Data Sensors (S-dot) operated by the Seoul Metropolitan Government (Figure 3). In this study, the universal kriging interpolation method was applied based on the Gaussian process regression model [30] to consider the variables of distance between measurement points, altitude, and distance to the river. Air temperature maps at 9 p.m. and 6 a.m. for eight days in the summer of 2021 on a 50 m × 50 m resolution grid were produced. Finally, the difference between the 9 p.m. and 6 a.m. air temperatures was calculated to create a mean urban air temperature reduction map.

Table 3. Time of analyzing urban air temperatures.

	Minimum Air Temperature (°C)	Maximum Air Temperature (°C)	Mean Air Temperature (°C)	Mean Wind Speed (m/s)	Mean Cloud Cover
9 June 2021	19.5	31.6	25.8	1.6	3.4
13 June 2021	20.9	29.7	24.8	2.0	3.5
1 July 2021	21.4	31.0	26.3	1.8	2.3
21 July 2021	25.3	35.3	30.5	1.7	3.5
23 July 2021	27.2	35.8	31.2	1.8	2.3
24 July 2021	26.9	36.5	31.7	1.7	3.0
28 July 2021	27.1	34.7	30.4	1.8	3.6
7 August 2021	23.4	32.3	28.0	2.0	3.3

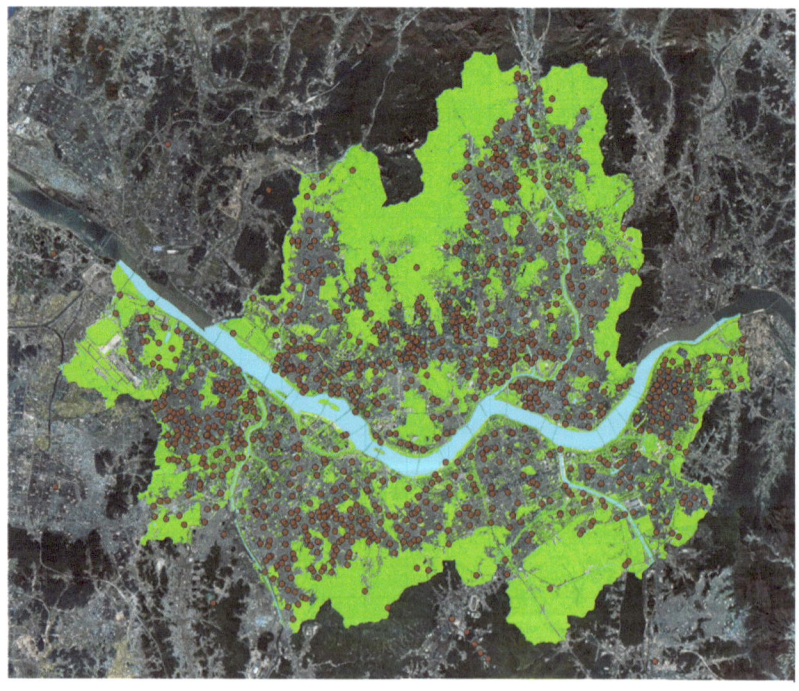

Figure 3. AWSs and S-dot sensors in the study area.

2.2. Analysis of the Relationship between the CAVF and the Amount of Urban Air Temperature Reduction

To examine the relationship between CAVF and the amount of air temperature reduction in the case study area, correlation analysis was conducted between CAVF and the amount of air temperature reduction. This is to confirm the contribution of CAF generated from mountainous and green areas in the study area to air temperature reduction.

2.3. Urban Spatial Categorization

A city made up of many different spatial factors will have different amounts of CAF and air temperature reduction depending on the characteristics of the spaces. Therefore, in order to categorize urban spaces, it is necessary to classify them according to the CAVF and amount of air temperature reduction. The results of the CAF modeling analysis and the urban air temperature reduction amount calculations were classified into tertile (high, medium, and low) using the Natural Breaks Jenk function (Figure 4). Consequently, the urban spaces were categorized into nine areas as follows: areas with the highest, medium, and lowest CAVF, and areas with the highest, medium, and lowest amount of urban air temperature reduction. Next, the characteristics of each space were identified based on the classified urban spatial categories.

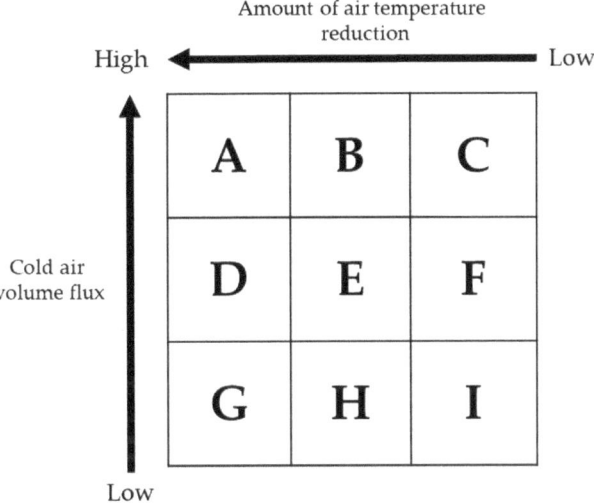

Figure 4. Urban spatial categorization method.

2.4. Identification of Urban Air Temperature Reduction Factors by Urban Spatial Type

Correlation analysis was performed to analyze the relationship between the CAVF and the amount of urban air temperature reduction by urban spatial type. Regression analysis was also performed to find the factors influencing CAF that affect air temperature reduction. The independent variables for the regression analysis are elevation, BH, WAI, and BA among the input variables of KLAM_21. The peripheral borders of the mountainous and green areas have cooler air temperatures than built-up areas, so CAF has a less significant effect on air temperature drops. Therefore, distance to green areas and the normalized difference vegetation index (NDVI) were added as independent variables to account for the effect of mountainous areas and green areas. Regression analysis was performed using the standardized values of the variables, and the amount of air temperature reduction was selected as the dependent variable to find the influential variables that affect air temperature reduction. Moreover, the amount of urban air temperature reduction utilizes air temperatures measured by the AWSs and s-dot; as spatial autocorrelation is generally known to exist, this can be problematic for spatial autocorrelation due to the first law of geography [31]. Therefore, spatial autoregressive models (spatial lag model (SLM) and spatial error model (SEM)), which are regression models that control for spatial autocorrelation, were applied.

3. Results

3.1. Analysis Results of the CAVF and the Amount of Urban Air Temperature Reduction

3.1.1. Analysis Results of the CAVF

KLAM_21 was utilized to analyze CAF for a 9 h period from 9 p.m. to 6 a.m., when cold air is generated after sunset (Figure A1). The unit of the CAVF is m^3/ms, which indicates the volume of cold air traveling a unit distance per unit area. As the time passes from after sunset to before sunrise, the CAF generated by the mountainous and green areas entering the city can be detected. Figure 5 is a CAF map that adds all the analysis results for 9 h, excluding water bodies (e.g., rivers and streams), mountains, and green areas of the study area. Many areas are experiencing an inflow of CAF, with a few exceptions. In the north and southeast, where large mountainous regions are located, a relatively smooth CAF can be determined. However, densely developed urban areas, including the southwestern part of the study area, were receiving relatively little CAF. This suggests that some spaces within the study area do not have a smooth CAF within the urban space.

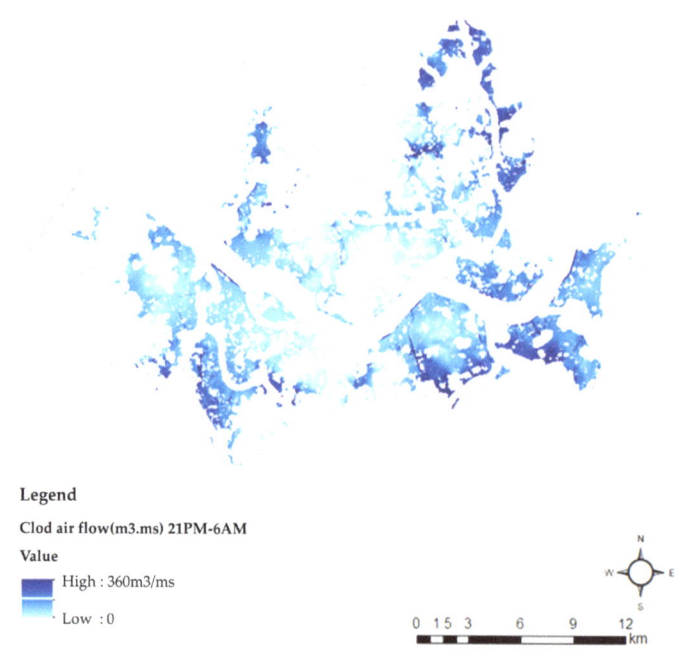

Figure 5. The total CAVF from 9 p.m. to 6 a.m. in the study area.

3.1.2. Analysis Results of Urban Air Temperature and Amount of Air Temperature Reduction

Figure 6 is a map of the air temperature at 9 p.m. and 6 a.m. in the study area. The air temperature at 9 p.m. ranged from 28.6 °C to 34.4 °C with a mean value of 32.8 °C, and the difference between the maximum and minimum temperatures was 5.8 °C. This confirms that the urban heat island phenomenon is severe in the study area. This phenomenon can also be seen in the air temperature distribution map. Higher air temperatures were found in urban areas with relatively high development densities and lower air temperatures were found in the north, east, and south, where large urban forests are predominantly located. At 6 a.m., the air temperature ranged from 25.4 °C to 29.9 °C with a mean value of 28.7 °C. The difference between the maximum and minimum air temperatures was 4.5 °C. The analysis of air temperature reduction between 9 p.m. and 6 a.m. showed that the north and east areas closest to the urban forests experienced a maximum nighttime air temperature reduction of 5.9 °C, while the lowest air temperature difference of 2.3 °C was found in the west area, which has a relatively high development density and no urban forests nearby (Figure 7).

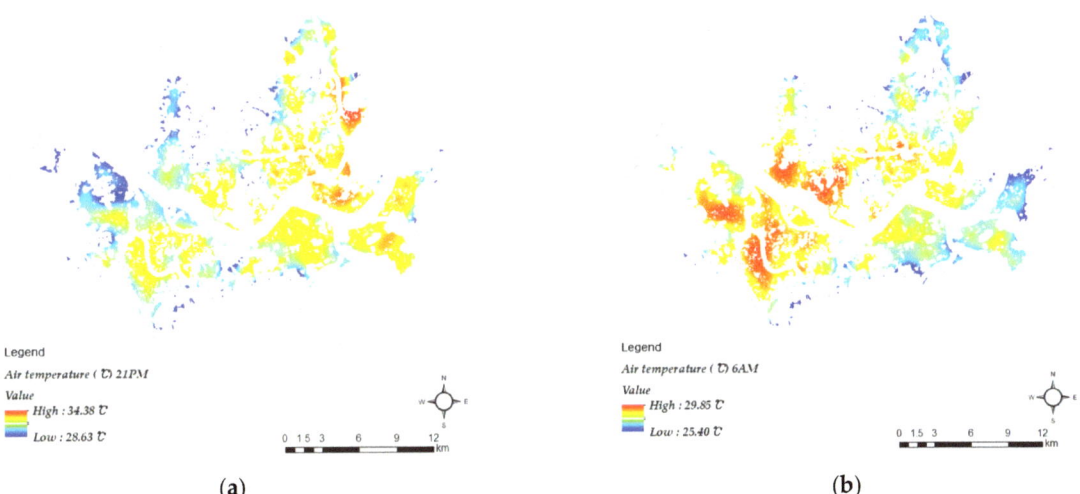

Figure 6. Air temperature analysis results: (**a**) left: 9 p.m.; (**b**) right: 6 a.m.

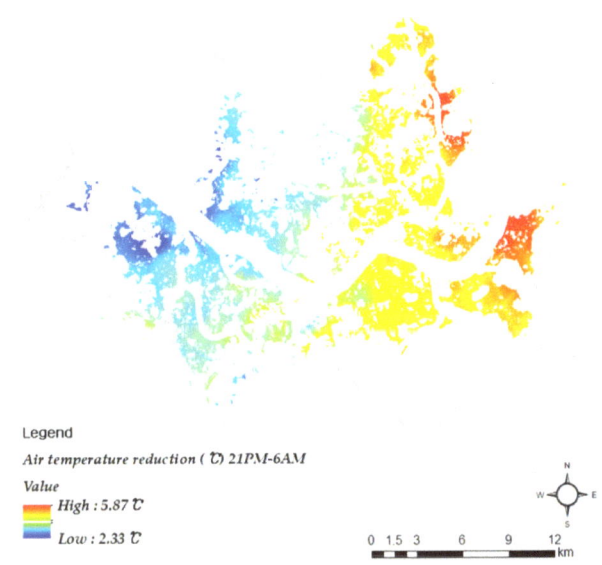

Figure 7. Air temperature reduction analysis (9 p.m.–6 a.m.).

3.2. Analysis Results of the Relationship between the CAVF and The Amount of Urban Air Temperature Reduction

Table 4 shows the results of the correlation analysis between the total CAVF and the amount of air temperature reduction in the study area: the higher the positive correlation coefficient, the greater the air temperature reduction effect of the CAVF. A positive correlation was shown across the study areas, and thus confirms that CAVF contributes to air temperature reduction. This is the result of the analysis of the entire study area. Additional analysis is needed by urban spatial type because the degree to which CAVF contributes to urban temperature reduction will vary depending on the characteristics of urban space.

Table 4. Analysis results of the correlation between the amount of overall CAF and the amount of air temperature reduction across study areas.

		CAVF (Seoul)
Amount of Air temperature Reduction	Pearson Correlation	0.394 **
	N	50,950

** Correlation is significant at the 0.01 level (two-sided).

3.3. Urban Spatial Categorization Results

Using the tertile analysis results of the CAVF and the amount of urban air temperature reduction, the urban spaces of the study areas were categorized into the following nine types: areas with high CAVF and high air temperature reduction (Type A), areas with high CAVF and medium air temperature reduction (Type B), areas with high CAVF and low air temperature reduction (Type C), areas with medium CAVF and high air temperature reduction (Type D), areas with medium CAVF and medium air temperature reduction (Type E), areas with medium CAVF and low air temperature reduction (Type F), areas with low CAVF and high air temperature reduction (Type G), areas with low CAVF and medium air temperature reduction (Type H), and areas with low CAVF and low air temperature reduction (Type I) (Figure 8 and Table 5). The spatial characteristics of each urban spatial type are shown in Table 6. The BH, WAI, and BA were all relatively low for Types A, B, and C, which have more CAF, while Types G, H, and I, which have less CAF, were classified as areas with relatively high BH, WAI, and BA. Except for Types H and G, which are located at high altitudes, the more the CAVF and the amount of air temperature reduction, the greater the distance from green areas, and the higher the BH and WAI. In the case of NDVI, it was similar for all urban spatial types because mountainous and green areas were excluded, which are the source spaces of generating CAF and classified urban spaces. In terms of land use, the lower the CAVF and the amount of air temperature reduction in relatively flat types (except for Types G and H, which have higher elevations and a higher proportion of high-rise residences), the lower the proportion of high-rise residences and commercial/office buildings with higher BH than low-rise residences. The high CAF types (Types A, B, and C) make up only ~12% of the region's total study area. However, Types G, H, and I, which make up about 48% of the study area, showed less CAF and therefore have more potential to utilize CAF to reduce air temperatures than other types. These types of urban spaces are relatively high-rise and densely developed areas. In sum, depending on the elevation and development density in the study area, it is possible to improve urban heat islands by actively considering factors influencing cold air in urban planning and design for a smooth CAF.

Table 5. Urban spatial categorization results.

	Type A (n = 4888)	Type B (n = 1167)	Type C (n = 256)	Type D (n = 11,475)	Type E (n = 4798)	Type F (n = 3769)	Type G (n = 5516)	Type H (n = 10,730)	Type I (n = 8351)
Elevation	−0.3298	−0.2955	−0.3453	−0.2449	−0.3003	−0.5688	0.4402	0.6715	−0.1429
NDVI	−0.0620	−0.1033	0.5304	−0.0561	−0.1,100	−0.0931	−0.1123	0.1787	0.0613
BH	−0.1288	−0.2675	−0.6371	0.0541	−0.0596	−0.1949	0.0437	0.1522	−0.0443
WAI	−0.1242	−0.3562	−0.8158	0.0788	−0.0863	−0.1624	0.0316	0.1409	−0.0398
BA	−0.2603	−0.2610	−0.5723	−0.1318	0.1296	0.0284	0.0522	0.0105	0.0665
Distance to green areas	0.0358	−0.0065	−0.4046	0.1074	0.0635	0.0450	0.2597	−0.3434	0.1151

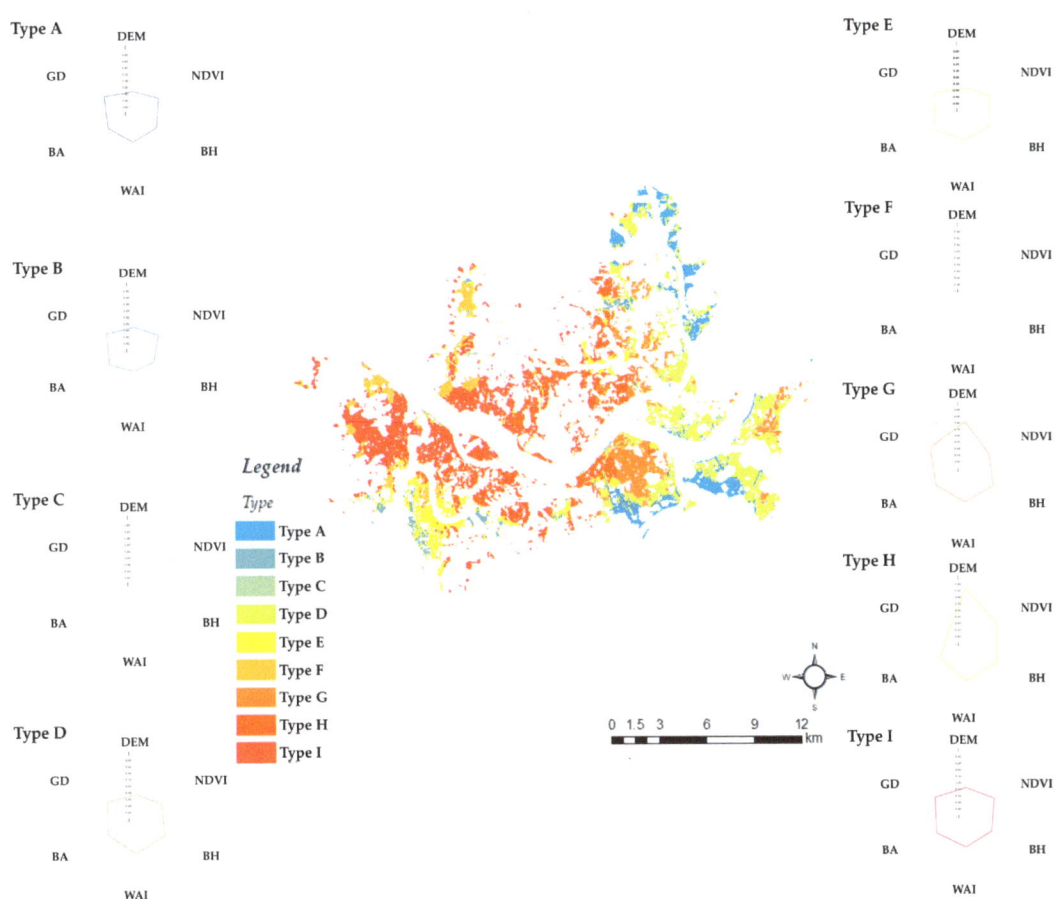

Figure 8. Urban spatial categorization results.

Table 6. Spatial characteristics by urban spatial type.

		CAVF	Amount of Air Temperature Reduction	Elevation	NDVI	BH	WAI	BA	Distance to Green Areas	Land Use (%)
Type A (n = 4888)	Max	359.99	5.85	81	0.42	165	24.95	1	1104.54	Low-rise residential: 34.39
	Min	63.82	4.43	10	0.01	0	0	0	0	Road: 23.85
	Mean	84.11	4.95	19.97	0.14	18.69	4.64	0.25	360.88	High-rise residential: 20.50
	S.D.	24.50	0.30	6.80	0.07	17.86	3.24	0.18	206.08	Commercial: 14.92
Type B (n = 1167)	Max	256.62	4.43	71	0.42	87	22.02	0.96	806.22	Low-rise residential: 28.61
	Min	63.82	3.59	0	0.01	0	0	0	0	Road: 24.65
	Mean	84.02	3.93	20.66	0.13	15.93	3.80	0.25	352.29	High-rise residential: 13.90
	S.D.	30.04	0.23	14.10	0.07	17.26	3.61	0.22	198.36	Commercial: 10.31
Type C (n = 256)	Max	289.02	3.59	81	0.43	48	10.97	0.92	860.23	Low-rise residential: 30.88
	Min	63.82	2.76	7	0.03	0	0	0	0	Road: 24.57
	Mean	82.39	3.24	19.66	0.16	8.58	2.13	0.20	271.37	Open space: 17.28
	S.D.	26.80	0.23	14.47	0.09	10.68	2.18	0.22	220.24	Commercial: 6.46

Table 6. Cont.

		CAVF	Amount of Air Temperature Reduction	Elevation	NDVI	BH	WAI	BA	Distance to Green Areas	Land Use (%)
Type D (n = 11,475)	Max	63.81	5.87	60	0.47	174	27.37	1	1204.16	Low-rise residential: 40.37
	Min	28.39	4.43	10	−0.01	0	0	0	0	High-rise residential: 21.82
	Mean	44.34	4.94	21.67	0.14	22.33	5.37	0.27	375.42	Road: 16.43
	S.D.	9.90	0.31	8.61	0.06	21.26	3.89	0.16	203.42	Open space: 12.25
Type E (n = 4798)	Max	63.81	4.43	82	0.47	123	21.54	1	860.23	Low-rise residential: 40.58
	Min	28.39	3.59	5	−0.02	0	0	0	0	Road: 17.73
	Mean	42.11	3.91	20.56	0.13	20.07	4.77	0.32	366.50	High-rise residential: 17.40
	S.D.	9.69	0.23	11.88	0.06	18.91	3.46	0.18	187.12	Commercial: 12.69
Type F (n = 3769)	Max	66.75	3.59	83	0.52	108	20.22	0.99	948.68	Low-rise residential: 41.76
	Min	28.39	2.53	7	−0.01	0	0	0	0	Road: 18.39
	Mean	40.13	3.21	15.17	0.13	17.38	4.50	0.31	362.75	High-rise residential: 15.44
	S.D.	8.96	0.21	8.16	0.06	14.99	2.88	0.17	197.90	Commercial: 9.82
Type G (n = 5516)	Max	28.39	5.68	129	0.49	162	23.87	1	1204.16	Low-rise residential: 40.57
	Min	0	4.43	11	−0.03	0	0	0	0	High-rise residential: 16.83
	Mean	17.26	4.76	35.42	0.13	22.12	5.20	0.31	406.38	Commercial: 16.77
	S.D.	8.54	0.26	16.68	0.07	19.28	3.61	0.16	246.66	Road: 13.36
Type H (n = 10,730)	Max	28.37	4.43	233	0.48	207	26.46	1	860.23	Low-rise residential: 40.88
	Min	0	3.59	6	−0.02	0	0	0	0	High-rise residential: 24.58
	Mean	17.08	3.99	40.06	0.15	24.28	5.59	0.30	283.81	Commercial: 12.24
	S.D.	9.18	0.25	27.45	0.07	22.28	3.94	0.17	160.26	Road: 13.64
Type I (n = 8351)	Max	28.38	3.59	293	0.49	207	26.15	1	1019.80	Low-rise residential: 44.38
	Min	0	2.33	6	−0.02	0	0	0	0	High-rise residential: 17.27
	Mean	14.90	3.22	23.72	0.14	20.37	4.94	0.31	377.00	Commercial: 13.09
	S.D.	8.08	0.25	24.34	0.07	18.28	3.23	0.17	207.46	Road: 15.94

3.4. Results of Identifying Urban Air Temperature Reduction Factors by Type of Urban Space

Table 7 shows the results of the correlation analysis between the CAVF and the amount of air temperature reduction by urban spatial type. The higher the positive correlation coefficient, the more the CAVF contributes to the air temperature reduction. A positive correlation was found across all urban spatial types. The higher the air temperature reduction (Type A and D), the stronger the positive correlation, while the lower the air temperature reduction or the lower the CAVF, the lower the positive correlation. Areas with a higher positive correlation tend to be located on flat land, far from green areas, and show a development form with relatively high BH and WAI.

Table 7. Correlation of the CAVF and the amount of air temperature reduction by urban spatial type.

		Type A	Type B	Type C	Type D	CAVF Type E	Type F	Type G	Type H	Type I
Amount of air temperature reduction	Pearson Correlation	0.481 *	0.407 *	0.326 *	0.511 *	0.491 *	0.366 *	0.333 *	0.344 *	0.350 *
	N	4888	1167	256	11,475	4798	3769	5516	10,730	8351

* Correlation is significant at the 0.01 level (two-sided).

Table A1 shows the regression analysis results of the amount of air temperature reduction and factors influencing CAVF by urban spatial type. Among the three regression models, the SEM model was found to have the highest log likelihood value, which indicates high suitability. In the SEM, all six independent variables were significant at the 1% level. Analysis results of SEM revealed the six variables can be categorized into natural and artificial factors based on whether they can be manipulated in urban planning and design.

The natural factors are elevation, NDVI, and distance to green areas, and the artificial factors are BH, WAI, and BA. The absolute values of the coefficients of the independent variables in all urban spatial types were higher in the order of elevation, distance to green areas, and NDVI for natural factors, and higher in the order of BH, WAI, and BA for artificial factors. However, the signs of the coefficients of the independent variables were analyzed differently for each type of urban space. Elevation, NDVI, and distance to green areas showed positive signs for all urban spatial types. BH showed a positive sign for all urban spatial types, and BA showed a negative sign for all urban spatial types. However, the effects of WAI vary depending on elevation and development density. Positive signs were found in areas with relatively low BA, such as Types A, B, C, and D, and in Types G and H with high elevation and BH, while negative signs were found in Types E, F, and I with relatively high BA (Tables 8 and 9).

Table 8. Spatial regression analysis results by urban spatial type.

	Type A	Type B	Type C	Type D	Type E	Type F	Type G	Type H	Type I
Elevation	0.015715	0.131069	0.050024	0.02743	0.040678	0.052277	0.04338	0.002747	0.00667
NDVI	0.00097	0.00289	0.005849	0.000958	0.00255	0.001553	0.001692	0.00017	0.00132
BH	0.00361	0.01382	0.092827	0.00785	0.00067	0.013152	0.008919	0.00045	0.002249
WAI	0.005989	0.009598	0.05023	0.005491	−0.00054	−0.01139	0.003234	0.00032	−0.00253
BA	−0.001165	−0.00317	−0.01934	−0.000825	−0.00028	−0.0012	−0.00105	−0.00032	−0.0008
Distance to green areas	0.01227	0.011781	0.005794	0.007555	0.007517	0.02466	0.034309	0.00559	0.012248
Constant term	1.05584	−0.33896	−1.18172	0.949379	−0.29672	−1.23183	0.885141	−0.30435	−1.2174
Log likelihood	5733.09	1211.93	168.69	18154.13	6677.75	5592.04	7799.59	17207.67	13935.50
R2	0.80	0.75	0.74	0.79	0.78	0.75	0.81	0.78	0.79

Table 9. Relationship between the factors influencing the amount of air temperature reduction and CAF.

	Type A	Type B	Type C	Type D	Type E	Type F	Type G	Type H	Type I
Elevation	+	+	+	+	+	+	+	+	+
NDVI	+	+	+	+	+	+	+	+	+
BH	+	+	+	+	+	+	+	+	+
WAI	+	+	+	+	−	−	+	+	−
BA	−	−	−	−	−	−	−	−	−
Distance to green areas	+	+	+	+	+	+	+	+	+

4. Discussion and Conclusions

The key findings of this study are the following. The analysis of the relationship between the CAVF and the amount of air temperature reduction by urban spatial type showed that the higher the amount of air temperature reduction, the more relevant it is to CAVF (Types A and D). Furthermore, it was found that Type E, which is located in a flat area and has a relatively high BH and WAI, is highly relevant. On the contrary, types with higher elevation (Types G and H) and types with lower proportions of high-rise residences in flat areas (Types F and I) were less relevant. Furthermore, in areas with low BA (Types A, B, C, and D), the relevance of the CAVF and the amount of air temperature reduction varied depending on BH and WAI. The results suggest that elevation, BH, WAI, and BA are the factors in urban planning and design that can reduce urban air temperatures by smoothly enhancing CAF.

Next, the mechanisms through which these factors affect CAF and affect air temperature reduction by urban spatial type were identified. As a result, in urban planning and design, it is necessary to prepare improvement measures for increasing the factors with CAF that positively (+) influence air temperature reduction and decreasing those with negative (−) influence according to the characteristics of urban spatial types. BH had a positive (+) influence on all urban spatial types, and the absolute influence of WAI and BA was a larger factor. BA had a negative (−) influence on all urban spatial types. The higher the height of the building and the lower the BA, the more unimpeded the CAF, maximizing the air temperature reduction effect. However, in the case of the WAI, it needs to be applied differently depending on the development density. Relatively low BA areas (Types A, B, C, and D) should focus on development that increases height to increase WAI, while relatively high BA areas (Types E and F) should focus on development that reduces BA to reduce WAI by targeting development types closer to the tower type. Even in neighborhoods with similar development densities, the impact of development density varies by elevation. Thus, it is possible to ensure smooth CAF as well as the continuity of cold air by providing the direction for improvement based on the characteristics of urban spatial types.

This study has quantitatively analyzed the relationship between the CAVF and the amount of air temperature reduction by urban spatial type, and proposed measures to improve the urban heat island through smooth CAF. The analysis showed that a combination of factors, such as elevation, BH, WAI, and BA, should be considered. The preceding research has mainly focused on the formation and movement of CAF due to natural factors [24,32–34]. However, this study has a significant advantage in focusing on the movement of CAF and temperature reduction caused by physical factors in urban spaces. Additionally, employing a statistical approach enables the confirmation of the influence of physical factors in urban spaces on temperature reduction in a more scientific and quantitative manner, providing practical information for urban planning and design. In particular, when making plans to mitigate the urban heat island phenomenon in large cities such as Seoul, where various urban spaces exist in a complex manner, it is possible to enhance the efficiency of such plans by identifying the factors that hinder CAF in detail through a spatially customized approach. If the smooth CAF can be maintained within the city while simultaneously increasing the inflow of CAF through the strengthening of mountainous and green areas around the city, which serve as resources for urban ecosystem services, a great synergy will be achieved in the circulating and cooling of heat in the city. These implications will provide useful information for the efficient use of CAF, a major element of urban ecosystem services.

This study has the following limitations. Among the physical characteristics of various urban spaces, the analysis focused on factors influencing CAF in KLAM_21. A more comprehensive consideration and analysis of other urban spatial physical characterization variables that were not considered in this study, such as sky view factor (SVF), height to road width (H/W) ratio, and porosity, is needed to determine the exact relationship between factors influencing the amount of urban air temperature reduction and CAF. Furthermore, since there is the limitation of not comprehensively considering both natural and physical factors in urban spaces that contribute to the formation and movement of CAF, future research should involve a more comprehensive and holistic investigation that takes into account both of these factors.

Author Contributions: This article is the result of joint work by all the authors. K.O. supervised and coordinated work on the paper. Conceptualization, K.O. and H.K.; methodology, K.O. and H.K.; validation, K.O. and H.K.; formal analysis, H.K.; data curation, H.K. and I.Y.; writing—original draft preparation, H.K.; writing—review and editing, K.O. and H.K.; visualization, H.K.; supervision, K.O.; project administration, K.O.; funding acquisition, K.O. All authors have read and agreed to the published version of the manuscript.

Funding: This research was funded by the Korea Ministry of Environment (MOE) grant number 2022003570004.

Data Availability Statement: The data presented in this study are available on request from the corresponding author. The data are not publicly available due to an ongoing study.

Acknowledgments: This work was supported by Korea Environment Industry & Technology Institute (KEITI) through "Climate Change R&D Project for New Climate Regime", funded by Korea Ministry of Environment (MOE) (2022003570004).

Conflicts of Interest: Author Ilsun Yoo was employed by the company Cheil Engineering Co., Ltd. The remaining authors declare that the research was conducted in the absence of any commercial or financial relationships that could be construed as a poten-tial conflict of interest.

Appendix A

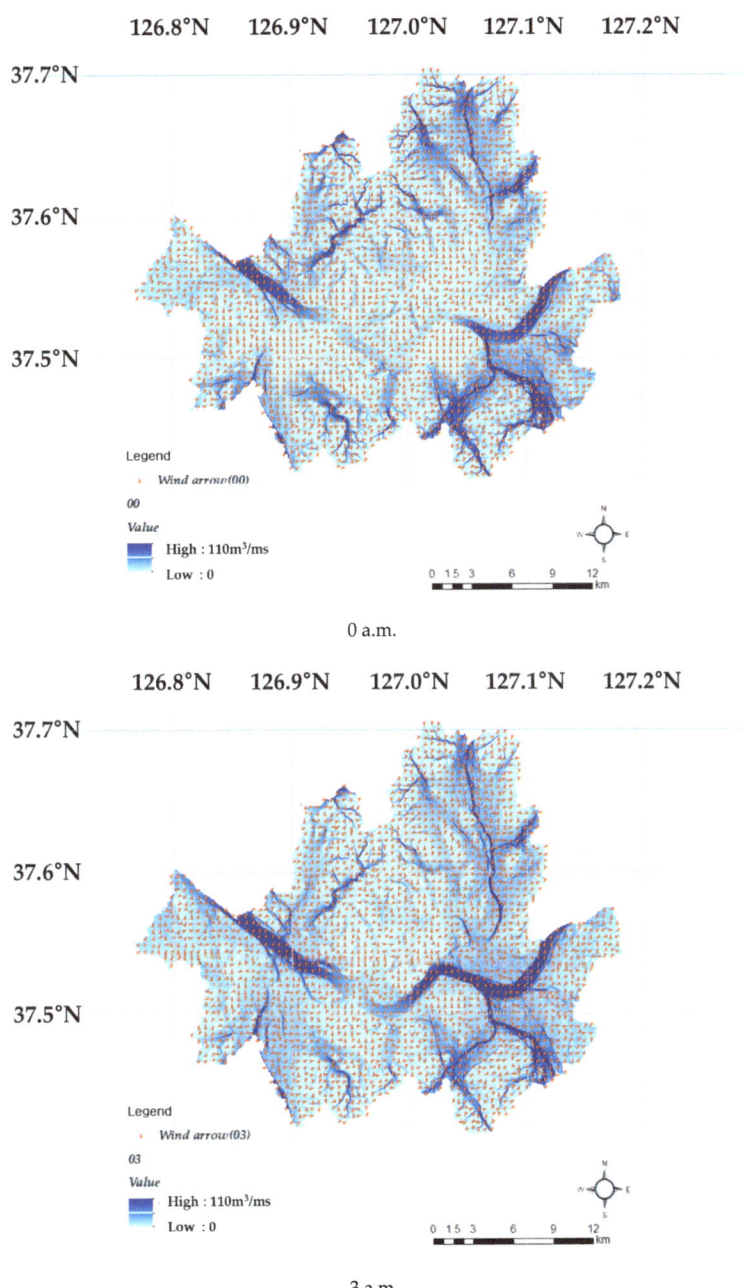

Figure A1. *Cont.*

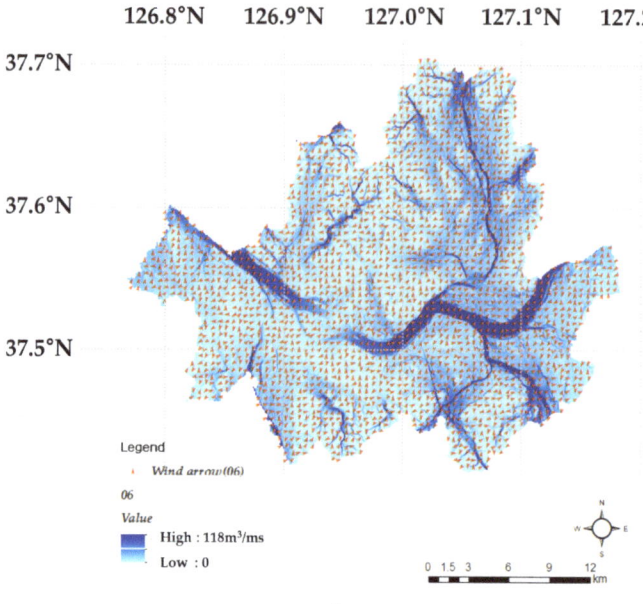

6 a.m.

Figure A1. The CAVF from 0 a.m., 3 a.m. and 6 a.m.

Table A1. Regression analysis results between air temperature reduction and the factors influencing CAF.

		OLS	SLM	SEM
Type A (n = 4888)	Elevation	0.009527	0.02162880987	0.015715
	NDVI	0.013	0.001732	0.00097
	BH	0.20824	0.04015	0.00361
	WAI	0.132445	0.021205	0.005989
	BA	0.001273	−0.0059	−0.001165
	Distance to green areas	0.03936	0.00844	0.01227
	Constant term	1.0414	0.177992	1.05584
	Log likelihood	−2423.76	2009.12	5733.09
	R2	0.26	0.78	0.80
Type B (n = 1167)	Elevation	0.138952	0.020185	0.131069
	NDVI	0.01851	0.00643	0.00289
	BH	0.16345	0.03621	0.01382
	WAI	0.1733	0.035978	0.009598
	BA	−0.01419	−0.00265	−0.00317
	Distance to green areas	0.05924	0.00775	0.011781
	Constant term	−0.29121	−0.05245	−0.33896
	Log likelihood	−83.25	897.92	1211.93
	R2	0.30	0.71	0.75
Type C (n = 256)	Elevation	0.005751	0.03205	0.050024
	NDVI	0.03912	0.03373	0.005849
	BH	0.201706	0.128547	0.092827
	WAI	0.02889	0.04067	0.05023
	BA	−0.047697	−0.030245	−0.01934
	Distance to green areas	0.025339	0.020728	0.005794
	Constant term	−1.12038	−0.80723	−1.18172
	Log likelihood	−20.19	35.69	168.69
	R2	0.32	0.62	0.74

Table A1. *Cont.*

		OLS	SLM	SEM
Type D (n = 11475)	Elevation	0.06705	0.00033	0.02743
	NDVI	0.00024	0.001853	0.000958
	BH	0.11839	0.00822	0.00785
	WAI	0.112607	0.006854	0.005491
	BA	0.033998	−0.000955	−0.000825
	Distance to green areas	0.01465	0.00146	0.007555
	Constant term	1.0155	0.090324	0.949379
	Log likelihood	−6396.11	8789.41	18154.13
	R^2	0.22	0.75	0.79
Type E (n = 4798)	Elevation	0.163369	0.00918	0.040678
	NDVI	0.03471	0.00443	0.00255
	BH	0.02464	0.00271	0.00067
	WAI	0.017313	−0.003426	−0.00054
	BA	−0.03298	−0.00161	−0.00028
	Distance to green areas	0.006601	0.002262	0.007517
	Constant term	−0.31871	−0.02243	−0.29672
	Log likelihood	−1035.19	5767.68	6677.75
	R^2	0.11	0.67	0.78
Type F (n = 3769)	Elevation	0.163246	0.009748	0.052277
	NDVI	0.04314	0.010399	0.001553
	BH	0.08502	0.01537	0.013152
	WAI	−0.06946	−0.00863	−0.01139
	BA	−0.005411	−0.00659	−0.0012
	Distance to green areas	0.056493	0.019646	0.02466
	Constant term	−1.22814	−0.35632	−1.23183
	Log likelihood	−490.66	1989.70	5592.04
	R^2	0.20	0.69	0.75
Type G (n = 5516)	Elevation	0.04289	0.01055	0.04338
	NDVI	0.050936	0.011022	0.001692
	BH	0.02006	0.00184	0.008919
	WAI	0.038631	0.002736	0.003234
	BA	−0.015738	−0.000262	−0.00105
	Distance to green areas	0.113154	0.010995	0.034309
	Constant term	0.780519	0.106907	0.885141
	Log likelihood	−1759.95	3749.93	7799.59
	R^2	0.25	0.73	0.81
Type H (n = 10730)	Elevation	0.033696	0.000781	0.002747
	NDVI	0.001823	0.00155	0.00017
	BH	−0.07323	0.00266	0.00045
	WAI	0.057249	0.003143	0.00032
	BA	−0.01682	−0.00106	−0.00032
	Distance to green areas	0.018694	0.001708	0.00559
	Constant term	−0.27792	−0.01038	−0.30435
	Log likelihood	−3502.54	15,245.60	17,207.67
	R^2	0.14	0.78	0.78
Type I (n = 8351)	Elevation	0.00788	0.00403	0.00667
	NDVI	0.010434	0.0037	0.00132
	BH	0.133405	0.018523	0.002249
	WAI	−0.08795	−0.00789	−0.00253
	BA	−0.07131	−0.00825	−0.0008
	Distance to green areas	0.045243	0.010169	0.012248
	Constant term	−1.31879	−0.18362	−1.2174
	Log likelihood	−2399.94	6189.14	13,935.50
	R^2	0.21	0.71	0.79

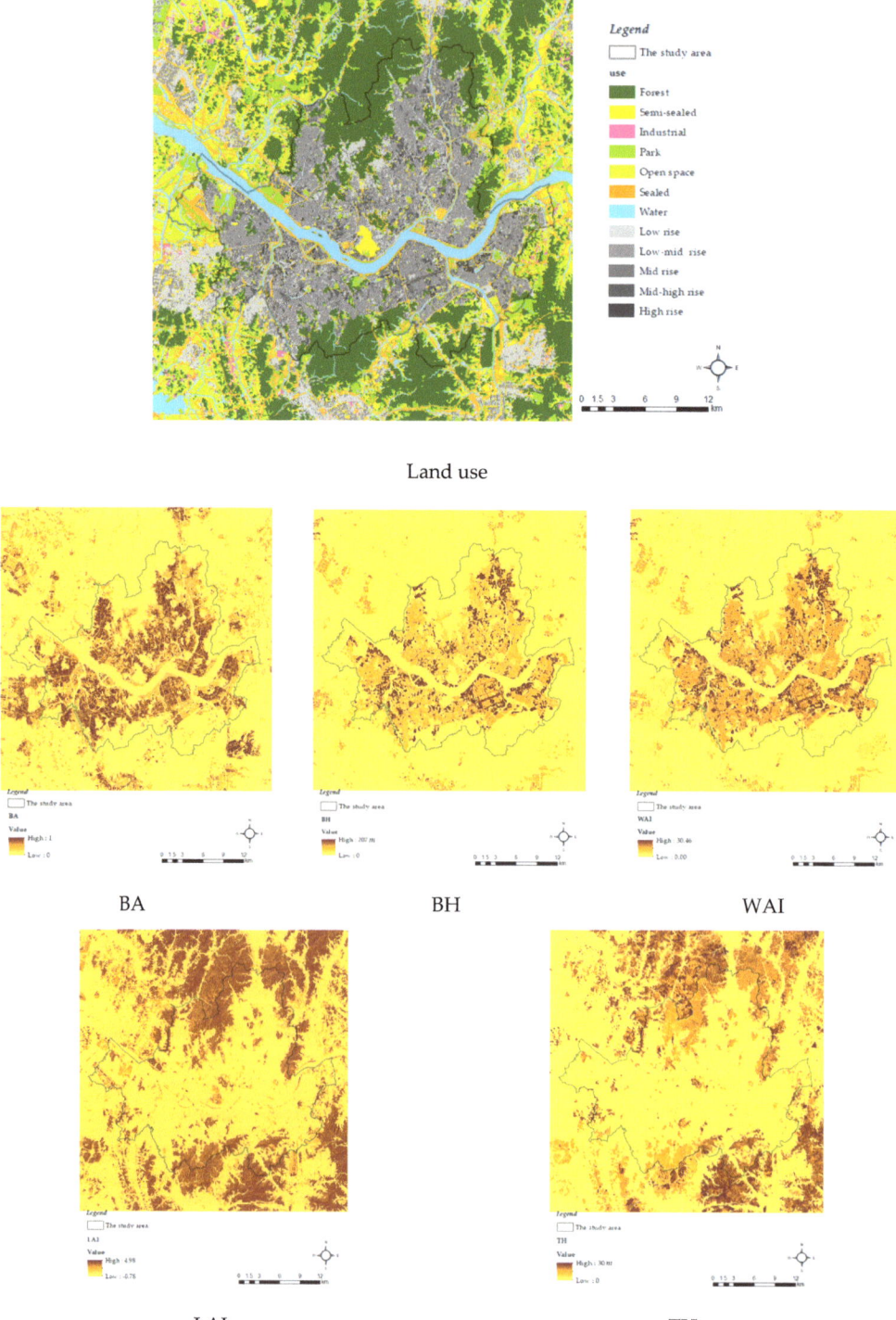

Figure A2. KLAM_21 modeling Physical Parameters results.

References

1. Landsberg, H.E. *The Urban Climate*; Elsevier Science: Amsterdam, The Netherlands, 1981.
2. Rizwan, A.M.; Dennis, L.Y.; Chunho, L. A review on the generation, determination and mitigation of Urban Heat Island. *J. Environ. Sci.* **2008**, *20*, 120–128. [CrossRef] [PubMed]
3. Oke, T.R. Towards better scientific communication in urban climate. *Theor. Appl. Climatol.* **2006**, *84*, 179–190. [CrossRef]
4. Acosta, M.P.; Vahdatikhaki, F.; Santos, J.; Hammad, A.; Dorée, A.G. How to bring UHI to the urban planning table? A data-driven modeling approach. *Sustain. Cities Soc.* **2021**, *71*, 102948. [CrossRef]
5. Parker, J. The Leeds urban heat island and its implications for energy use and thermal comfort. *Energy Build.* **2021**, *235*, 110636. [CrossRef]
6. Ng, E.; Chen, L.; Wang, Y.; Yuan, C. A study on the cooling effects of greening in a high-density city: An experience from Hong Kong. *Build. Environ.* **2012**, *47*, 256–271. [CrossRef]
7. Kwok, Y.T.; de Munck, C.; Lau, K.K.-L.; Ng, E. To what extent can urban ventilation features cool a compact built-up environment during a prolonged heatwave? A mesoscale numerical modelling study for Hong Kong. *Sustain. Cities Soc.* **2022**, *77*, 103541. [CrossRef]
8. Gedzelman, S.D.; Austin, S.; Cermak, R.J.; Stefano, N.; Partridge, S.D.; Quesenberry, S.; Robinson, D.A. Mesoscale aspects of the Urban Heat Island around New York City. *Theor. Appl. Climatol.* **2003**, *75*, 29–42. [CrossRef]
9. Mayer, H. Bestimmung von stadtklimarelevanten Luftleitbahnen. *UVP-Report* **1994**, *5*, 265–267.
10. Priyadarsini, R.; Hien, W.N.; David, C.K.W. Microclimatic modeling of the urban thermal environment of Singapore to mitigate urban heat island. *Sol. Energy* **2008**, *82*, 727–745. [CrossRef]
11. Yim, S.H.L.; Fung, J.C.H.; Ng, E.Y.Y. An assessment indicator for air ventilation and pollutant dispersion potential in an urban canopy with complex natural terrain and significant wind variations. *Atmos. Environ.* **2014**, *94*, 297–306. [CrossRef]
12. Hsieh, C.-M.; Huang, H.-C. Mitigating urban heat islands: A method to identify potential wind corridor for cooling and ventilation. *Comput. Environ. Urban Syst.* **2016**, *57*, 130–143. [CrossRef]
13. Wang, Z.-H.; Li, Q. Thermodynamic characterisation of urban nocturnal cooling. *Heliyon* **2017**, *3*, e00290. [CrossRef] [PubMed]
14. Barlag, A.-B.; Kuttler, W. The significance of country breezes for urban planning. *Energy Build.* **1991**, *15*, 291–297. [CrossRef]
15. Ministerium, F.V.; Baden-Württemberg, I. *Städtebauliche Klimafibel. Hinweise für die Bauleitplanung*; Ministerium für Verkehr und Infrastruktur Baden-Württemberg: Stuttgart, Germany, 2012.
16. Blumen, W.; Grossman, R.; Piper, M. Analysis of heat budget, dissipation and frontogenesis in a shallow density current. *Bound.-Layer Meteorol.* **1999**, *91*, 281–306. [CrossRef]
17. Ingenieure, V.D. Environmental meteorology-local cold air. In *VDI Guideline 3787, Part 5*; Beuth: Berlin, Germany, 2003; Volume 85, pp. 14–29.
18. Sandeepan, B.; Rakesh, P.; Venkatesan, R. Numerical simulation of observed submesoscale plume meandering under nocturnal drainage flow. *Atmos. Environ.* **2013**, *69*, 29–36. [CrossRef]
19. King, E.; Wetterdienst, D. *Untersuchungen über Kleinräumige Änderungen des Kaltluftflusses und der Frostgefährdung durch Strassenbauten: Mit 5 Tab*; Dt. Wetterdienst, Zentralamt: Offenbach, Germany, 1973.
20. Röckle, R.; Richter, C.; Höfl, H.; Steinicke, W.; Streifeneder, M.; Matzarakis, A. *Klimaanalyse Stadt Freiburg*; Auftraggeber Stadtplanungsamt der Stadt Freiburg: Freiburg, Germany, 2003.
21. Pypker, T.; Unsworth, M.H.; Lamb, B.; Allwine, E.; Edburg, S.; Sulzman, E.; Mix, A.; Bond, B. Cold air drainage in a forested valley: Investigating the feasibility of monitoring ecosystem metabolism. *Agric. For. Meteorol.* **2007**, *145*, 149–166. [CrossRef]
22. Sachsen, T.; Ketzler, G.; Knörchen, A.; Schneider, C. Past and future evolution of nighttime urban cooling by suburban cold air drainage in Aachen. *J. Geogr. Soc. Berl.* **2013**, *144*, 274–289.
23. Sievers, U.; Kossmann, M. The cold air drainage model KLAM_21-Model formulation and comparison with observations. *Weather. Clim.* **2016**, *36*, 2–24. [CrossRef]
24. Grunwald, L.; Kossmann, M.; Weber, S. Mapping urban cold-air paths in a Central European city using numerical modelling and geospatial analysis. *Urban Clim.* **2019**, *29*, 100503. [CrossRef]
25. Scherer, D.; Fehrenbach, U.; Beha, H.-D.; Parlow, E. Improved concepts and methods in analysis and evaluation of the urban climate for optimizing urban planning processes. *Atmos. Environ.* **1999**, *33*, 4185–4193. [CrossRef]
26. Katzschner, L. The urban climate as a parameter for urban development. *Energy Build.* **1988**, *11*, 137–147. [CrossRef]
27. Son, J.M.; Eum, J.H.; Kim, D.P.; Kwon, J. Management strategies of thermal environment in urban area using the cooling function of the mountains: A case study of the Honam Jeongmaek areas in South Korea. *Sustainability* **2018**, *10*, 4691. [CrossRef]
28. Gu, K.; Fang, Y.; Qian, Z.; Sun, Z.; Wang, A. Spatial planning for urban ventilation corridors by urban climatology. *Ecosyst. Health Sustain.* **2020**, *6*, 1747946. [CrossRef]
29. He, B.-J. Potentials of meteorological characteristics and synoptic conditions to mitigate urban heat island effects. *Urban Clim.* **2018**, *24*, 26–33. [CrossRef]
30. Lee, D.; Oh, K.; Jung, S. Classifying Urban Climate Zones (UCZs) Based on Spatial Statistical Analyses. *Sustainability* **2019**, *11*, 1915. [CrossRef]
31. Tobler, W.R. A computer movie simulating urban growth in the Detroit region. *Econ. Geogr.* **1970**, *46*, 234–240. [CrossRef]
32. Song, B.; Park, K. Mountain valley cold air flow interactions with urban morphology: A case study of the urban area of Changwon, South Korea. *Landsc. Urban Plan.* **2023**, *233*, 104703. [CrossRef]

33. Son, J.-M.; Eum, J.-H.; Kim, S. Wind corridor planning and management strategies using cold air characteristics: The application in Korean cities. *Sustain. Cities Soc.* **2022**, *77*, 103512. [CrossRef]
34. Grunwald, L.; Schneider, A.-K.; Schroeder, B.; Weber, S. Predicting urban cold-air paths using boosted regression trees. *Landsc. Urban Plan.* **2020**, *201*, 103843. [CrossRef]

Disclaimer/Publisher's Note: The statements, opinions and data contained in all publications are solely those of the individual author(s) and contributor(s) and not of MDPI and/or the editor(s). MDPI and/or the editor(s) disclaim responsibility for any injury to people or property resulting from any ideas, methods, instructions or products referred to in the content.

Review

Decision Support Systems in Forestry and Tree-Planting Practices and the Prioritization of Ecosystem Services: A Review

Neelesh Yadav [1], Shrey Rakholia [1] and Reuven Yosef [2,*]

1. Bioinformatics Center, Forest Research Institute, Dehradun 248006, Uttarakhand, India; neelesh_yadav@icfre.org (N.Y.); rakholias@gmail.com (S.R.)
2. Eilat Campus, Ben Gurion University of the Negev, P.O. Box 272, Eilat 881020, Israel
* Correspondence: ryosef60@gmail.com

Abstract: In this study, tree-selection/plantation decision support systems (DSSs) were reviewed and evaluated against essential objectives in the available literature. We verified whether existing DSSs leverage multiple data sources and available online resources such as web interfaces. We compared the existing DSSs, and in this study mainly focused on five main objectives that DSSs can consider in tree selection, including (a) climate resilience, (b) infrastructure/space optimization, (c) agroforestry, (d) ecosystem services, and (e) urban sustainability. The climate resilience of tree species and urban sustainability are relatively rarely taken into account in existing systems, which can be integrated holistically in future DSS tools. Based on this review, deep neural networks (DNNs) are recommended to achieve trade-offs between complex objectives such as maximizing ecosystem services, the climate resilience of tree species, agroforestry conservation, and other benefits.

Keywords: decision support system; climate resilience; ecosystem services; deep neural networks; sustainability

Citation: Yadav, N.; Rakholia, S.; Yosef, R. Decision Support Systems in Forestry and Tree-Planting Practices and the Prioritization of Ecosystem Services: A Review. *Land* **2024**, *13*, 230. https://doi.org/10.3390/land13020230

Academic Editors: Alessio Russo and Giuseppe T. Cirella

Received: 16 December 2023
Revised: 26 January 2024
Accepted: 9 February 2024
Published: 12 February 2024

Copyright: © 2024 by the authors. Licensee MDPI, Basel, Switzerland. This article is an open access article distributed under the terms and conditions of the Creative Commons Attribution (CC BY) license (https:// creativecommons.org/licenses/by/ 4.0/).

1. Introduction

The global climate is changing and is predicted to change even faster in the near future [1]. The importance of planting trees for climate change adaptation and mitigation is increasing, as forests act as carbon sinks [2,3]. This is particularly true in areas with desertification and complex environmental problems that require robust processes that allow the ever-growing human population to benefit from the environment's ecosystem services [4,5]. Furthermore, in many cases, ecosystem services cannot be easily quantified in monetary terms, are taken for granted, and often involve moral and ethical principles [6]. The rapid growth of tree planting and land-use conversion from grassland to forests directly impacts ecosystem services, resulting in increased regulation and service provision [7]. However, further planning is needed to ensure that local environmental concerns and cultural values are internalized and that additional ecosystem services such as timber availability, water quality, biodiversity enrichment, and carbon sequestration are enabled [8].

However, this type of land-use change does not always lead to an improvement in ecosystem services, as grassland biomes are often considered to have potential for forest restoration and planting. Biodiversity and ecosystem services are typically reduced once these grasslands/savannahs are converted, resulting in significant protective measures to plantation strategies, and thus separate ecosystem services need to be identified for forest and grassland biomes [9]. Furthermore, identifying suitable tree species for adaptability is crucial for future climate scenarios, especially in urban areas, as changing climates lead to the loss of tree species, which can lead to a reduction in ecosystem services such as Urban Heat Island (UHI) mitigation, which can pose a challenge to adaptation and mitigation strategies for human-caused climate change [10].

Using deep neural networks (DNNs), a decision support system (DSS) can be trained to learn from a large dataset of tree data, including information about tree species, climate, soil conditions, and other factors that influence tree growth and survival. This is because the use of neural networks was proposed three decades ago to solve forest management problems by integrating forest knowledge with artificial intelligence (AI; [11]). AI greatly benefits sustainability and the preservation of ecosystem values, as increasing disruptions in a changing world can only be managed beyond human intelligence [12]. Furthermore, despite the various DSSs and AI systems used, the appointment of appropriate project managers is crucial to the execution and subsequent success of a project [8].

Our study examines various DSSs and compares them based on their objectives and applications. In addition, we provide a literature review focusing on the need for an ecosystem-services-focused DSS and discuss the potential applications of DNNs for these systems.

2. Review of Existing DSS Tools for Tree Selection and Plantations

2.1. Review of the Existing Literature

One of the earliest DSSs for tree plantations in forestry was developed at the University of Canterbury: a framework-based system coded in Prolog. The focus was on knowledge-based decision support by linking to the Forest Management Information System (FMIS) or Geographic Information System (GIS) databases, enabling location-based access to information about the field microenvironment such as soils, climate, elevation, and earlier land/crop use and current conditions, along with multiple management options for optimization [13].

Further efforts to develop a DSS for tree plantations began in the 2000s using a GIS with a focus on street and neighborhood tree plantations, while attempting to address management aspects such as DSS-based strategies to reduce energy, fuel, and pesticides/fertilizers for plantation management [14]. In addition, the focus was also expanded to include aspects such as soil-property-based tree planting, feasibility of the planting area, tree age, species diversity, shade, and canopy cover [15]. It is also important to conduct an existing urban tree cover (UTC) analysis prior to tree-planting decisions, using object-oriented satellite image analysis to identify existing vegetation cover and land-use types [16].

Mitigating a region's hydrological problems also requires selecting appropriate species, prioritizing sites for re-vegetation, and simulating different hydro-climatological conditions annually. These aspects were incorporated into China's bilingual GUI decision support tool for re-vegetation programs, ReVegIH, which could also reduce sediment load release through afforestation modeling [17].

A multilingual programming (C++ and Fortran)-based DSS known as the Motti Simulator, developed by the Natural Resources Institute Finland (Luke), has also been used for tree selection based on detailed forest stand dynamics and incorporating tree growth and yield models [18]. Additionally, simplified open-source and open-code DSSs such as PT^2 (Prairie and Tree Planting Tool) have allowed users to explore and delineate areas of interest for tree/prairie planting or management using scaled dimensional drawing tools, and then select seeds/woody plants for the various soils with a drop-down menu. This also enabled the selection and calculation of financial costs and long-term management options [19].

Nevertheless, advances in machine learning in recent years have enabled the selection of tree species taking into account climate variability using MaxEnt to determine the suitability and resilience of trees in different climate scenarios. A recent example is the online platform "Which Plant Where" in Australia, which was developed using Python, Django, and PostgreSQL [20]. In addition, others use tree-selection tools developed by the United States Department of Agriculture (USDA) such as the Tree Advisor and the Woody Plant Selection Tool for multi-functional purposes, using MySQL and the Drupal framework [21]. In addition, a spatio-temporal urban tree DSS was developed using ensemble CAD and GIS tools. This integrates detailed 3D trees into urban design, allowing the testing of tree placement, species selection, solar exposure, etc. Valuable elements of computational botany and lighting engineering technology make this possible [22].

Although tree-planting decision support systems have addressed tree-selection ecosystem services such as UHI mitigation, only simple filtering techniques with limited variables that filter the attributes from tree databases have been used [23]. In addition, ensemble models that use higher-resolution datasets to infer the potential suitability and realized distribution of tree species through batch generalization are also proposed. This is a boosting method that uses random forest (RF), gradient-booster trees (GBTs), and generalized linear models (GLMs), which are further processed by the meta-learner [24].

2.2. Methods

In this section, reviews and analyses of existing DSSs for tree selection and plantations are reviewed and analyzed using obvious keywords such as 'DSS', 'Decision Support tool', 'Tree selection,', etc., via a Google Scholar search (Figure 1). Keywords such as 'ecosystem services', 'agroforestry', 'urban', 'climate', etc., were also frequently searched for in the literature texts. We compare different DSSs for tree selection/planting based on their objectives, their programming language framework and software (Table 1), and their comparison with the main objectives (Figure 2). Table 1 summarizes DSSs for urban tree plantations, agroforestry, etc., which show prevailing trends of using R and Python tools. This comparison is crucial as it highlights the core concept of tree planting, as the DSS for tree selection/planting is based on a basic structure that includes species, location, and value. Data sources include research articles from Google Scholar, DSS web interfaces, and the gray literature; the practical use of DSS web interfaces was crucial in determining the capabilities and objectives of various DSSs. During the review process, text comparison revealed important patterns and themes in the literature. The DSS findings include recommendations with critical objectives and advocate for advanced techniques such as deep neural networks (DNNs) to improve decision accuracy in tree selection and planting, thereby providing more informed and insightful guidance.

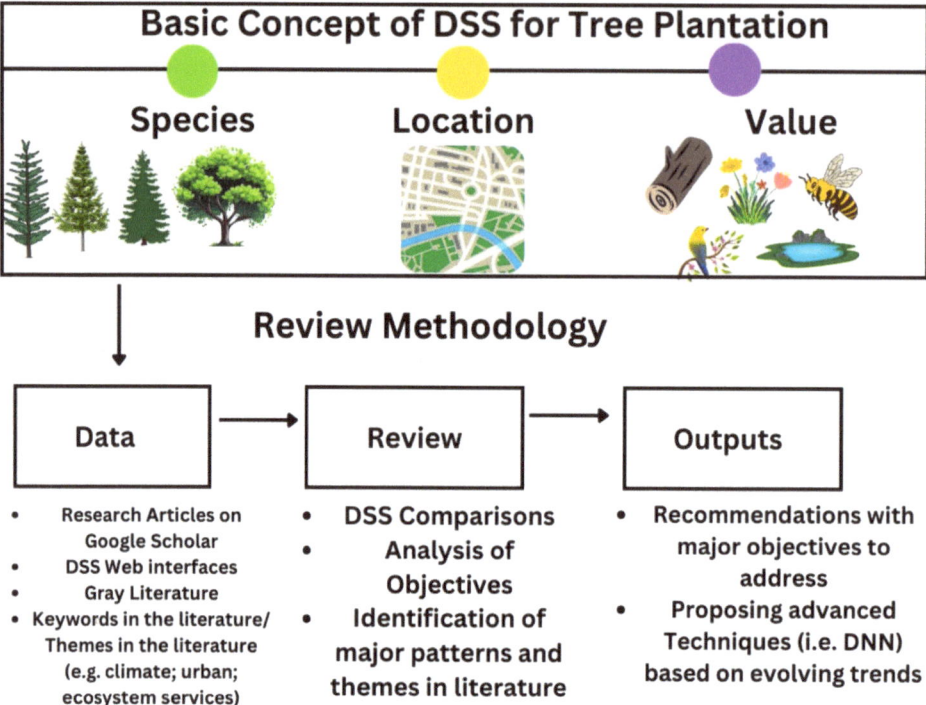

Figure 1. Flow chart of the purpose of this study highlighting the key concepts and objectives of DSSs.

Table 1. Comparison of various DSSs developed for tree selection/plantation.

	DSS Name (Provisional)	Software/Language/Framework	Objective Type	Reference
1	Knowledge-based DSS	Prolog	Forest plantation DSS	[13]
2	Prototype Decision Support System	SMODT; ArcTrees; Treemodules ǀ Visual Basic Analysis (VBA)	Urban tree plantation suitability	[15]
3	ReVegIH Decision Support Tool	C#, Visual Basic, C++, .NET	Tree species selection with ecohydrological modelling	[17]
4	Prototype Decision Support System (Randall)	ArcView GIS Extension ǀ Avenue	Neighborhood greening	[14]
5	Decision Support Tool—Precision Forestry	HprAnalys, ArcGIS, Motti stand simulator	Tree species selection with stand dynamics	[18]
6	Virginia UTC Assessment Process	ERDAS; ISODATA	Object-oriented classification of urban tree canopy analysis	[16]
7	Right Place, Right Tree—Boston	R packages—shinydashboard; leaflet; tigris; DT	Tree plantation DSS for UHI mitigation	[23]
8	Which Plant Where?	Python; Django; PostgreSQL	Plant selection tool for climate resilience and sustainability	[20]
9	Tree Advisor USDA	MySQL; Drupal	Woody plant selection tool for multifunctional objectives	[21]
10	Plant-Best	R	Plant selection tool for slope protection	[25]
11	Spatio-Temporal Decision Support System for Street Trees	QGIS/ArcGIS; exlevel GrowFX; Autodesk; AutoCAD; ForestPro	Detailed 3D trees for urban design	[22]
12	Florida Agroforestry Decision Support System (FADSS)	Delphi; SQL	Agroforestry planning and tree selection	[26]
13	PT^2 (Prairie and Tree Planting Tool)	HTML; CSS; Javascript	Prairie and tree planting selection and financial cost estimation	[19]
14	Diversity for Restoration (D4R)	JavaScript, Python, and R.	Ecosystem restoration and agroforestry	[27]
15	Citree	PHP; MariaDB server	Tree selection for urban areas in temperate climate	[28]
16	i-tree USDA	Java; Javascript; Python	Multi-module suite for urban tree structures and ecosystem service evaluation	[29]
17	Unique DSS for Agroforestry Systems	R; HTML	Decision support tool for coffee and cocoa agroforestry systems	[30]

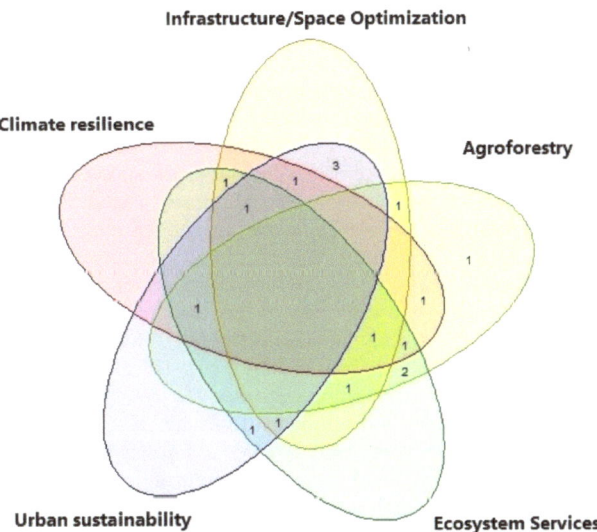

Figure 2. A Venn diagram of DSSs (the number represents the number of DSSs that fit into the categories) and their main goals to show similarities and differences in DSS goals. Details can be found in Table 2. Numbers in the VENN diagram ellipses represent the number of DSS fitting the categories.

Table 2. The DSS reviews and the relevant objectives they address. CR = climate resilience, I/SO = infrastructure/space optimization, AF = agroforestry, ES = ecosystem services, US = urban sustainability.

#	DSS	CR	I/SO	AF	ES	US	Ecosystem Services
1	Knowledge-based DSS	No	No	Yes	No	No	-
2	Prototype Decision Support System	No	Yes	No	No	Yes	-
3	ReVegIH Decision Support Tool	Yes	No	Yes	No	No	-
4	Prototype Decision Support System (Randall)	No	Yes	No	No	Yes	-
5	Decision Support Tool—Precision Forestry	No	Yes	Yes	No	No	-
6	Virginia UTC Assessment Process	No	Yes	No	Yes	Yes	Air quality; flood mitigation; UHI mitigation
7	Right Place, Right Tree—Boston	No	No	No	Yes	Yes	UHI mitigation
8	Which Plant Where?	Yes	Yes	No	No	Yes	-
9	Tree Advisor USDA	No	No	Yes	Yes	No	Extensive ecosystem services
10	Plant-Best	Yes	Yes	No	Yes	No	Slope protection (landslide prevention)
11	Spatio-Temporal Decision Support System for Street Trees	No	Yes	No	No	Yes	-
12	Florida Agroforestry Decision Support System (FADSS)	Yes	Yes	Yes	Yes	No	Runoff reduction; erosion control; timber provisioning, etc.
13	PT^2 (Prairie and Tree Planting Tool)	No	Yes	Yes	Yes	No	Biodiversity (wildlife and pollinator habitat); water quality
14	Diversity for Restoration (D4R)	Yes	No	Yes	Yes	No	Extensive ecosystem services
15	Citree	Yes	Yes	No	Yes	Yes	Air quality; bird feeding; provisioning (honey and edibles)
16	i-tree	Yes	No	Yes	Yes	Yes	Air quality; runoff reduction; Carbon sequestration; Cooling effect, etc.
17	Unique Decision Support Tool for Cocoa and Coffee	No	No	Yes	Yes	No	Microclimate buffering; soil fertility; pest/weed suppression; provisioning (timber, food and fuelwood), etc.

Furthermore, in this study, we focus on five main objectives that DSSs can consider in tree selection, including (a) climate resilience, (b) infrastructure/space optimization, (c) agroforestry, (d) ecosystem services, and (e) urban sustainability (Figure 2). The goal of infrastructure/space optimization has been addressed by some DSSs mentioned above (Table 1), and includes aspects such as tree selection to optimize shading, infrastructure constraints, tree placement, spatial considerations, etc. [15,19,20,22].

Climate resilience was also addressed in some DSSs, covering aspects such as drought tolerance, heat resistance, resilience of trees, etc., to extreme events in different climate change scenarios [20,25–28]. In addition, the main DSSs aimed at agroforestry include DSS Nos. 9, 12, 13, 14, and 17, which also more or less internalize ecosystem services, since the relevant literature contains the term 'ecosystem services' and a range of ecosystem services are explicitly mentioned (Table 1). The specific ecosystem services are listed in the Section 2.2 (Table 2). Many DSSs are specifically focused on urban sustainability related to tree planting, including DSS Nos. 2, 6, 7, 8, 15 and 16, as these DSSs even emphasized the word 'urban' and subsequently sustainability in urban areas in their published articles (Table 1). DSSs specifically targeted at forestry were very rare, as DSSs generally referred to agroforestry, mainly because farmers and planters were key stakeholders rather than stand-alone forestry applications. However, the stand-alone forestry DSSs also included DSS Nos. 1 and 5, which also highlighted the term 'forestry' in the literature, particularly in the article abstracts (Table 1). Therefore, the independent forestry objective was not taken into account in this analysis.

In addition, the Venn diagram of DSS comparisons reflects how existing DSSs have combinations and commonalities of objectives, such as infrastructure/space optimization with urban sustainability, which are described in the Section 2.3.

2.3. Results

Table 1 summarizes the DSSs focusing on urban tree plantations, agroforestry, etc., and shows a prevailing trend of using R and Python tools. However, the technologies used span a wide range and include languages such as C#, C++, .NET, Python, R, and Java, as well as web development tools such as HTML, CSS, and JavaScript. The emphasis on these languages suggests a shared recognition within the community of their effectiveness in tackling data-driven tasks and facilitating interdisciplinary collaboration in environmental decision making [29,30].

Since DSSs have historically paid the least attention to the goal of climate resilience, it is important to focus on climate-tolerant planting strategies by increasing the functional diversity of trees, as this ensures the maintenance of ecosystem services by preventing tree death [31]. Interestingly, the trade-offs between climate resilience and ecosystem services are particularly embedded in the concept of climate adaptation, as in this context the focus on the use of regulating ecosystem services is important [32].

According to the Venn diagram, the commonalities in the system in terms of the goals they address include infrastructure/space optimization and urban sustainability (n = 6), followed by ecosystem services and agroforestry (n = 6; Figure 2). This shows that existing DSS developers in particular have placed emphasis on spatial optimization in tree selection in urban areas, as well as maximizing ecosystem services in agroforestry ecosystems. Furthermore, climate resilience and the urban sustainability of trees are the least considered, while infrastructure/space optimization and ecosystem services are relatively more considered in many DSS tools (Figure 2). Nevertheless, the existing DSSs address all issues at different times, but not comprehensively. This is evident from the analysis (Table 2).

DSSs aimed at urban sustainability often includes the regulation of ecosystem services such as UHI mitigation (or cooling effect) and air quality. Other regulating ecosystem services such as water quality, runoff reduction, and pollination were also included. In addition, some existing DSSs have extensively addressed a wide range of ecosystem services, for example, DSS Nos. 9, 14, 15, and 17. Provision services are also included, such as in DSS Nos. 12, 13, 15, 17, etc., which include wood and non-timber forest products. Some supporting ecosystem services such as soil fertility were also included (Table 2).

3. The Need for an Ecosystem-Services-Focused DSS

It is crucial to understand the ecosystem services received from trees during selection and planting, as trees provide various regulatory (carbon sequestration, air pollution reduction) and provisioning services (timber, tree crops). Non-market values sometimes exceed commercial values and threats, such as forest fires and pests, and this must be taken into account for resilience [33]. Additionally, models such as the Natural Capital Protocol can be applied to improve agroforestry decision making and evaluation at the farm level. They describe the connection between a natural capital, its condition, the resulting ecosystem services, and the benefits that people derive from these services. Better benefit representation can also promote the public benefits of agroforestry at the farm level [34].

One of the most important ecosystem services is flood protection, which can be improved by riparian forests as part of agroforestry (e.g., riparian buffers), providing the same benefits at almost 30% of the cost compared to an engineered protection structure, as shown in a study in Germany [35]. Satellite datasets and IDF-based (Intensity, Duration and Frequency) flood models can provide valuable information about the flooding and water logging situation in regions experiencing monsoons and persistent floods. The areas affected by flooding and erosion can be identified using flood depth and flow velocity forecasts for 25-, 50- and 100-year return periods [36]. Therefore, the selection of tree species adapted to this water logging must be assessed based on the literature that evaluates parameters such as stomatal conductance and net photosynthesis, since some tree species show a reduction in these two processes after flooding [37]. In addition, trees such as poplars in riparian zones are very tolerant of flooding because nitrogen metabolism is not affected by flooding compared to species such as oak and beech, which are sensitive to successive flooding, and the depth and duration of flooding must also be taken into account in detail [38].

It is important to understand the dynamics of the UHI effect. There are regional and zonal differences, including in urban areas, because although trees are effective in reducing air temperature in areas with high building density, they are ineffective in built-up areas with low building density, and therefore high-density trees with taller trunks are recommended for built-up areas [39]. Changes in land use and land cover can influence local surface temperatures. For example, as previously irrigated croplands and forests transform into built-up urban areas over time, this can lead to increases in air and land surface temperatures (LSTs). Conversely, a transition from bare land cover to urban areas could reduce the average LST for semi-arid regions [40,41]. This highlights the significant influence of both vegetation and urban development on LSTs at the local scale. The vegetation has a cooling effect through transpiration, shading, and rainwater retention.

Similarly, urbanized zones contribute more to temperature reduction than regions with exposed ground or rocky terrain due to their surface properties and materials that promote convection more effectively [42]. There is a unique approach to UHI mitigation that involves creating a regional Heat Vulnerability Index (HVI) that incorporates socioeconomic (family income, age, building density) and environmental data (e.g., LST, vegetation) for decision making [43], which helps to increase urban tree canopy cover with the most suitable tree species. To mitigate UHI, urban areas need to be divided into high- and low-density areas because land use and tree availability are limited in cities.

Furthermore, nature-based solutions (NbSs) to air pollution can be implemented zone-wise by involving plantations. Air-pollution-tolerant species such as *Shorea robusta*, *Ficus religiosa*, and *Mangifera indica* have high tolerance to pollutants and high metal accumulation capacity in industrial areas. Dust removal and deposition are excellent in residential areas in *Azadirachta indica*, *Dalbergia sissoo*, and *Ficus religiosa* [44]. Tools for slope protection and landslide mitigation include Plant-Best, which was developed in the statistical programming language R [25].

Many factors influence tree plantation, including the value and placement of trees, particularly in urban areas. This includes public lands, parks, and roadsides, as well as private land, i.e., residential properties [45]. Kirkpatrick et al. [46] suggested that small fruit trees on private property were more aesthetically pleasing and practical. A study on agroforestry found that the management of forests involves significant uncertainty regarding future timber prices, tree growth, and the impact of climate-related changes on tree growth. Because most forest owners prefer to avoid risk and tree growth and timber prices are unpredictable, the study suggests the following implication: longer rotations should be compared to recommended guidelines. There may be a greater preference for mixed stands than deterministic calculations suggest; the concentration of timber revenues should be less focused on the final harvest, as currently recommended. The consistent retention of multiple timber assortments in the inventory is advantageous, which indicates that more uneven stand structures should be pursued [47].

Therefore, the suitability process must include mixed stands and not just monoculture recommendations. However, this may not be the case for all tree species as agarwood monoculture plantations could also be favorable in terms of growth, as they are endangered [48]. Nevertheless, plantation agriculture in tropical countries must be managed on the basis of polyculture systems and not monocultures since the ecosystem services provided by the former are much better, as they include the improvement of biodiversity, pollination, and biological pest control even in the context of small-scale silviculture [49].

Hirsch et al. [50] found species-specific tolerance to drought and traffic pollution in urban areas, suggesting the use of certain tree species along roads and in residential areas.

DSSs such as the FADSS (Florida Agroforestry DSS) dealt with economic and environmental services and used GIS databases that contained important datasets on tree attributes, infrastructure, climate, soil, and cropping, including critical levels such as key agroforestry management practices [26]. It is also important to include soil datasets on soil pH, sand content, etc., for tree species distribution models (SDMs) as soil variables are strong predictors of habitat suitability [51]. Soil datasets are often neglected in many SDMs, so these datasets should be some of the core variables in decision support systems. Finally, recent developments in tree selection DSSs include the Diversity for Restoration (D4R) tool, which allows users to make multiple selections from a menu for restoration objectives, ecosystem services, seeding zones, climate, and other environmental data on decision-making for individual and combined tree species selection [27].

Therefore, by incorporating rich ecosystem services, DSS tools are enriched with more data-driven and knowledge-driven capabilities, introducing complexities in these systems that can then be addressed and improved through the implementation of DNNs, as explained in the following section.

4. Proposed Use of DNN in DSS for Tree Selection/Plantation

In order to improve decision making in urban forestry for sustainable and livable cities, AI has been increasingly used as part of smart technology in recent years [52]. However, only half of the studies using AI manage to take into account aspects such as the limitations of methods, including robustness and lack of precision in some datasets, the combined use of discrete and continuous data variables, overfitting, collinearity, etc. [53]. The application of AI in forestry can be improved by incorporating the XAI (Explainable Artificial Intelligence), LTNL (Learning To Not Learn), and FUL (Feature Unlearning) methods which allow the qualitative and quantitative comparison of model accuracy and explanations through the use of predefined annotation matrices, i.e., expert knowledge that can improve these deep learning models. Therefore, the combination of XAI, FUL, and expert knowledge can improve the understanding of how the model works, instead of only obtaining simple model results [54].

In addition, the use of CNNs (convolutional neural networks) is increasing significantly with a large number of applications in agriculture/agroforestry DSSs generally based on frameworks such as *Keras*, *Tensorflow*, *Tensorflow-Keras*, *PyTorch*, *Tensorflow-PyTorch*, and *Deeplearning 4j* [55]. In addition, the applications of DNNs for intelligent geographic data analysis in DSSs in agriculture have shown promising results, especially when Back-Propagation Neural Network (BPNN)-based prediction models are used to predict agricultural indicator values [56,57]. In addition, DNN-based species distribution models show better results than traditional models, including DNNs built using bootstraps to improve the prediction performance of species distribution. These can be built in the Python environment using the *Scikit-learn* package with bootstrapping aggregation (bagging) performed in the R statistics package *boot* to train the DNN [58]. Regardless, CNN-based SDMs offer broader advantages, including better learning of non-linear environmental descriptors, compelling distribution predictions of environmental descriptors, and the use of high-dimensional data, enabling an improved collection of information about environmental landscapes structured on tensors, rather than local values of environmental factors [59]. Likewise, the ecosystem service component of a tree plantation DSS can be better understood and improved through these tensors [60], i.e., different functions of multiple vectors (as ecosystem services include multiple services and complex relationships, such as between the existing environment and land use) can be considered in one vector. Ecosystem services can be viewed as multi-linear functions of the vector [61].

As explained in the Section 2.3, a trend of DSS frameworks over the years is towards the use of scripting and data analysis languages such as R and Python (Table 1). This trend now also brings with it the possibility of using deep neural networks to solve complexities and improve automation processes, as DNNs can be developed with R [62] and Python much more extensively [63–66].

TensorFlow uses the term "Tensor" to denote the primary data structure used in deep learning algorithms. This "Tensor" represents a multi-dimensional array of numerical values [67]. In addition, deep neural networks have been widely used in recent years. This rise in popularity of deep learning models is due to TensorFlow, an open-source deep learning framework, as this framework offers users the ability to rely on pre-defined, network-trained deep learning classification (and regression) models while enabling the customizable training of their personalized or custom datasets [68].

The TensorFlow Deep Neural Network (TF-DNN) is used in the Python environment as the primary model of this study because TF-DNN has been applied in GIS studies that have shown higher spatial prediction accuracy than other techniques such as random forest (RF), support vector machines (SVMs), and logistic regression (LR) [69]. The TF-DNN can be applied with semi-supervised learning with a multivariate multilayer perceptron with training datasets, where the soil, climate, and landscape environmental layers can be used to determine the land suitability of the plant species in the study, with the results providing continuously better decision-making potential when validated through K-fold cross-validation [70].

For the proper implementation of the TF-DNN, it is important to use multiple libraries, including *TensorFlow, Keras, NumPy*, and *Matplotlib*. *Keras* is used as a backend to build and implement the TF-DNN algorithm, while *TensorFlow* acts as a numerical computing library. The *Numpy* library is useful for many mathematical functions that operate on arrays, and *Matplotlib* is similarly used to visualize statistical outputs [71].

Therefore, the use of DNNs is crucial for improving the precision and effectiveness of DSSs and contributes to sustainable and informed tree selection and plantation strategies in both urban and regional environments.

5. Conclusions

This study not only highlights existing DSSs developed for the purpose of tree selection and plantation, but also highlights the evolving trends and goals that DSSs address. It outlines various goals that are commonly addressed in the existing literature and notes the lack of a comprehensive DSS that takes into account all of these goals as well as future challenges such as climate resilience and sustainable urban spaces.

Based on this review, it is important to focus on increasing the selection of climate-resilient trees in DSSs, along with urban sustainability requirements, to maximize ecosystem services in urban environments. Given the evolving trend of using scripting and data analysis languages such as R and Python, incorporating DNNs can also improve decision making when considering multiple ecosystem services and the benefits of agroforestry, especially when the goal is better predictive modeling capabilities in the context of tree plantation.

In addition, the main objectives set out in this review must be addressed simultaneously and taken into account and included in the reviewed DSSs. The application of DNNs in future DSS tools will enable the internalization of these challenging goals, especially when it comes to finding a balance between complex trade-offs such as maximizing ecosystem services, the climate resilience of tree species, and maintaining the benefits of agroforestry. This study will provide future DSS developers with an important comparison to address some of the objectives not previously considered in DSSs. The future implementation of DNNs will improve decision making under challenging climate change conditions and the resilience of ecosystem services.

Author Contributions: Conceptualization, S.R., R.Y. and N.Y.; methodology, N.Y.; software, S.R.; validation, R.Y. and N.Y.; formal analysis, S.R.; data curation, S.R.; writing—original draft preparation, S.R. and R.Y.; writing—review and editing, S.R. and R.Y.; visualization, S.R.; supervision, N.Y.; funding acquisition, N.Y. All authors have read and agreed to the published version of the manuscript.

Funding: This research received no external funding.

Data Availability Statement: This is a review of historical and existing platforms, and no new data were created.

Conflicts of Interest: The authors declare no conflicts of interest.

Abbreviations

AI, Artificial Intelligence; BPNN, Back-Propagation Neural Network; CNNs, Convolutional Neural Networks; DNNs, Deep Neural Networks; DSS, Decision Support System; D4R, Diversity for Restoration; FADSS, Florida Agroforestry Decision Support System; FMIS, Forest Management Information System; FUL, Feature Unlearning; GBTs, Gradient Booster Trees; GISs, Geographic Information Systems; GLMs, Generalized Linear Models; HVI, Heat Vulnerability Index; IDF, Intensity, Duration and Frequency; IPCC, International Panel for Climate Change; LST, Land Surface Temperature; LTNL, Learning To Not Learn; NbSs, Nature-based Solutions; RF, Random Forests; SDM, Species Distribution Modelling; SVM, Support Vector Machine; TF-DNN, TensorFlow Deep Neural Network; UHI, Urban Heat Island; USDA, United States Department of Agriculture; UTC, Urban Tree Cover; XAI, Explainable Artificial Intelligence.

References

1. IPCC 2022. *Climate Change 2022: Impacts, Adaptation, and Vulnerability*; Cambridge University Press: Cambridge, UK; New York, NY, USA, 2022; p. 3056. Available online: https://www.ipcc.ch/report/sixth-assessment-report-working-group-ii/ (accessed on 16 December 2023).
2. Nunes, L.J.; Meireles, C.I.; Pinto Gomes, C.J.; Almeida Ribeiro, N.M. Forest contribution to climate change mitigation: Management oriented to carbon capture and storage. *Climate* **2020**, *8*, 21. [CrossRef]
3. Yang, H.; Ciais, P.; Frappart, F.; Li, X.; Brandt, M.; Fensholt, R.; Fan, L.; Saatchi, S.; Besnard, S.; Deng, Z.; et al. Global increase in biomass carbon stoc dominated by growth of northern young forests over past decade. *Nat. Geosci.* **2023**, *16*, 886–892. [CrossRef]
4. Poch, M.; Comas, J.; Rodriguez-Roda, I.; Sanchez-Marri, M.; Cortes, U. Ten years of experience in Designing and Building real Environmental Decision Support Systems. In Proceedings of the What Have We Learnt?—1st International Congress on Environmental Modelling and Software, Lugano, Switzerland, 24–27 June 2002; p. 168. Available online: https://scholarsarchive.byu.edu/iemssconference/2002/all/168 (accessed on 10 December 2023).
5. Prince, S.; Safriel, U. Land use and degradation in a desert margin: The Northern Negev. *Remote Sens.* **2021**, *13*, 2884. [CrossRef]
6. Kiker, G.A.; Bridges, T.S.; Varghese, A.; Seager, T.P.; Linkov, I. Application of multi criteria decision analysis in environmental decision making. *Integr. Environ. Assess. Manag.* **2009**, *1*, 95–108. [CrossRef] [PubMed]
7. Vihervaara, P.; Marjokorpi, A.; Kumpula, T.; Walls, M.; Kamppinen, M. Ecosystem services of fast-growing tree plantations: A case study on integrating social valuations with land-use changes in Uruguay. *For. Policy Econ.* **2012**, *14*, 58–68. [CrossRef]
8. Muller, R.; Turner, R. The influence of project managers on project success by type of project. *Eur. Manag. J.* **2007**, *25*, 298–309. [CrossRef]
9. Veldman, J.W.; Overbeck, G.E.; Negreiros, D.; Mahy, G.; Le Stradic, S.; Fernandes, G.W.; Durigan, G.; Buisson, E.; Putz, F.E.; Bond, W.J. Where tree planting and forest expansion are bad for biodiversity and ecosystem services. *BioScience* **2015**, *65*, 1011–1018. [CrossRef]
10. Lanza, K.; Stone, B., Jr. Climate adaptation in cities: What trees are suitable for urban heat management? *Landsc. Urban Plan.* **2016**, *153*, 74–82. [CrossRef]
11. Kourtz, P. Artificial intelligence: A new tool for forest management. *Can. J. For. Res.* **1990**, *20*, 428–437. [CrossRef]
12. Silvestro, D.; Goria, S.; Sterner, T.; Antonelli, A. Improving biodiversity protection through artificial intelligence. *Nat. Sustain.* **2022**, *5*, 415–424. [CrossRef]
13. Mason, E.G. Planning forest establishment operations with a computerised decision-support system: A case study analysis of decision-making over a full rotation. *NZJ For. Sci* **1996**, *26*, 222–234.
14. Randall, T.A.; Churchill, C.J.; Baetz, B.W. A GIS-based decision support system for neighbourhood greening. *Environ. Plan. B Plan. Des.* **2003**, *30*, 541–563. [CrossRef]
15. Kirnbauer, M.C.; Kenney, W.A.; Churchill, C.J.; Baetz, B.W. A prototype decision support system for sustainable urban tree planting programs. *Urban For. Urban Green.* **2009**, *8*, 3–19. [CrossRef]
16. McGee, J.A., III; Day, S.D.; Wynne, R.H.; White, M.B. Using geospatial tools to assess the urban tree canopy: Decision support for local governments. *J. For.* **2012**, *110*, 275–286. [CrossRef]
17. McVicar, T.R.; Li, L.; Van Niel, T.G.; Zhang, L.; Li, R.; Yang, Q.; Zhang, X.P.; Mu, X.M.; Wen, Z.M.; Liu, W.Z.; et al. Developing a decision support tool for China's re-vegetation program: Simulating regional impacts of afforestation on average annual streamflow in the Loess Plateau. *For. Ecol. Manag.* **2007**, *251*, 65–81. [CrossRef]
18. Saksa, T.; Uusitalo, J.; Lindeman, H.; Häyrynen, E.; Kulju, S.; Huuskonen, S. Decision support tool for tree species selection in forest regeneration based on harvester data. *Forests* **2021**, *12*, 1329. [CrossRef]

19. Tyndall, J. Prairie and Tree Planting Tool—PT 2 (1.0): A Conservation Decision Support Tool for Iowa, USA. *Agrofor. Syst.* **2021**, *96*, 49–64. [CrossRef]
20. Tabassum, S.; Beaumont, L.J.; Shabani, F.; Staas, L.; Griffiths, G.; Ossola, A.; Leishman, M.R. Which Plant Where: A Plant Selection Tool for Changing Urban Climates. *Arboric. Urban For.* **2023**, *49*, 190–210. [CrossRef]
21. Bentrup, G.; Dosskey, M.G. Tree Advisor: A Novel Woody Plant Selection Tool to Support Multifunctional Objectives. *Land* **2022**, *11*, 397. [CrossRef]
22. White, M.R.; Langenheim, N. A spatio-temporal decision support system for designing with street trees. In *Geospatial Intelligence: Concepts, Methodologies, Tools, and Applications*; IGI Global: Hershey, PA, USA, 2019; pp. 533–560.
23. Werbin, Z.R.; Heidari, L.; Buckley, S.; Brochu, P.; Butler, L.J.; Connolly, C.; Bloemendaal, L.H.; McCabe, T.D.; Miller, T.K.; Hutyra, L.R. A tree-planting decision support tool for urban heat mitigation. *PLoS ONE* **2020**, *15*, e0224959. [CrossRef]
24. Bonannella, C.; Hengl, T.; Heisig, J.; Parente, L.; Wright, M.N.; Herold, M.; De Bruin, S. Forest tree species distribution for Europe 2000–2020: Mapping potential and realized distributions using spatiotemporal machine learning. *PeerJ* **2022**, *10*, e13728. [CrossRef] [PubMed]
25. Gonzalez-Ollauri, A.; Mickovski, S.B. Plant-Best: A novel plant selection tool for slope protection. *Ecol. Eng.* **2017**, *106*, 154–173. [CrossRef]
26. Ellis, E.A.; Nair, P.K.R.; Linehan, P.E.; Beck, H.W.; Blanche, C.A. A GIS-based database management application for agroforestry planning and tree selection. *Comput. Electron. Agric.* **2000**, *27*, 41–55. [CrossRef]
27. Fremout, T.; Thomas, E.; Taedoumg, H.; Briers, S.; Gutiérrez-Miranda, C.E.; Alcázar-Caicedo, C.; Lindau, A.; Kpoumie, H.M.; Vincenti, B.; Kettle, C.; et al. Diversity for Restoration (D4R): Guiding the selection of tree species and seed sources for climate-resilient restoration of tropical forest landscapes. *J. Appl. Ecol.* **2022**, *59*, 664–679. [CrossRef]
28. Vogt, J.; Gillner, S.; Hofmann, M.; Tharang, A.; Dettmann, S.; Gerstenberg, T.; Schmidt, C.; Gebauer, H.; Van de Riet, K.; Berger, U.; et al. Citree: A database supporting tree selection for urban areas in temperate climate. *Landsc. Urban Plan.* **2017**, *157*, 14–25. [CrossRef]
29. Nowak, D.J.; Maco, S.; Binkley, M.J.A.C. i-Tree: Global tools to assess tree benefits and risks to improve forest management. *Arboric. Consult.* **2018**, *51*, 10–13.
30. Van der Wolf, J.; Jassogne, L.; Gram, G.I.L.; Vaast, P. Turning local knowledge on agroforestry into an online decision-support tool for tree selection in smallholder's farms. *Exp. Agric.* **2019**, *55*, 50–66. [CrossRef]
31. Wood, S.L.R.; Dupras, J. Increasing functional diversity of the urban canopy for climate resilience: Potential tradeoffs with ecosystem services? *Urban For. Urban Green.* **2021**, *58*, 126972. [CrossRef]
32. Munang, R.; Thiaw, I.; Alverson, K.; Liu, J.; Han, Z. The role of ecosystem services in climate change adaptation and disaster risk reduction. *Curr. Opin. Environ. Sustain.* **2013**, *5*, 47–52. [CrossRef]
33. Cavender-Bares, J.M.; Nelson, E.; Meireles, J.E.; Lasky, J.R.; Miteva, D.A.; Nowak, D.J.; Pearse, W.D.; Helmus, M.R.; Zanne, M.E.; Fagan, W.F.; et al. The hidden value of trees: Quantifying the ecosystem services of tree lineages and their major threats across the contiguous US. *PLOS Sustain. Transform.* **2022**, *1*, e0000010. [CrossRef]
34. Marais, Z.E.; Baker, T.P.; O'Grady, A.P.; England, J.R.; Tinch, D.; Hunt, M.A. A natural capital approach to agroforestry decision-making at the farm scale. *Forests* **2019**, *10*, 980. [CrossRef]
35. Barth, N.C.; Döll, P. Assessing the ecosystem service flood protection of a riparian forest by applying a cascade approach. *Ecosyst. Serv.* **2016**, *21*, 39–52. [CrossRef]
36. Quirogaa, V.M.; Kurea, S.; Udoa, K.; Manoa, A. Application of 2D numerical simulation for the analysis of the February 2014 Bolivian Amazonia flood: Application of the new HEC-RAS version 5. *Ribagua* **2016**, *3*, 25–33. [CrossRef]
37. Anderson, P.H.; Pezeshki, S.R. Effects of flood pre-conditioning on responses of three bottomland tree species to soil waterlogging. *J. Plant Physiol.* **2001**, *158*, 227–233. [CrossRef]
38. Kreuzwieser, J.; Fürniss, S.; Rennenberg, H. Impact of waterlogging on the N-metabolism of flood tolerant and non-tolerant tree species. *Plant Cell Environ.* **2002**, *25*, 1039–1049. [CrossRef]
39. Aboelata, A.; Sodoudi, S. Evaluating the effect of trees on UHI mitigation and reduction of energy usage in different built up areas in Cairo. *Build. Environ.* **2020**, *168*, 106490. [CrossRef]
40. Yosef, R.; Rakholia, S.; Mehta, A.; Bhatt, A.; Kumbhojkar, S. Land Surface temperature regulation ecosystem service: A case study of Jaipur, India, and the urban island of Jhalana Forest Reserve. *Forests* **2022**, *13*, 1821. [CrossRef]
41. Sahdev, S.; Yosef, R.; Rakholia, S.; Mehta, A.; Yadav, N. Bidecadal Analysis of Ecosystem Services and Urbanization Effects in the Western Himalayas: A Case Study from Haldwani, Uttarakhand. *Remote Sens. Appl. Soc. Environ.* **2023**. *accepted for publication*.
42. Rasul, A.; Balzter, H.; Smith, C. Applying a normalized ratio scale technique to assess influences of urban expansion on land surface temperature of the semi-arid city of Erbil. *Int. J. Remote Sens.* **2017**, *38*, 3960–3980. [CrossRef]
43. Johnson, D.P.; Stanforth, A.; Lulla, V.; Luber, G. Developing an applied extreme heat vulnerability index utilizing socioeconomic and environmental data. *Appl. Geogr.* **2012**, *35*, 23–31. [CrossRef]
44. Menon, J.S.; Sharma, R. Nature-based solutions for co-mitigation of air pollution and urban heat in Indian cities. *Front. Sustain. Cities* **2021**, *3*, 705185. [CrossRef]
45. Kowarik, I. The Mediterranean tree *Acer manspessulanum* invades urban green spaces in Berlin. *Dendrobiology* **2023**, *89*, 20–26. [CrossRef]

46. Kirkpatrick, J.B.; Davison, A.; Daniels, G.D. Resident attitudes towards trees influence the planting and removal of different types of trees in eastern Australian cities. *Landsc. Urban Plan.* **2012**, *107*, 147–158. [CrossRef]
47. Pukkala, T.; Kellomäki, S. Anticipatory vs adaptive optimization of stand management when tree growth and timber prices are stochastic. *For. Int. J. For. Res.* **2012**, *85*, 463–472. [CrossRef]
48. Nath, P.C.; Nath, A.J.; Sileshi, G.W.; Das, A.K. Growth and Coppicing Ability of the Critically Endangered Agarwood (*Aquilaria malaccensis* Lam.) Tree in Monoculture and Polyculture in North East India. *J. Sustain. For.* **2022**, *42*, 947–959. [CrossRef]
49. Yahya, M.S.; Syafiq, M.; Ashton-Butt, A.; Ghazali, A.; Asmah, S.; Azhar, B. Switching from monoculture to polyculture farming benefits birds in oil palm production landscapes: Evidence from mist netting data. *Ecol. Evol.* **2017**, *7*, 6314–6325. [CrossRef] [PubMed]
50. Hirsch, M.; Boddeker, H.; Albrecht, A.; Saha, S. Drought tolerance differs between urban tree species but is not affected by the intensity of traffic pollution. *Trees* **2023**, *37*, 111–131. [CrossRef]
51. Henderson, A.F.; Santoro, J.A.; Kremer, P. Impacts of spatial scale and resolution on species distribution models of American chestnut (*Castanea dentata*) in Pennsylvania, USA. *For. Ecol. Manag.* **2023**, *529*, 120741. [CrossRef]
52. Nitoslawski, S.A.; Galle, N.J.; Van Den Bosch, C.K.; Steenberg, J.W. Smarter ecosystems for smarter cities? A review of trends, technologies, and turning points for smart urban forestry. *Sustain. Cities Soc.* **2019**, *51*, 101770. [CrossRef]
53. de Lima Araújo, H.C.; Martins, F.S.; Cortese, T.T.P.; Locosselli, G.M. Artificial intelligence in urban forestry—A systematic review. *Urban For. Urban Green.* **2021**, *66*, 127410. [CrossRef]
54. Cheng, X.; Doosthosseini, A.; Kunkel, J. Improve the deep learning models in forestry based on explanations and expertise. *Front. Plant Sci.* **2022**, *13*, 902105. [CrossRef] [PubMed]
55. Altalak, M.; Ammad Uddin, M.; Alajmi, A.; Rizg, A. Smart agriculture applications using deep learning technologies: A survey. *Appl. Sci.* **2022**, *12*, 5919. [CrossRef]
56. Zeng, C.; Zhang, F.; Luo, M. A deep neural network-based decision support system for intelligent geospatial data analysis in intelligent agriculture system. *Soft Comput.* **2022**, *26*, 10813–10826. [CrossRef]
57. Araujo, S.O.; Peres, R.S.; Filipe, L.; Manta-Costa, A.; Lidon, F.; Ramalho, J.C.; Barata, J. Intelligent data-driven decision support for agricultural systems-ID3SAS. *IEEE Access* **2023**, *11*, 115798–115815. [CrossRef]
58. Rew, J.; Cho, Y.; Hwang, E. A robust prediction model for species distribution using bagging ensembles with deep neural networks. *Remote Sens.* **2021**, *13*, 1495. [CrossRef]
59. Deneu, B.; Servajean, M.; Bonnet, P.; Botella, C.; Munoz, F.; Joly, A. Convolutional neural networks improve species distribution modelling by capturing the spatial structure of the environment. *PLoS Comput. Biol.* **2021**, *17*, e1008856. [CrossRef]
60. Unawong, W.; Yaemphum, S.; Nathalang, A.; Chen, Y.; Domec, J.-C.; Torngern, P. Variations in leaf water status and drought tolerance of dominant tree species growing in multi-aged tropical forests in Thailand. *Sci. Rep.* **2022**, *12*, 6882. [CrossRef]
61. Zhang, P.; Ren, H.; Dong, X.; Wang, X.; Liu, M.; Zhang, Y.; Zhang, Y.; Huang, J.; Dong, S.; Xiao, R. Understanding and Applications of Tensors in Ecosystem Services: A Case Study of the Manas River Basin. *Land* **2023**, *12*, 454. [CrossRef]
62. Ciaburro, G.; Venkateswaran, B. *Neural Networks with R: Smart Models Using CNN, RNN, Deep Learning, and Artificial Intelligence Principles*; Packt Publishing Ltd.: Birmingham, UK, 2017.
63. Zaccone, G.; Karim, M.R. *Deep Learning with TensorFlow: Explore Neural Networks and Build Intelligent Systems with Python*; Packt Publishing Ltd.: Birmingham, UK, 2018.
64. Vasilev, I.; Slater, D.; Spacagna, G.; Roelants, P.; Zocca, V. *Python Deep Learning: Exploring Deep Learning Techniques and Neural Network Architectures with Pytorch, Keras, and TensorFlow*; Packt Publishing Ltd.: Birmingham, UK, 2019.
65. Rothman, D. *Transformers for Natural Language Processing: Build Innovative Deep Neural Network Architectures for NLP with Python, PyTorch, TensorFlow, BERT, RoBERTa, and More*; Packt Publishing Ltd.: Birmingham, UK, 2021.
66. Chollet, F. *Deep Learning with Python*; Simon and Schuster: New York, NY, USA, 2021.
67. Singh, P.; Manure, A. *Learn TensorFlow 2.0: Implement Machine Learning and Deep Learning Models with Python*; Apress: New York, NY, USA, 2019.
68. Pally, R.J.; Samadi, S. Application of image processing and convolutional neural networks for flood image classification and semantic segmentation. *Environ. Model. Softw.* **2022**, *148*, 105285. [CrossRef]
69. Truong, T.X.; Nhu, V.H.; Phuong, D.T.N.; Nghi, L.T.; Hung, N.N.; Hoa, P.V.; Bui, D.T. A New Approach Based on TensorFlow Deep Neural Networks with ADAM Optimizer and GIS for Spatial Prediction of Forest Fire Danger in Tropical Areas. *Remote Sens.* **2023**, *15*, 3458. [CrossRef]
70. Bhullar, A.; Nadeem, K.; Ali, R.A. Simultaneous multi-crop land suitability prediction from remote sensing data using semi-supervised learning. *Sci. Rep.* **2023**, *13*, 6823. [CrossRef]
71. Osah, S.; Acheampong, A.A.; Fosu, C.; Dadzie, I. Deep learning model for predicting daily IGS zenith tropospheric delays in West Africa using TensorFlow and Keras. *Adv. Space Res.* **2021**, *68*, 1243–1262. [CrossRef]

Disclaimer/Publisher's Note: The statements, opinions and data contained in all publications are solely those of the individual author(s) and contributor(s) and not of MDPI and/or the editor(s). MDPI and/or the editor(s) disclaim responsibility for any injury to people or property resulting from any ideas, methods, instructions or products referred to in the content.

Article

Identifying the Optimal Area Threshold of Mapping Units for Cultural Ecosystem Services in a River Basin

Ye Li [1], Junda Huang [1] and Yuncai Wang [1,2,*]

[1] Department of Landscape Architecture, College of Architecture and Urban Planning, Tongji University, Shanghai 200092, China; ly0201@tongji.edu.cn (Y.L.); 23310120@tongji.edu.cn (J.H.)
[2] Joint Laboratory of Ecological Urban Design (Research Centre for Land Ecological Planning, Design and Environmental Effects, International Joint Research Centre of Urban-Rural Ecological Planning and Design), College of Architecture and Urban Planning, Tongji University, Shanghai 200092, China
* Correspondence: wyc1967@tongji.edu.cn

Abstract: Mapping cultural ecosystem services (CES) in river basins is crucial for spatially identifying areas that merit conservation due to their significant CES contributions. However, precise quantification of the appropriate area of mapping units, which is the basis for CES assessment, is rare in existing studies. In this study, the optimal area threshold of mapping units (OATMU) identification, consisting of a multi-dimensional indicator framework and a methodology for validation, was established to clarify the boundary and the appropriate area of the mapping units for CES. The multi-dimensional indicator framework included geo-hydrological indicator (GI), economic indicator (EI) and social management indicator (SMI). The OATMU for each indicator was determined by seeking the inflection point in the second-order derivative of the power function. The minimum value of the OATMU for each indicator was obtained as the OATMU for CES. Finally, the OATMU for CES was validated by comparing it with the area of administrative villages in the river basin. The results showed the OATMU for CES was 3.60 km^2. This study adopted OATMU identification, with easy access to basic data and simplified calculation methods, to provide clear and generic technical support for optimizing CES mapping.

Keywords: watershed landscapes; mapping units; CES assessment; minimal unit; multi-dimensional indicators

1. Introduction

The benefits provided by ecosystems and appropriated by humans are called Ecosystem Services (ES) [1]. According to the Millennium Ecosystem Assessment, ESs are classified into four categories: provisioning, regulating, supporting, and cultural services, each contributing uniquely to various aspects of human well-being [2]. Cultural ecosystem services (CES) are the non-material benefits obtained from ecosystems and are classified into 10 main subcategories: recreation and tourism, aesthetics, spiritual values, education, inspirational values, cultural diversity, knowledge systems, sense of place and identity, social relations, and cultural heritage values [2]. CES are fundamentally shaped and sustained by ecological, economic, and social factors [2–4]. However, the rapid pace of urbanization poses significant challenges, leading to ecosystem fragmentation and functional degradation, which in turn diminishes the CES supply capacity [5]. As a key to society, CES must be more thoroughly acknowledged as crucial for enhancing human health and demand increased attention from policymakers [6].

Through a long history of human interaction with nature, river basins have become densely populated human settlements and indispensable birthplaces of civilization [7]. The role of river basins in providing CES—such as cultural heritage, recreation and ecotourism, and aesthetic appreciation—is widely recognized [8]. These areas are considered the foundational units for ecosystem conservation planning and are deemed appropriate scales

Citation: Li, Y.; Huang, J.; Wang, Y. Identifying the Optimal Area Threshold of Mapping Units for Cultural Ecosystem Services in a River Basin. *Land* **2024**, *13*, 346. https://doi.org/10.3390/land13030346

Academic Editors: Alessio Russo and Giuseppe T. Cirella

Received: 24 January 2024
Revised: 4 March 2024
Accepted: 6 March 2024
Published: 8 March 2024

Copyright: © 2024 by the authors. Licensee MDPI, Basel, Switzerland. This article is an open access article distributed under the terms and conditions of the Creative Commons Attribution (CC BY) license (https://creativecommons.org/licenses/by/4.0/).

for the management of CES [9]. CES mapping in river basins is primarily used to concretely visualize the value of an area, allowing for an intuitive analysis of whether the supply and demand of CES are balanced [9]. Areas with a high CES supply, often characterized by abundant recreational activities or cultural values, contribute significantly to human health and are therefore critical to identify and prioritize for enhanced conservation efforts [6]. Given the relatively high value of CES attributed to aquatic areas [10], it is of paramount importance to conduct CES mapping studies in river basins.

However, without a clear understanding of the mapping units, the protection and development of CES will be hampered [11]. To divide these units, Campos-Campos et al. [12] analyzed variations in landscape units through the Supervised Classification method, while Alvioli et al. [13] introduced an automated approach utilizing the r.slopeunits v1.0 software for delineation. Currently, the mapping units for CES in river basins mainly consist of existing grid cells [14], land-use units [15], homogeneous landscape units [16], and administrative units [17]. In addition, different scales of catchments are applied to measure CES supply and demand in river basins [18,19]. Recognized as units that encapsulate ecological, social, economic, and cultural dimensions [11], catchments are extensively applied as mapping units in the fields of Hydrology and Geography [13,20,21]. There is a well-established and systematic method for delineating catchment boundaries using the ArcGIS 10.8 software hydrological analysis tool [21,22]. With clear boundaries that match the natural geographic extent of river basins, catchments avoid the problem of incompatibility between the boundaries of the mapping units and the study area.

The selection of mapping units profoundly affects the conclusions drawn from the CES assessment because the area of these units determines the credibility of the spatial images [11]. However, the determination of the appropriate area of mapping units has been largely neglected [23], leading to inadequate representation of the study area's characteristics and compromised mapping accuracy [24]. Particularly when the precision of mapping units falls below that of the socio-practical system, there is a notable deficiency in motivation or comprehensive guidelines for their application [11]. Therefore, with the establishment of appropriate units, CES can be accurately quantified [25]. Several studies in environmental hydraulics have used multiple catchment area thresholds for selecting the optimal area thresholds to obtain relatively homogeneous and refined mapping units [20,22,26]. To enhance mapping precision alongside computational efficiency, this study introduced the concept of area threshold, offering both a theoretical framework and a practical approach for accurately determining the appropriate area of mapping units. The area threshold of mapping units (ATMU) is defined as the minimum area of mapping units. The optimal area threshold of mapping units (OATMU) is defined as the ATMU that most effectively fulfills the objectives of the study, thereby enhancing the precision of the outcomes and accurately reflecting the characteristics of the study area [22]. The value of OATMU, which determines the most appropriate area of mapping units, varies according to the research objectives and is highly related to the indicator selection. For instance, in the pursuit of accurately delineating actual river networks and catchment boundaries, drainage density serves as a key indicator for determining the OATMU [22].

This study mainly focused on (1) clarifying the boundary and the most appropriate area of the mapping units for CES; (2) determining the indicators affecting the area of the mapping units and methods of their calculation; and (3) testing the applicability of OATMU identification methods for CES in river basins. As a cornerstone in the research on CES spatial mapping, this study has constructed a novel, streamlined, and transferable method that leverages available spatial data to identify the optimal area for mapping units with high precision. By tackling the earlier challenges of subjective selection of mapping unit types and the absence of quantitative assessments of their areas, this study significantly advanced the precise application of CES quantification and supply demand research outcomes in real-world spaces.

2. Materials and Methods

2.1. Study Area

The Qiantang River Basin is located in southeastern China, covering an area of 42,770.05 km^2 (116°52′ to 120°56′ E, 27°56′ to 30°59′ N) (Figure 1). The whole river, with a length of 668.10 km, is the longest river in Zhejiang Province. As of the end of 2020, the basin boasted a population of 12,667,200 and a GDP of approximately 1.65 trillion dollars [27]. As one of the top ten practices of Nature-Based Solutions jointly published by the Chinese Ministry of Natural Resources and the World Conservation Union (IUCN), the Qiantang River Basin serves as a pilot location for developing and demonstrating a replicable and scalable model of key ecological functional zones across the nation [27].

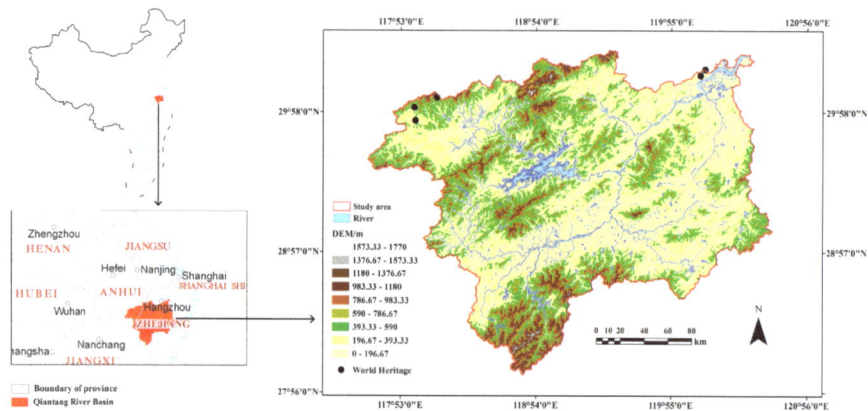

Figure 1. The location and elevation of study area.

There are 3 reasons for choosing the Qiantang River Basin as the study area: (1) Diverse Geomorphology for CES Spatial Analysis. The Qiantang River Basin is a representative region featuring a wide range of catchment geomorphic types, including mountains, basins, and plains [28]. This diversity in topographic features across different catchments yields a variety of CES [8]. (2) Rich and Emblematic Cultural Heritage. Serving as a cradle of ancient Chinese civilization, the basin is closely linked with four World Heritage sites: Huangshan Mountain, West Lake, the Grand Canal, and ancient villages in southern Anhui province, such as Xidi and Hongcun. It is also home to the renowned Liangzhu Culture (5300–4200 cal. a BP) and Hui Culture (1121–1911), making it a region of significant cultural value [28]. (3) Challenges Posed by Rapid Urbanization. As one of the most densely populated and economically advanced areas in China, the Qiantang River Basin faces threats from fast-paced urbanization, which leads to the encroachment and fragmentation of CES areas. Notably, the construction of hydroelectric power plants has contributed to the loss of numerous historical sites. This underscores the critical need for a systematic and accurate approach to CES space optimization within the basin.

2.2. Research Framework

As a foundational study of CES mapping, the OATMU identification method was constructed to identify the boundary and the most appropriate area of the mapping units. Distinct from other ESs, the assessment of CES leans heavily on human perceptions, cultural values, and demands [29]. Consequently, incorporating indicators that reflect economic and social dimensions is crucial. Based on ecological, economic and social aspects, a multi-dimensional indicator framework with specific indicator calculation methods and the validation of the final results were established in this study (Figure 2).

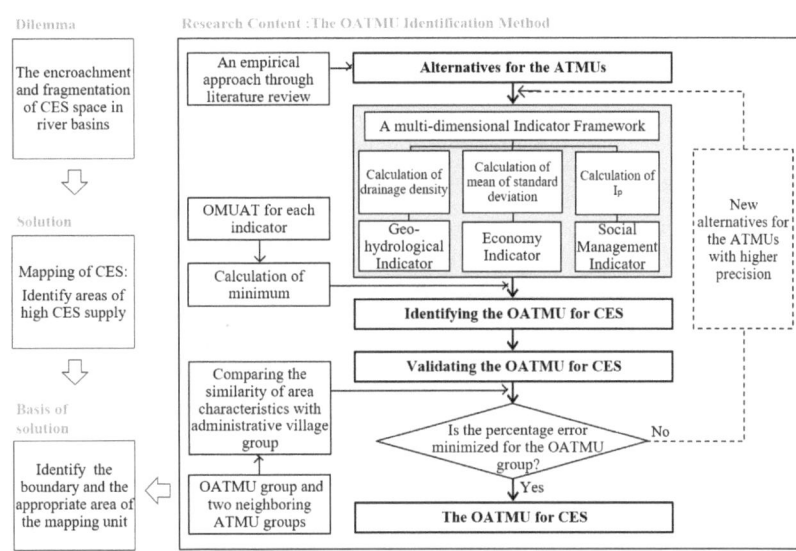

Figure 2. Research framework.

2.3. Data Sources and Preprocessing

Five data sets were used to calculate the OATMU: the digital elevation model (DEM), China population spatial distribution data, China GDP spatial distribution data, China administrative boundary data and the point of interest (POI) (Table 1).

Table 1. Data sources.

Data	Time	Scale/Format/Resolution	Source
Digital elevation Model (DEM)	2022	The Qiantang River Basin, grid, 30 m	Geospatial Data Cloud (https://www.gscloud.cn/, accessed on 26 May 2023)
China population spatial distribution data	2019	China, gird, 1 km	Resource and Environmental Science and Data Center of the Chinese Academy of Sciences (https://www.resdc.cn/, accessed on 29 May 2023)
China GDP spatial distribution data	2019	China, gird, 1 km	Resource and Environmental Science and Data Center of the Chinese Academy of Sciences (https://www.resdc.cn/, accessed on 29 May 2023)
China administrative boundary data	2021	The Qiantang River Basin, shape, shape	Resource and Environmental Science and Data Center of the Chinese Academy of Sciences (https://www.resdc.cn/, accessed on 5 May 2023)
Point of interest (POI)	2022	The Qiantang River Basin, point, point	Amap Company (https://www.amap.com/, accessed on 26 May 2023)

Among these, the China population spatial distribution data and China GDP spatial distribution data have not been updated in recent years; thus, the most recent data from 2019 were used. At Amap Company, one of China's most popular mapping services, POI data were categorized into 23 types using classification codes. For this study, POI data for government organizations and social groups within the study area were collected, amounting to 57,644 records.

The projection coordinate system for the aforementioned data was converted to WGS 1984 UTM Zone 50N. Spatial data processing was conducted using ArcGIS 10.8.

2.4. Alternatives for ATMUs

The ATMUs are typically determined by the number of grids with different DEM accuracies [22] or by the limitations of available computing memory [30]. This study established the ATMUs based on the following criteria: (1) Integration of ecological and social research findings. For social ecosystems, the average area of mapping units is 3.50 km^2 [23]. Consistent with studies of river basins of comparable size, the OATMUs range from 1.41 to 7.20 km^2 at the ecological and social levels. (2) Accuracy Assurance. Available studies offer a range of 6 to 15 alternatives (Table 2). Finer ATMUs and a greater number of alternatives lead to more accurate results [22,31]. To ensure precision, this study adopts the maximum number of alternatives for calculations. (3) Maximizing the OATMU range of values. The upper limit was increased to the maximum reported value of 40.5 km^2 in existing studies (Table 2). (4) Grid Multiplicity [22]. The ATMU is an integer multiple of the number of grids. In this study, with a DEM data accuracy of 30 m and each grid area being 30 m × 30 m, the ATMU is an integer multiple of 0.0009 km^2.

Table 2. Research results on the optimal catchment area threshold.

Study Area (km^2)	Range of Alternative (km^2)	Number of Alternatives	Optimal Catchment Area Threshold (km^2)	Identification Method
1350	0.45~40.5	15	7.2	Drainage density method [32]
772.6	0.0081~6.4800	10	4.05	Box dimension method [22]
Multi-basins	0.5~10	6	5	Coefficient of line correspondence [33]
1700.61	0.9~18	7	7.2	Drainage density method [34]
95,400	0.078~3.125	7	1.5625	Box-Counting Method [26]
95,400	0.1562~3.90625	13	1.40625	Multifractal Method [26]

In summary, following the research of Zhang [32], this study selected 15 alternatives with 500, 1000, 1500, 2000, 4000, 6000, 8000, 10,000, 15,000, 20,000, 25,000, 30,000, 35,000, 40,000, and 45,000 grids. The calculation formula for each alternative is as follows:

$$ATMU_j = A_{grid} \times Q_j (1 \leq j \leq 15) \quad (1)$$

where $ATMU_j$ is the ATMU of jth alternative, A_{grid} is the area of grid, Q_j is the number of grids of jth alternatives.

Consequently, the 15 alternatives of ATMUs were determined to be 0.45 km^2, 0.90 km^2, 1.35 km^2, 1.80 km^2, 3.60 km^2, 5.40 km^2, 7.20 km^2, 9.00 km^2, 13.50 km^2, 18.00 km^2, 22.50 km^2, 27.00 km^2, 31.50 km^2, 36.00 km^2, 40.50 km^2.

Utilizing the four-step process of terrain preprocessing, flow direction identification, ATMU setting, and catchment extraction in the Hydrological Analysis module of ArcGIS 10.8 software [22], 15 catchment alternatives were generated. Patches smaller than the ATMU were automatically eliminated to optimize the ATMU alternatives.

2.5. Indicator Framework

The following criteria were utilized to select the OATMU for CES: (1) Covering ecological, economic and social aspects [3,4]; (2) Data availability, accessibility and reliability; (3) Representativeness of indicators for core characteristics. In summary, 3 indicators were selected to identify the OATMU for CES: the geo-hydrological indicator (GI), economy indicator (EI), and the social management indicator (SMI).

2.5.1. Geo-Hydrological Indicator

The drainage density was used as a proxy for the geomorphologic and hydrologic characteristics of river basins [35]. The drainage density, which shows a strong correlation with surface roughness, vegetation index, and water storage changes, is better modeled and more generalizable [36]. Hence, the drainage density was selected to characterize GI. Its calculation formula is as follows:

$$D_d = \frac{\sum_{i=1}^{n} Li}{\sum_{i=1}^{n} Ai} \tag{2}$$

where D_d is the drainage density, i is the ith mapping unit, n is the maximum number of mapping units, Li is the river length in the ith mapping unit, and Ai is the area of the ith mapping unit.

2.5.2. Economy Indicator

GDP per capita is frequently utilized as a significant measurement of economic standards [37]. The level of economic development is similar in each mapping unit; therefore, the economic homogeneity within the mapping unit can be reflected by the degree of dispersion of GDP per capita.

The standard deviation is often utilized to depict the extent of data dispersion or aggregation uniformity in space [38]. Hence, the mean value of the standard deviation of GDP per capita was selected to represent EI. The formula is as follows:

$$MSD = \frac{\sum_{i=1}^{n} S_i}{n} \tag{3}$$

where MSD is the mean standard deviation of GDP per capita, i is the ith mapping unit, n is the maximum number of mapping units, S_i is the standard deviation of GDP per capita in the ith mapping unit.

2.5.3. Social Management Indicator

As one of the 3 dimensions of CES, social aspect includes a societal or shared interpretation at stake, as in social process, social scale, social problem, etc. [4]. Social management is crucial in the social aspect, as it often determines the specific preferences for management programs and the implementation of management decisions. Government organizations and social groups are the mainstays of social management. The greater the concentration of government organizations and social groups, the more likely it is to be a complete spatial unit with control, management, supervision and service functions [39]. Traditional methods such as the kernel density method use POI as a single data source [40], which failed to combine the mapping unit variables to reflect the differences among alternatives. The Index of Patchiness (I_p) is used in population ecology to measure the intensity of population aggregation by calculating patches occupied by the number of individuals [41]. Due to the need to clarify the spatial extent of study individuals and distribution patches, this study selected I_p of POI for government organizations and social groups to illustrate SMI. The formula is as follows:

$$I_p = \frac{I_c}{P} \tag{4}$$

where I_c is the index of mean crowding, P is the average number of POI for government organizations and social groups within each mapping unit. Where P is as follows:

$$P = \frac{\sum_{i=1}^{n} P_i}{n} \tag{5}$$

where i is the ith mapping unit, n is the maximum number of mapping units, P_i is the number of POI for government organizations and social groups in the ith mapping unit. I_c is as follows:

$$I_c = p + (\frac{s^2}{p} - 1) \tag{6}$$

where S^2 is the variance of the number of POI for government organizations and social groups in the mapping units from i to n.

2.6. Identification of the OATMU

The OATMU is typically determined by the inflection points on the relationship curve that illustrates the correlation between the indicator and ATMUs [42]. This method is not limited to calculating optimal area thresholds corresponding to drainage density but is also widely used to calculate optimal area thresholds for slope [43], socio-economic [44], and so on. Second-order derivative analysis is commonly used as a research method for extracting inflection points from relationship curves [45], with the advantages of computational simplicity and ascertainable results. The inflection point in this method is where the second derivative approaches 0 and no longer changes afterwards [46]. In this study, the second-order derivative of the fitting function curve was chosen to detect particular inflection points.

Through the Matlab R2020a software, the relationship between the 3 indicators and ATMUs was curve-fitted using the power function [47], the second-order derivatives of the power function were obtained, and the inflection point was identified [46]. The ATMU corresponding to the inflection point is the OATMU for each indicator.

When there are different OATMUs for each indicator, it is generally believed that the smallest OATMU is more conducive to fine-scale surface research [30,31]. To prevent the loss of spatial information during calculation, this study selected the minimum OATMU value from the 3 indicators as the OATMU for CES in river basins. The formula is as follows:

$$O = min\{O_{GI}, O_{EI}, O_{SMI}\} \tag{7}$$

where O_{GI}, O_{EI}, O_{SMI} is the OATMU corresponding to the inflection points of the GI, EI, and SMI, O is the OATMU for CES in the basin.

2.7. Validation of the OATMU for CES

As the practical units of a basin ecosystem, catchments do not coincide with the boundaries of administrative units [18], but often have strong correlations with them [48]. Administrative villages are the smallest administrative units and also have relatively complete social-ecological systems, often appearing as the basic units of cultural landscapes [49]. In this study, the validation of the OATMU for CES was carried out by measuring the similarity of area data between the mapping units and administrative villages. The mapping unit dataset with the OATMU for CES was defined as the OATMU group; the mapping unit dataset with the left ATMUs adjacent to the OATMU was defined left-OATMU group; the mapping unit dataset with the right ATMU adjacent to the OATMU was defined as the right-OATMU group; and the administrative village dataset within the Qiantang River Basin was defined as the administrative village group. By comparing the mean area, area quartile distance, and area maximum interval of the 3 ATMU groups and the administrative village group [26], the group with the minimum percent error of all three is the OATMU group for CES, and the corresponding ATMU is the OATMU for CES in the river basin. Validation was conducted using descriptive statistics and box plots in IBM SPSS Statistics 26 software.

3. Results

3.1. Characteristics of the Indicators

Fifteen alternatives for ATMUs were obtained through extraction and optimization. The mean area of mapping units for each alternative was concentrated between 1.05 and 92.62 km² (Figure 3).

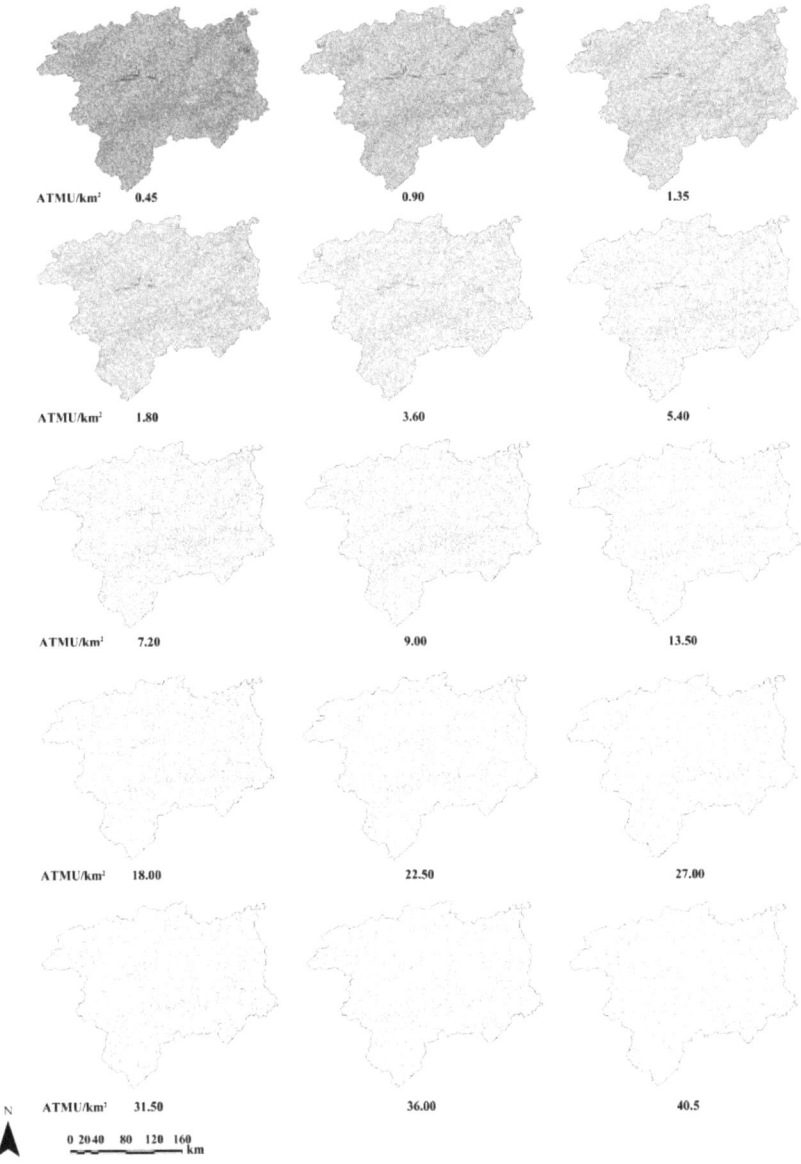

Figure 3. Fifteen alternatives for ATMUs in the Qiantang River Basin.

For GI, among the 15 alternatives, the minimum area of mapping units was 0.45 km^2, and the maximum area was 421.52 km^2 (Figure 4A). The average river length of 15 alternatives was concentrated at 0.45~3.93 km, with the shortest length of 0.03 km and the maximum length of 51.39 km (Figure 4B). For EI, GDP per capita varied considerably within each alternative, resulting in more extreme values. The minimum value of the standard deviation of GDP per capita within each mapping unit was 0, and the maximum value was 16.00 (Figure 4C). For SMI, the minimum number of POIs for government organizations and social groups in each mapping unit was 0, with the maximum reaching 3471 (Figure 4D).

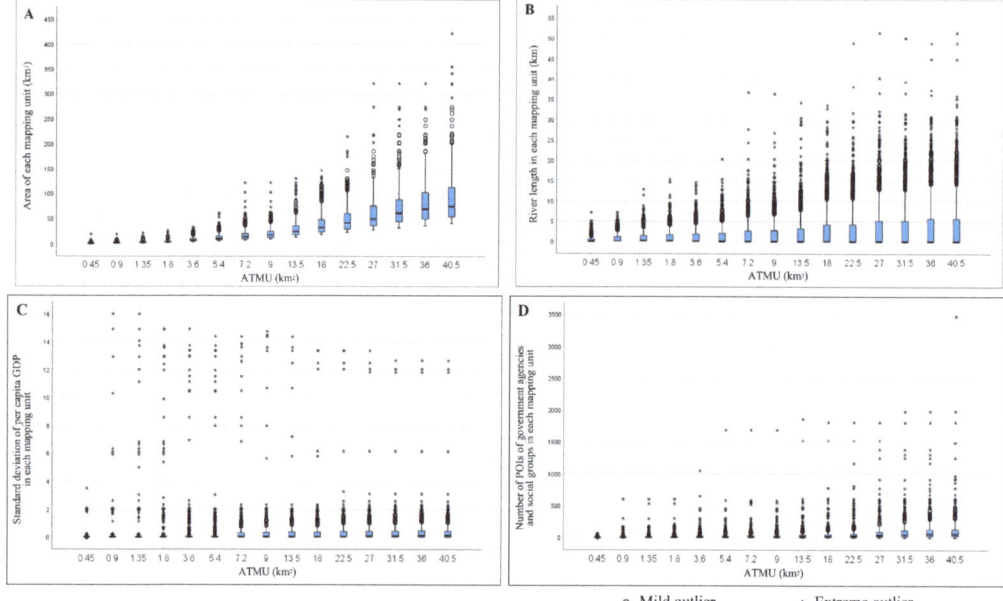

Figure 4. (**A**) Area of each mapping unit of 15 alternatives for ATMUs; (**B**) River length in each mapping unit of 15 alternatives for ATMUs; (**C**) Standard deviation of per capita GDP in each mapping unit of 15 alternatives for ATMUs; (**D**) Number of POIs for government organizations and social groups in each mapping unit of 15 alternatives for ATMUs.

3.2. Evaluation of the OATMU for CES

(1) The OATMU for GI

EI was used as the vertical axis, and ATMU was used as the horizontal axis. The power function used for curve fitting was $y = 0.8058x^{-0.4586}$, with a goodness-of-fit $R^2 = 1$ (Figure 5A). The second derivative of power function fitting for RE was $y = 0.539010868968x^{-2.4586}$ (Figure 5B). The curve image plotted by Matlab showed the ATMU corresponding to the inflection point was 3.60 km². For GI, the OATMU in the Qiantang River Basin was 3.60 km².

(2) The OATMU for EI

EI was used as the vertical axis, and ATMU was used as the horizontal axis. The power function used for curve fitting was $y = 0.07288x^{0.4725}$, with a goodness-of-fit $R^2 = 0.991$ (Figure 5C). The second derivative of power function fitting for SEI was $y = -0.0181648845x^{-1.5275}$ (Figure 5D), and the ATMU corresponding to the inflection point was 9.00 km². For EI, the OATMU in the Qiantang River Basin was 9.00 km².

(3) The OATMU for SMI

SMI was used as the vertical axis, and ATMU was used as the horizontal axis. The power function used for curve fitting was $y = 13.56x^{-0.3133}$, with a goodness-of-fit $R^2 = 0.9888$ (Figure 5E). The second derivative of power function fitting for SMI was $y = 5.5793554284x^{-2.3133}$ (Figure 5F), and the ATMU corresponding to the inflection point was 3.60 km². For SMI, the OATMU in the Qiantang River Basin was 3.60 km².

(4) The OATMU for CES

According to Equation (7), the OATMU for CES in the Qiantang River Basin was determined to be 3.60 km². When the ATMU was 3.60 km², the GI was 0.4592, the EI was 0.1383 and the SMI was 6.6596. There were 4910 mapping units in the OATMU dataset, of which the area of mapping units smaller than 6.37 km² accounted for 46.74% of the total. The mapping units between 6.38 km² and 10.11 km² accounted for 29.04%. The mapping units between 10.12 km² and 15.27 km² accounted for 15.58% (Figure 6).

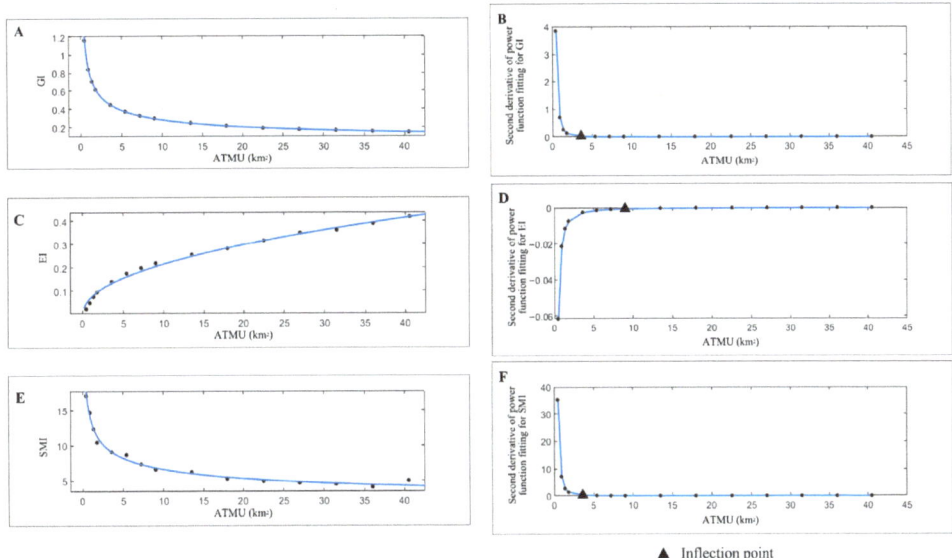

Figure 5. (**A**) Fitting power function for GI; (**B**) Second derivative of power function fitting for GI; (**C**) Fitting power function for EI; (**D**) Second derivative of power function fitting for EI; (**E**) Fitting power function for SMI; (**F**) Second derivative of power function fitting for SMI.

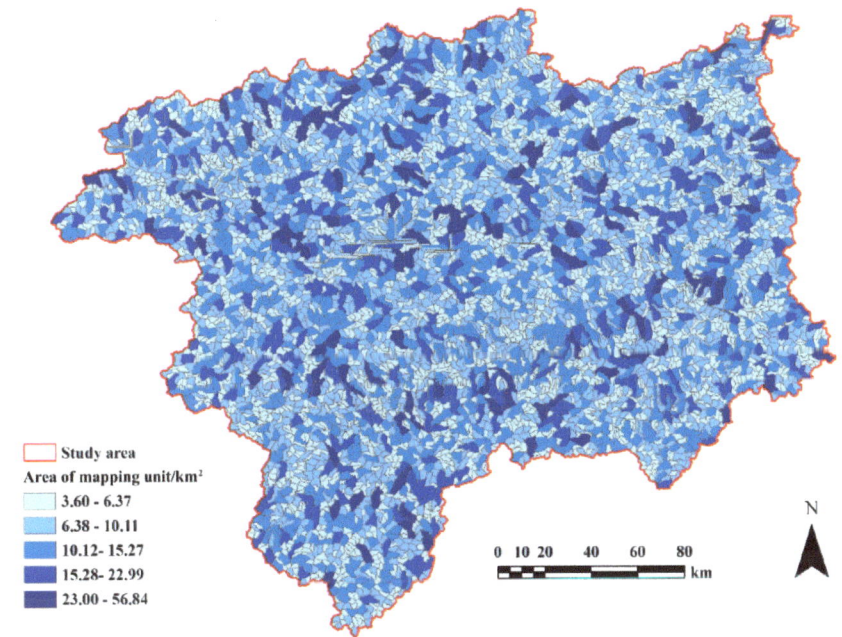

Figure 6. The dataset of OATMU for CES in the Qiantang River Basin.

3.3. Validation of ATMUs Groups and Administrative Village Group

The OATMU group (3.60 km^2), the left-OATMU group (1.80 km^2), and the right-OATMU group (5.40 km^2) were selected for validation with the administrative village group. The mean area of the administrative village group was 6.29 km^2, with an interquartile range of 4.68 km². The mean area of the left-OATMU group was 4.19 km² with an error of 33.39% from the mean area of the administrative village group and an interquartile range of 2.64 km² with an error of 43.59%. The mean area of the OATMU group was 8.23 km² with an error of 30.84%, and the interquartile range was 5.16 km² with an error of 10.26%. The mean area of the right OATMU group was 12.63 km² with an error of 100.79% and the interquartile range was 8.19 km² with an error of 75%. Compared to the left-OATMU and the right-OATMU groups, the boxplots for the administrative village group and the OATMU group were the most similar, with both sets of boxes clustered in the 0–10 km², with the lower edges clustered in the 0–10 km² and the upper edges clustered in the 10–20 km² (Table 3).

Table 3. Validation of the OATMU for CES in the Qiantang River Basin.

	Administrative Village Group	Left-OATMU Group	OATMU Group	Right-OATMU Group
Mean area	6.29	4.19	8.23	12.63
Range	63.42	24.84	53.24	62.72
Minimum	0.34	1.80	3.60	5.40
Maximum	63.75	26.64	56.84	68.12
Lower quartile	2.25	2.42	4.79	7.28
Median	4.13	3.41	6.66	10.14
Upper quartile	6.93	5.06	9.95	15.47
Boxplot				Area (km²)

The median area of the left-OATMU group was closer to that of the administrative village group. However, the range of the maximum value was 24.84 km², which was significantly smaller than that of the administrative village group (63.42 km²). There is a risk that a large number of basic cultural landscape units could be divided excessively resulting in data redundancy. The minimum and mean values of the area for the right-OATMU group were much larger than the values of the administrative village group. For the right-OATMU group, the area of the mapping units was too large, resulting in decreased accuracy of the results. The percentage error of the mean area and quartile distance of the OATMU group was the smallest. In addition, the range and degree of data dispersion were most similar to those of the administrative village group, and it avoided the data redundancy caused by excessive delineation of basic cultural landscape units. Therefore, the OATMU for CES in the Qiantang River Basin was determined to be 3.60 km².

4. Discussion

4.1. Applicability of the Multi-Dimensional Indicators and Calculation Methods

This study systematically identified the OATMU for CES by constructing a multi-dimensional indicator framework that includes GI, EI, and SMI. They were derived from Zhou et al., MEA, Scholte et al., Barnes and Hamylton, and Heasley et al. [2,3,36,41,50]. In contrast to previous studies that constructed a GI framework for river basin management [51], this study integrated EI and social SMI, adding two measurement dimensions (Figure 4) that yielded more systematic results.

For calculating GI, the drainage density method proves to be more effective for simulating areas with large topographic relief compared to the fitness index and fractal theory methods [52]. The difference in elevation in the Qiantang River basin is 1770 m (Figure 2). Combined with the actual situation of the study area, it is more reasonable to choose the drainage density method. Through the drainage density method, Zheng et al. [53] found that many parallel river networks were formed in areas with little change in elevation. This aligns with the results of the four alternatives for ATMUs, which take values between 0.45 km² and 3.60 km² (Figure 3). Additionally, as the ATMU becomes larger, the slope network chain is gradually eliminated, revealing the main rivers [53]. This is similar to the findings of this study (Figure 4B). The study areas have different locations and cover a wide range of scales (Table 2), which demonstrates the universality of the drainage density method for calculating GI.

For EI, the variability of GDP per capita between mapping units can be calculated by absolute differences, including standard deviations [50]. In this study, the mean value of the standard deviation of GDP per capita was used to reflect the differences for the EI of the mapping units, which facilitates cross-sectional comparisons of alternatives when there is only one value for each ATMU. In the absence of finer data sources, the standard deviation of GDP per capita within many mapping units is 0 for ATMUs smaller than or equal to 9 km² (Figure 4C), which is consistent with the study of Liang et al. [54]. The large degree of data dispersion (Figure 4C) indicated that the economic distribution of the Qiantang River Basin is not uniform and that there are large differences in economic development between different mapping units, which is consistent with the study of Zhou et al. and Wang et al. [55,56].

For SMI, I_p was introduced to explore the possibility that the clustering pattern of POIs varies with the spatial scale of the mapping unit. The distribution of POI for government organizations and social groups, whose primary function is management and accessibility, is geospatially relevant. The 15 ATMUs from 0.45–40.5 km² all have a number of POIs within a single mapping unit of 0 (Figure 4D), due to the fact that those mapping units are located in large bodies of water or in mountainous areas that are not easily accessible (Figure 2). The presence of large ecological reserves with geomorphological diversity limits the spatial distribution of POIs, which is consistent with Zhen et al. [57]. The maximum value of the number of POIs within a single mapping unit reaches 3471, and the minimum value is 0 (Figure 4D). Such a huge difference in the extreme value of the number for POI reflects the imbalance of the social management situation in the Qiantang River Basin, which is in line with the study of Li [58].

4.2. Identification and Comparison of the OATMUs for Each Indicator

Based on the conclusion from the existing literature, this study applied the method that accurately identified the OATMU by fitting the second-order derivative of the power function [46]. Some studies exacted the OATMU by calculating the rate of reduction for the first-order derivative of the power function relationship [53] or the rate of change for the relation curve [26]. However, in the fitted curves for the three indicators of GI, EI, and SMI (Figure 5A,C,E), the rate of change for the relationship curves has been decreasing with the increase in ATMUs, and it becomes challenging to determine the corresponding OATMUs from the appropriate rate of change. Tang et al. used Matlab to calculate the OATMU by interpolating the data with a one-in-ten-thousand threshold [59]. However, since the area ratio of the study area to the mapping unit in this study is sufficiently large, refining the accuracy to 0.0009 km² would result in a large amount of redundant computational data. Considering precision and ease of operation, this method, which was identified by fitting the second-order derivative of the power function, was used in this study for calculation.

In the Qiantang River Basin, the OATMU for GI was 3.60 km², which was similar to the studies of Wu et al. [22] and Olsen et al. [33]. Wu et al. calculated very close results using three DEM datasets to evaluate the reasonableness of OATMU [22]. An ATMU of 3.60 km² represents the first interval change from 0.45 km² to 0.90 km². The rule by which the OATMU is more likely to be identified falls between the first interval change of ATMU, consistent with the results of Zhang et al. [46]. Wu et al. [22] found three phases of the relationship curve with a rapid decline, a flat fluctuation and a convergence to a fixed value, which is consistent with the findings of this study (Figure 5A). The OATMU of GI and SMI were consistent, suggesting that the catchments naturally generated by the geographic environment are highly similar to the mapping units with social management functions, which was consistent with the study of Martín-López et al. [48]. The OATMU for EI was 9.00 km^2, which was larger than the value of the other indicators. This is due to the fact that EI is usually plotted for larger mapping units, such as town administrative units [54]. Furthermore, as the last ATMU before the interval became larger, 10,000 grids corresponding to 9.00 km^2 were more likely to be an OATMU. This pattern is also consistent with the results of Zhang et al. [26].

4.3. Limitations

Despite the important contributions of this study, it also has limitations and uncertainties. First, due to the lack of finer data sources for the GDP data in the Qiantang River Basin, the OATMU for EI in the Qiantang River Basin did not have much influence on the final OATMU for CES. However, if the scale of the study area is larger or if there is a finer GDP per capita data source, the EI will affect the calculation process and the value of the final results. Second, differences in factors, such as the ratio of the relationship curves to the axes, may lead to errors in the inflection points of the curve images plotted by Matlab. To address this issue, this study selected the two neighboring ATMUs to validate the OATMU for CES. CES is comprehensively influenced by multiple factors [21], and there are trade-off or synergy relationships between the indicators [60]. Therefore, the small number of corresponding indicators may also affect the accuracy of the final OATMU identification, despite the fact that the study added two key indicators in both dimensions.

In addition, the alternatives for ATMUs were set based on the literature, which may affect their accuracy. In the relevant literature, the area ratio of the study area to the mapping unit ranges from 10^2:1 to 10^3:1 [22,26]. In this study, the ratio was raised to 10^4:1, which split the mapping units sufficiently to ensure accuracy. Even if there were more appropriate area thresholds within each alternative, the errors would likely have minimal impact on the results. Due to the need for more precise segmentation and the large amount of data involved in the subdivision of catchments, Python and Matlab can be utilized to assist with computation in the future.

5. Conclusions

Based on the OATMU identification method, this study identified the OATMU for CES in the Qiantang River Basin from a multi-dimensional perspective. The method included four main steps: finalizing alternatives for ATMUs, establishing a multi-dimensional indicator framework for identifying the OATMU for CES, calculating the OATMU for each indicator, and calculating and validating the OATMU for CES.

Mapping units were appropriately divided according to the CES characteristics of the study area, avoiding division of the basic cultural landscape units or computational redundancy. Fifteen alternatives of ATMUs were established to control the calculation and selection of the OATMU, which improved the accuracy of the quantitative study for mapping units. The results showed that for the Qiantang River Basin, the OATMU for GI was 3.60 km^2, the OATMU for EI was 9.00 km^2, the OATMU for SMI was 3.60 km^2, and the OATMU for CES was 3.60 km^2.

The indicators and calculation methods proposed here can be spatially replicated and can be applied to various river basins, providing CES. Despite the limitations in terms of data accuracy and alternative settings, the research method of this study helped to clarify the optimal mapping units to demonstrate the spatial distribution of CES more scientifically and accurately when calculating supply and demand. For future urban planning in the Qiantang River Basin, it is recommended that the mapping units delineated by the OATMU of 3.60 km² be considered as homogenized units with closely similar CES characteristics, which can be utilized in planning practice to facilitate CES conservation and development. The study's reasonably evaluated units, which reflect the spatial variation of CES most realistically, can provide a basis for ecosystem service valuation and ecological compensation transfer payments.

Author Contributions: Conceptualization, Y.L. and Y.W.; methodology, Y.L.; software, Y.L.; validation, Y.L.; formal analysis, Y.L.; investigation, Y.L.; resources, Y.L.; data curation, Y.L.; writing—original draft preparation, Y.L.; writing—review and editing, Y.L., J.H. and Y.W.; visualization, Y.L.; supervision, J.H. and Y.W.; project administration, J.H. and Y.W.; funding acquisition, Y.W. All authors have read and agreed to the published version of the manuscript.

Funding: This research was funded by Key Project of the National Natural Science Foundation of China "Theory and Method of Landscape Ecological Planning for Livable Urban-rural Areas: Taking the Mountainous Region of Southwest China as Example" (No. 52238003).

Data Availability Statement: The original contributions presented in the study are included in the article, further inquiries can be directed to the corresponding author.

Conflicts of Interest: The authors declare no conflicts of interests.

References

1. Bachi, L.; Ribeiro, S.C.; Hermes, J.; Saadi, A. Cultural Ecosystem Services (CES) in landscapes with a tourist vocation: Mapping and modeling the physical landscape components that bring benefits to people in a mountain tourist destination in southeastern Brazil. *Tour. Manag.* **2020**, *77*, 104017. [CrossRef]
2. MEA. *Ecosystems and Human Well-Being*; Island Press: Washington, DC, USA, 2005.
3. Scholte, S.S.K.; van Teeffelen, A.J.A.; Verburg, P.H. Integrating socio-cultural perspectives into ecosystem service valuation: A review of concepts and methods. *Ecol. Econ.* **2015**, *114*, 67–78. [CrossRef]
4. Breyne, J.; Dufrêne, M.; Maréchal, K. How integrating 'socio-cultural values' into ecosystem services evaluations can give meaning to value indicators. *Ecosyst. Serv.* **2021**, *49*, 101278. [CrossRef]
5. Liu, Z.; Liu, Z.; Zhou, Y.; Huang, Q. Distinguishing the Impacts of Rapid Urbanization on Ecosystem Service Trade-Offs and Synergies: A Case Study of Shenzhen, China. *Remote Sens.* **2022**, *14*, 4604. [CrossRef]
6. Kalinauskas, M.; Bogdzevič, K.; Gomes, E.; Inácio, M.; Barcelo, D.; Zhao, W.; Pereira, P. Mapping and assessment of recreational cultural ecosystem services supply and demand in Vilnius (Lithuania). *Sci. Total Environ.* **2023**, *855*, 158590. [CrossRef]
7. Nie, X.; Xie, Y.; Xie, X.; Zheng, L. The characteristics and influencing factors of the spatial distribution of intangible cultural heritage in the Yellow River Basin of China. *Herit. Sci.* **2022**, *10*, 121. [CrossRef]
8. Thiele, J.; Albert, C.; Hermes, J.; von Haaren, C. Assessing and quantifying offered cultural ecosystem services of German river landscapes. *Ecosyst. Serv.* **2020**, *42*, 101080. [CrossRef]
9. Meng, S.; Huang, Q.; Zhang, L.; He, C.; Inostroza, L.; Bai, Y.; Yin, D. Matches and mismatches between the supply of and demand for cultural ecosystem services in rapidly urbanizing watersheds: A case study in the Guanting Reservoir basin, China. *Ecosyst. Serv.* **2020**, *45*, 101156. [CrossRef]
10. Xiong, L.; Li, R. Assessing and decoupling ecosystem services evolution in karst areas: A multi-model approach to support land management decision-making. *J. Environ. Manag.* **2024**, *350*, 119632. [CrossRef]
11. Shen, J.; Chen, C.; Wang, Y. What are the appropriate mapping units for ecosystem service assessments? A systematic review. *Ecosyst. Health Sust.* **2021**, *7*, 1888655. [CrossRef]
12. Campos-Campos, C.; Cruz-Cárdenas, G.; Aquino, R.J.C.; Moncayo-Estrada, R.; Machuca, M.A.V.; Meléndez, L.A.Á. Historical Delineation of Landscape Units Using Physical Geographic Characteristics and Land Use/Cover Change. *Open Geosci.* **2018**, *10*, 45–57. [CrossRef]
13. Alvioli, M.; Marchesini, I.; Pokharel, B.; Gnyawali, K.; Lim, S. Geomorphological slope units of the Himalayas. *J. Maps* **2022**, *18*, 300–313. [CrossRef]
14. Ocelli Pinheiro, R.; Triest, L.; Lopes, P.F.M. Cultural ecosystem services: Linking landscape and social attributes to ecotourism in protected areas. *Ecosyst. Serv.* **2021**, *50*, 101340. [CrossRef]
15. Zhang, S.; Muñoz Ramírez, F. Assessing and mapping ecosystem services to support urban green infrastructure: The case of Barcelona, Spain. *Cities* **2019**, *92*, 59–70. [CrossRef]
16. Aalders, I.; Stanik, N. Spatial units and scales for cultural ecosystem services: A comparison illustrated by cultural heritage and entertainment services in Scotland. *Landsc. Ecol.* **2019**, *34*, 1635–1651. [CrossRef]
17. Wu, M.; Che, Y.; Lv, Y.; Yang, K. Neighbourhood-scale urban riparian ecosystem classification. *Ecol. Indic.* **2017**, *72*, 330–339. [CrossRef]
18. Sun, R.; Jin, X.; Han, B.; Liang, X.; Zhang, X.; Zhou, Y. Does scale matter? Analysis and measurement of ecosystem service supply and demand status based on ecological unit. *Environ. Impact Asses* **2022**, *95*, 106785. [CrossRef]

19. Wu, J.; Guo, X.; Zhu, Q.; Guo, J.; Han, Y.; Zhong, L.; Liu, S. Threshold effects and supply-demand ratios should be considered in the mechanisms driving ecosystem services. *Ecol. Indic.* **2022**, *142*, 109281. [CrossRef]
20. Chen, M.; Cui, Y.; Gassman, P.; Srinivasan, R. Effect of Watershed Delineation and Climate Datasets Density on Runoff Predictions for the Upper Mississippi River Basin Using SWAT within HAWQS. *Water* **2021**, *13*, 422. [CrossRef]
21. Yuan, J.; Li, R.; Huang, K. Driving factors of the variation of ecosystem service and the trade-off and synergistic relationships in typical karst basin. *Ecol. Indic.* **2022**, *142*, 109253. [CrossRef]
22. Wu, H.; Liu, X.; Li, Q.; Hu, X.; Li, H. The Effect of Multi-Source DEM Accuracy on the Optimal Catchment Area Threshold. *Water* **2023**, *15*, 209. [CrossRef]
23. Neumann, A.; Kim, D.; Perhar, G.; Arhonditsis, G.B. Integrative analysis of the Lake Simcoe watershed (Ontario, Canada) as a socio-ecological system. *J. Environ. Manag.* **2017**, *188*, 308–321. [CrossRef]
24. García-Álvarez, D.; Camacho Olmedo, M.T.; Paegelow, M. Sensitivity of a common Land Use Cover Change (LUCC) model to the Minimum Mapping Unit (MMU) and Minimum Mapping Width (MMW) of input maps. *Comput. Environ. Urban Syst.* **2019**, *78*, 101389. [CrossRef]
25. Syrbe, R.; Walz, U. Spatial indicators for the assessment of ecosystem services: Providing, benefiting and connecting areas and landscape metrics. *Ecol. Indic.* **2012**, *21*, 80–88. [CrossRef]
26. Zhang, H.; Loáiciga, H.A.; Feng, L.; He, J.; Du, Q. Setting the Flow Accumulation Threshold Based on Environmental and Morphologic Features to Extract River Networks from Digital Elevation Models. *ISPRS Int. J. Geo-Inf.* **2021**, *10*, 186. [CrossRef]
27. Kong, F.B.; Cao, L.D.; Xu, C.Y. Measurement of Carbon Budget and Type Partition of Carbon Comprehensive: Compensation in the Qiantang River Basin. *Econ. Geogr.* **2023**, *43*, 150–161.
28. Zhang, Y.; Zeng, M.; Zhang, Q.; Ye, D.; Yang, C.; Xu, P.; Ren, X.; Jiang, R.; Li, F. Holocene Spatiotemporal Distribution of Sites and Its Response to Environmental Changes in Qiantang River Basin. *Resour. Environ. Yangtze Basin* **2022**, *31*, 2022–2034.
29. Hale, R.L.; Cook, E.M.; Beltrán, B.J. Cultural ecosystem services provided by rivers across diverse social-ecological landscapes: A social media analysis. *Ecol. Indic.* **2019**, *107*, 105580. [CrossRef]
30. Lin, P.; Pan, M.; Wood, E.F.; Yamazaki, D.; Allen, G.H. A new vector-based global river network dataset accounting for variable drainage density. *Sci. Data* **2021**, *8*, 28. [CrossRef]
31. Yang, K.; Smith, L.C. Internally drained catchments dominate supraglacial hydrology of the southwest Greenland Ice Sheet. *J. Geophys. Res. Earth Surf.* **2016**, *121*, 1891–1910. [CrossRef]
32. Zhang, J.X.; Tang, L.; Xie, T.; Peng, Q. Determination of catchment area threshold for extraction of digital river-network. *Water Resour. Hydropower Eng.* **2016**, *47*, 1–4. [CrossRef]
33. Olsen, N.R.; Tavakoly, A.A.; McCormack, K.A.; Levin, H.K. Effect of User Decision and Environmental Factors on Computationally Derived River Networks. *J. Geophys. Res. Earth Surf.* **2023**, *128*, e2022JF006873. [CrossRef]
34. Chen, J.M.; Lin, G.F.; Yang, Z.H.; Chen, H.Y. The Relationship between DEM Resolution, Accumulation Area Threshold and Drainage Network Indices. In Proceedings of the 2010 18th International Conference on Geoinformatics, Beijing, China, 18–20 June 2010.
35. Kim, S.; Yoon, S.; Choi, N. Evaluating the Drainage Density Characteristics on Climate and Drainage Area Using LiDAR Data. *Appl. Sci.* **2023**, *13*, 700. [CrossRef]
36. Heasley, E.L.; Clifford, N.J.; Millington, J.D.A. Integrating network topology metrics into studies of catchment-level effects on river characteristics. *Hydrol. Earth Syst. Sci.* **2019**, *23*, 2305–2319. [CrossRef]
37. Valente, D.; Pasimeni, M.R.; Petrosillo, I. The role of green infrastructures in Italian cities by linking natural and social capital. *Ecol. Indic.* **2020**, *108*, 105694. [CrossRef]
38. Montana, C.G.; Winemiller, K.O.; Sutton, A. Intercontinental comparison of fish ecomorphology: Null model tests of community assembly at the patch scale in rivers. *Ecol. Monogr.* **2014**, *84*, 91–107. [CrossRef]
39. Deng, D.C. Composite Politics: Governance Logic of Natural Unit and Administrative Unit. *Southeast Acad. Res.* **2017**, *6*, 25–37.
40. Lan, F.; Lin, Z.Y.; Huang, X. Evolution characteristics of multi-center spatial structure in Xi'an city anddriving factors: A POI data-based analysis. *J. Arid. Land Resour. Environ.* **2023**, *37*, 57–66. [CrossRef]
41. Barnes, R.S.K.; Hamylton, S.M. Isometric scaling of faunal patchiness: Seagrass macrobenthic abundance across small spatial scales. *Mar. Environ. Res.* **2019**, *146*, 89–100. [CrossRef]
42. Militino, A.; Moradi, M.; Ugarte, M. On the Performances of Trend and Change-Point Detection Methods for Remote Sensing Data. *Remote Sens.* **2020**, *12*, 1008. [CrossRef]
43. Colombo, R.; Vogt, J.V.; Soille, P.; Paracchini, M.L.; de Jager, A. Deriving river networks and catchments at the European scale from medium resolution digital elevation data. *Catena* **2007**, *70*, 296–305. [CrossRef]
44. Yang, G.; Deng, F.; Wang, Y.; Xiang, X. Digital Paradox: Platform Economy and High-Quality Economic Development—New Evidence from Provincial Panel Data in China. *Sustainability* **2022**, *14*, 2225. [CrossRef]
45. Villez, K.; Rosén, C.; Anctil, F.; Duchesne, C.; Vanrolleghem, P.A. Qualitative Representation of Trends (QRT): Extended method for identification of consecutive inflection points. *Comput. Chem. Eng.* **2013**, *48*, 187–199. [CrossRef]
46. Zhang, X.J.; Jiao, Y.F.; Liu, J.; Li, W.L.; Li, C.Z. Study on Method of Sub-Basin Partition of Daqing River Based on DEM. *Yellow River* **2020**, *42*, 13–17.
47. Zhang, W.; Li, W.; Loaiciga, H.A.; Liu, X.; Liu, S.; Zheng, S.; Zhang, H. Adaptive Determination of the Flow Accumulation Threshold for Extracting Drainage Networks from DEMs. *Remote Sens.* **2021**, *13*, 2024. [CrossRef]

48. Martín-López, B.; Palomo, I.; García-Llorente, M.; Iniesta-Arandia, I.; Castro, A.J.; García Del Amo, D.; Gómez-Baggethun, E.; Montes, C. Delineating boundaries of social-ecological systems for landscape planning: A comprehensive spatial approach. *Land Use Policy* **2017**, *66*, 90–104. [CrossRef]
49. Per, A.; Laine, B.; Grzegorz, M.; Ulf, S.; Anders, W. Assessing Village Authenticity with Satellite Images: A Method to Identify Intact Cultural Landscapes in Europe. *AMBIO J. Hum. Environ.* **2003**, *32*, 594–604. [CrossRef]
50. Zhou, Y.C.; Qi, Q.W.; Feng, C.F. Characteristics of dynamic variation of the inter-provincial economic difference in China in recent ten years. *Geogr. Res.* **2002**, *21*, 781–790.
51. González Del Tánago, M.; Gurnell, A.M.; Belletti, B.; García De Jalón, D. Indicators of river system hydromorphological character and dynamics: Understanding current conditions and guiding sustainable river management. *Aquat. Sci.* **2016**, *78*, 35–55. [CrossRef]
52. Cheng, Z.Y.; Zhang, X.N.; Fang, Y.H. Application and comparison of identification methods for critical catchment area threshold in Jialing River Basin. *Yangtze River* **2017**, *48*, 25–29. [CrossRef]
53. Zheng, Y.; Yu, C.; Zhou, H.; Xiao, J. Spatial Variations and Influencing Factors of River Networks in River Basins of China. *Int. J. Environ. Res. Public Health* **2021**, *18*, 11910. [CrossRef]
54. Liang, H.; Guo, Z.; Wu, J.; Chen, Z. GDP spatialization in Ningbo City based on NPP/VIIRS night-time light and auxiliary data using random forest regression. *Adv. Space Res.* **2020**, *65*, 481–493. [CrossRef]
55. Zhou, M.; Deng, J.; Lin, Y.; Zhang, L.; He, S.; Yang, W. Evaluating combined effects of socio-economic development and ecological conservation policies on sediment retention service in the Qiantang River Basin, China. *J. Clean. Prod.* **2021**, *286*, 124961. [CrossRef]
56. Wang, D.; Xie, H.M. Study on the Kuznets Effect of County Ecosystem Service Value within the Basin: Taking Qiantang River Basin as an Example. *Ecol. Econ.* **2021**, *37*, 147–152.
57. Zhen, Z.; Lei, T.; Ke, N. Ecological welfare performance and its convergence under the evolution of Qiantang River Basin ecological protection policies. *China Popul. Resour. Environ.* **2022**, *32*, 198–207.
58. Li, F.M. Study on the Path and Effects of Hangzhou's Strategy of "Embracing the River". *China Collect. Econ.* **2018**, *7*, 34–35.
59. Tang, J.; Cheng, Q.M.; Chen, Y. Determination Method of Optimal Catchment Area Threshold Based on ArcGIS-Matlab. *Water Resour. Power* **2021**, *39*, 46–48.
60. Shen, J.; Guo, X.; Wang, Y. Identifying and setting the natural spaces priority based on the multi-ecosystem services capacity index. *Ecol. Indic.* **2021**, *125*, 107473. [CrossRef]

Disclaimer/Publisher's Note: The statements, opinions and data contained in all publications are solely those of the individual author(s) and contributor(s) and not of MDPI and/or the editor(s). MDPI and/or the editor(s) disclaim responsibility for any injury to people or property resulting from any ideas, methods, instructions or products referred to in the content.

Article

Spatial and Temporal Changes in Supply and Demand for Ecosystem Services in Response to Urbanization: A Case Study in Vilnius, Lithuania

Giedrius Dabašinskas and Gintarė Sujetovienė *

Department of Environmental Sciences, Vytautas Magnus University, Universiteto 10, Akademija, 53361 Kaunas, Lithuania; giedrius.dabasinskas@vdu.lt
* Correspondence: gintare.sujetoviene@vdu.lt

Abstract: Intensification of urbanization is changing the supply capacities and demand levels of ecosystem services (ESs), and their mismatch has become a major problem for the sustainable development of urban areas. In this study, spatiotemporal changes of three ecosystem services (food provision, C sequestration, recreation) were quantified and imbalances between their supply and demand were identified in Vilnius County (Lithuania) in 2000–2020. The most significant land use transformation was the increase in forest and urbanized land at the expense of agricultural land. The lowest supply and the highest demand for food, carbon sequestration, and outdoor recreation were in the urban center. The urban land ratio had a negative impact on the provision of ecosystems' services during the study period, most notably affecting food supply. Urbanization indicators—population density and urban land area—showed a negative relationship with the provision of ecosystem services. The balance of supply and demand changed during the 2000–2020 period—the growth of suburbs led to the distance of the supply areas from the city, and the area of the intense demand increased. The results of the study highlight the importance of spatial scale in determining the impact of urbanization on ecosystem functions.

Keywords: ecosystem services; supply; demand; urbanization; spatiotemporal changes; land cover

Citation: Dabašinskas, G.; Sujetovienė, G. Spatial and Temporal Changes in Supply and Demand for Ecosystem Services in Response to Urbanization: A Case Study in Vilnius, Lithuania. *Land* **2024**, *13*, 454. https://doi.org/10.3390/land13040454

Academic Editors: Alessio Russo and Giuseppe T. Cirella

Received: 7 February 2024
Revised: 22 March 2024
Accepted: 27 March 2024
Published: 2 April 2024

Copyright: © 2024 by the authors. Licensee MDPI, Basel, Switzerland. This article is an open access article distributed under the terms and conditions of the Creative Commons Attribution (CC BY) license (https://creativecommons.org/licenses/by/4.0/).

1. Introduction

The overall increase in the world's population has been accompanied by an increase in the number of people living in cities. This relatively new phenomenon in modern human history is the main driver of many environmental changes [1]. Urban land development is a dynamic process that irreversibly and rapidly changes land cover and/or land use from natural ecosystems to artificial and built-up areas [2]. As the population has become more concentrated in urban centers, rural areas have become less dense. On the contrary, agricultural land in the suburban areas is being converted into built-up areas. These land cover changes have significant impacts on ecosystem functioning and represent a challenge for optimal land use management and biodiversity conservation [3].

Population growth and densification increase the scale and nature of supply and demand of ecosystem- and non-ecosystem-services (socioeconomic) [4]. The supply of ecosystem services is the ability of the ecosystem to provide a certain service for human well-being, and the demand is the need for these services. In such case, there was a high risk of oversupply and insufficient demand due to the identified differences in the capacity of ecosystems to provide a service compared to its demand [5]. A mismatch between the supply and demand of ecosystem services was in urbanized areas where demand is higher due to high population concentration [6,7]. The more densely buildings and other urban constructions cover the surface, the lower the ecosystem's capacity to provide human well-being [8]. As urban centers grow, the need for recreation opportunities increases. Only rural areas in rural areas or remaining fragments of green urban areas can meet the demands of the urban population. This creates a gradient of supply and demand in rural and urban areas, and the need for recreational areas in the urban center challenges urban planners [9,10].

Urbanization indicators clearly show a negative relationship with the various welfare benefits that people derive from ecosystems. The direct and effective impact of land urbanization on ecosystem functions was demonstrated by the negative linear relationship between land urbanization and total ecosystem services [11]. As urban land area increased, the provision of ecosystem services such as carbon sequestration, grain production, and habitat quality was negatively associated with urbanization indicators such as the night index, GDP, or population [12]. The potential for ecosystem services, especially regulating, supporting, and provisioning services [7,13–15] and health [16], has been greatly reduced by urban sprawl. However, not all ecosystem services have declined with urban expansion. For example, water yield increased with increasing levels of urbanization [12,15] as urban land reduced water retention, evaporation, and infiltration, resulting in a higher total water yield. Some ecosystem services are not directly related to urbanization and reflect different forms of associations. The relationship between provision of ecosystem service and land cover varied greatly across space and time, being both positive and negative [17]. The relationship between food provisioning and urbanization represented an inverted U shape [15]. In some cases, even a very high level of urbanization does not affect the provision of some ecosystem services—even a megalopolis like Shanghai met the desired amount of PM_{10} removal service [7]. Urban planning and sustainable development play an important role in assessing the ES's losses from urban sprawl.

Studies have shown inconsistent and conflicting findings on the relationships, so there is a need for further research on the relationship between the level of urbanization and the provision of ecosystem services. As most studies have focused on megacities e.g., [7,15,18] there is still a lack of research on the effects of a medium-level urbanization process. Metropolitan areas play an increasing role globally, becoming the centers of population and economic growth, but also having a significant impact on ecosystems and resource management. As is common to the post-socialist Central and Eastern Europe, since 1990, Lithuania has experienced rapid suburbanization, where the expansion of urban sprawl is very pronounced. Despite increases in urban spawl, another interesting fact was that since the 1990s, Lithuania lost almost a quarter of its population, and some regions within the country lost more than 50% of their residents [19]. The population has declined in almost the entire country, except for the areas around the largest cities, where metropolitan growth through urbanization has been observed since the early 1990s. Despite population decline across the country, immigration in major urban centers has generally been higher than emigration. Recent studies raised concern about the rapid urban sprawl of the Vilnius urban zone and its consequences, such as deteriorating ecosystem health [16,20]. However, a more detailed analysis of the spatial distribution ecosystem services, its change and supply–demand of ES in the wider Vilnius metropolitan area, including the suburbs, is needed. The aim of this study was to quantitatively assess the spatial and temporal dynamics of three ecosystem services (food, carbon sequestration, recreation) in the Vilnius metropolitan area from 2000 to 2020 and identify the mismatches between their supply and demand.

2. Materials and Methods

2.1. Study Area

Lithuania is located in a mid-latitude climate zone and belongs to the southwestern sub-region of the Atlantic continental forest zone (Figure 1). The average annual temperature is 6.1–6.8 and the average annual precipitation is 610–700 mm. The study area is Vilnius County located in the east of the country around the capital city of Vilnius. Vilnius County is located around the capital of Lithuania, Vilnius, which is the largest of Lithuania's 10 counties in terms of both area and population (30% of the population of Lithuania). The county consists of Vilnius city, its suburbs, and rural areas located farther away from

Vilnius. There are 8 municipalities which are divided into 105 elderships. We used the elderships with census data as the most comprehensive level in our study.

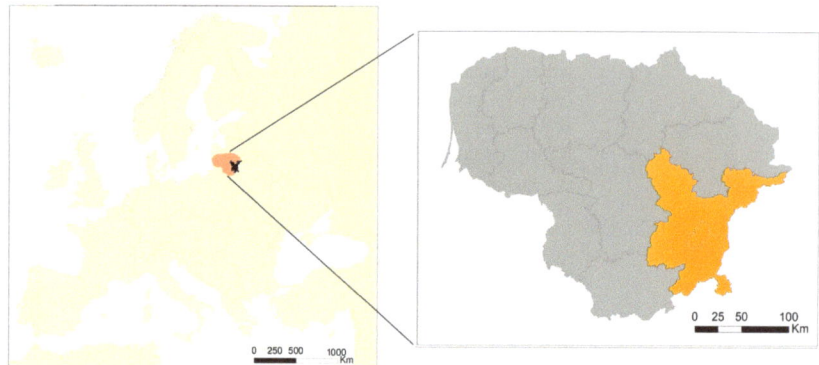

Figure 1. Location of the study area—Vilnius County (orange), Lithuania.

It is the fastest growing city in the region, and it is the only one of Lithuania's major cities where immigration has been higher than emigration and is the second youngest city in the Nordic European countries. The population in the study area decreased by 4.6%, from 850,064 in 2001 and to 810,797 inhabitants in 2021. However, the change was very different in certain regions of the county (Figure 2). Spatial population changes showed that in suburban areas, the share of population has decreased, while in the city center, it has increased, in some cases by more than 150%. This period experienced a large increase in suburban area around the city of Vilnius [21]. At the same time, remote rural areas have lost a significant part of their population, thus inflicting changes in land use as well as ES flows in the county.

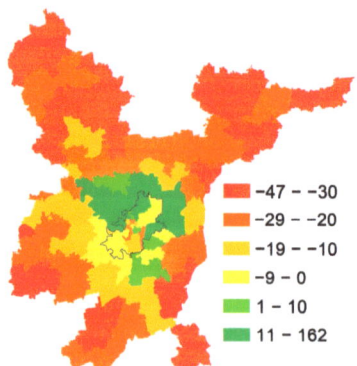

Figure 2. Population change (%) in elderships of Vilnius County, Lithuania from 2001 to 2021.

2.2. Mapping LULC Changes

Land use/land cover (LULC) was obtained from the 2000 and 2018 CORINE datasets. The CORINE dataset is the most comprehensive European land use database which uses remote sensing to classify different LULC types [22]. Its precision is 25 ha. There are 44 LULC classes in the dataset, which were grouped into 5 main classes for this study: urban; agricultural; forest; wetland; water (Figure 3). The dynamics of each LULC type was calculated as the difference of a certain LULC area at the start and the end of the study period, expressed in hectares. The percentage change of LULC type area from the total study area (%) was also calculated.

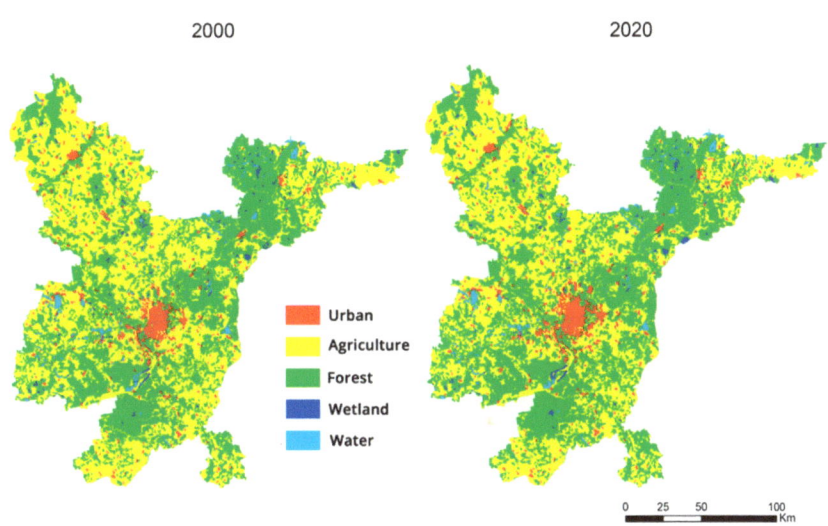

Figure 3. LULC in Vilnius County in 2000 and 2020.

2.3. Mapping Ecosystem Services

Three ecosystem services—food, carbon, and recreation—were quantified out at the eldership level in 2000 and 2020. These services were selected to represent different types of ecosystem services: provisioning (food), regulating (carbon), and cultural (recreation). The data required for the assessment of ecosystem services (population, CO_2 emissions, grain production) were obtained from the Lithuanian official statistics portal (https://osp.stat.gov.lt (accessed on 1 February 2024)). The data for 2000 reflects the post-independence situation, when societal and economic changes began. The year 2020 was chosen to reflect more recent changes such as increasing urbanization.

Since the capacity of an ecosystem to provide ecosystem services depends on its use, we used land use patterns to assess the supply and demand for ecosystem services. The main data we used to assess the ESs, together with their data sources and descriptions, are presented in Table 1.

Table 1. Data and methods used to map supply and demand for ecosystem services.

Services	Year 2000		Year 2020	
	Supply	Demand	Supply	Demand
Food	Eldership LULC (CORINE 2000) grain production in the municipality (2000): recalculated from municipality to eldership according to arable land area	Grain consumption in Lithuania (2001): per eldership according to its population	Eldership LULC (CORINE 2018) grain production in the municipality (2018): recalculated from municipality to eldership according to arable land area	Grain consumption in Lithuania (2021): per eldership according to its population
Carbon sequestration	Eldership LULC (CORINE 2000)	CO_2 emissions per capita (2005) in Lithuania: per eldership according to its population	Eldership LULC (CORINE 2018)	CO_2 emissions per capita (2020) in Lithuania: per eldership according to its population
Recreation	Eldership LULC (CORINE 2000)	Population density of the eldership (2001)	Eldership LULC (CORINE 2018)	Population density of the eldership (2021)

2.3.1. Mapping Food Supply and Demand

The total grain yield of each eldership in Vilnius County was used as its food supply. Food demand was estimated as the per capita consumption of grains in Vilnius County in a given year [18]. The supply and demand of the food ecosystem service (FES) was calculated as follows:

$$S_i^{FES} = I_i \quad (1)$$

$$D_i^{FES} = POP_i \times A \quad (2)$$

where S_i^{FES}—the supply of FES in eldership i, I_i—the grain production in eldership i, D_i^{FES}—the demand of FES in eldership i, POP_i—the number of residents in eldership i, and A represents the per capita consumption of grains in a given year.

2.3.2. Mapping Carbon Sequestration Supply and Demand

Carbon sequestration was quantified from the LULC data based on the carbon storage capacity of each land use type [3]. Carbon sequestration demand was estimated based on the carbon emissions data for each eldership of Vilnius County for the years 2000 and 2020. We assumed that larger emissions equate to a higher demand for carbon sequestration. The supply and demand for carbon sequestration ecosystem service (CSES) was calculated as follows:

$$S_i^{CSES} = \sum CS_{LULC} \times AREA \quad (3)$$

$$D_i^{CSES} = E_i \quad (4)$$

where S_i^{CSES} is the supply of CSEC in eldership i, CS_{LULC} is the carbon sequestration capacity of each LULC type, AREA is the area of each LULC type in the eldership i, D_i^{CSES}—the demand of CSES in eldership i, E_i represents the annual carbon emissions in the eldership i.

2.3.3. Mapping Outdoor Recreation Supply and Demand

The capacity of ecosystems to provide recreational opportunities was considered a service provided by outdoor recreation. Outdoor recreation ecosystem service supply was calculated by LULC type. First, LULC classes were categorized by their recreational potential, then total area of LULC with a coefficient of 7 or above was calculated for each eldership [3]. We defined the potential demand for outdoor recreation as everyone's basic right to connect with nature. For that purpose, we used the optimal area of 50 m² of green space per capita which is considered an ideal amount of urban green space per individual [23]. Outdoor recreation demand was calculated by multiplying the population of each eldership with the recreational optimal area. Supply and demand for carbon sequestration ecosystem service (RES) were calculated as follows:

$$S_i^{RES} = \sum RS_{LULC} \times AREA \quad (5)$$

$$D_i^{RES} = POP_i \times 0.005 \quad (6)$$

where S_i^{RES} is the supply of carbon sequestration ecosystem service in eldership i, RS_{LULC} is LULC type considered suitable for recreation, D_i^{RES} represents the demand for outdoor recreation ecosystem service in eldership i, POP_i represents population in the eldership i.

2.3.4. Ecological Supply and Demand Mismatch and Ratio

The mismatch between the supply and demand of ecosystem services was calculated by subtracting the value of estimated ES supply from the demand value at the eldership level. Positive values indicate that the demand for ES in the eldership was met, while negative values indicated that the ecosystem services provided did not satisfy the demand in the eldership. After evaluating the mismatches between supply and demand, it allowed us to determine their differences during the studied period (2000 and 2020).

We used the ecological supply and demand ratio (ESDR) and comprehensive supply–demand ratio (CESDR) to indicate surplus or deficit of a given ecosystem service and determine the status of all ESs considered, respectively. The ratios were calculated as follows [7]:

$$ESDR = \frac{S - D}{(S_{max} - D_{max})/2} \quad (7)$$

$$CESDR = \frac{1}{n}\sum_{i=1}^{n} ESDR_i \quad (8)$$

where S and D are the actual supply and demand for a given ES, respectively; S_{max} and D_{max} refers to the maximum value of supply and demand for a particular ES and year, respectively; n is the number of ESs (n = 3); $ESDR_i$—ecological supply–demand ratio of a given ES.

2.4. Data Sources and Analysis

Statistical data such as population, grain production, GDP, and CO_2 emissions were obtained from the State Data Agency of Lithuania. Census data from 2001 and 2021 were used to estimate the population in the municipality. Population density was measured as the number of inhabitants per unit area of the ward (km^2). The distance from the Vilnius city center to the center of each eldership was calculated as the length of a straight line between these points (km).

Regression analysis was used to indicate trendlines between the supply and demand ratio (ESDR, CESDR) and a certain land use ratio (urban and forest land ratio) for 2000 and 2020. Urban/forest land ratios were calculated as area of urban/forest lands divided by the total land area. Pearson correlation analysis was used to represent relationships between ecosystem services and urbanization indicators (population, natural capital, and urban land ratios). Statistical analyses were carried out using the R statistical software version 4.1. Spatial data were analyzed using ArcMap 10.7.

3. Results
3.1. Land Use Change

Between 2000 and 2020, the area of agricultural land (cropland and grassland) decreased by 49,469 ha (5.1% of the total area, Table 2). The rate of decline in agricultural land was 2473 ha per year. Forest and urban areas increased by 43,432 ha (4.5%) and 5192 ha (0.5%), respectively. The rates of change in forest and urban areas were 2172 ha and 260 ha per year, respectively. The area of wetlands increased by 6% over the period under study (Table 2).

Table 2. Land use changes in Vilnius County from 2000 to 2020.

LULC	Area (ha)		% of Total Area		Change from 2000 to 2020	
	2000	2020	2000	2020	ha	%
Urban	34,296	39,488	3.5	4.1	5192	0.5
Agriculture	496,372	446,903	51.1	46.0	−49,469	−5.1
Forest	412,848	456,280	42.5	47.0	43,432	4.5
Wetland	8397	8909	0.9	0.9	512	0.1
Water	19,785	20,118	2.0	2.1	332	0.0

The most significant land use transformation involved the conversion of agricultural land to forest and a relatively small area of urbanized land (Figure 4). Most of the forest area has been converted to agricultural land. A similar proportion of forest was converted to wetlands and, conversely, a small proportion was converted from wetlands to forest.

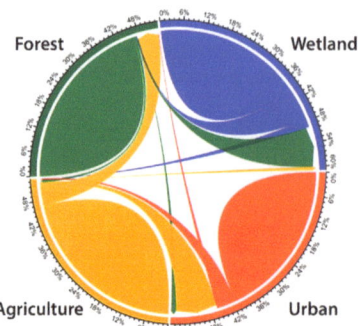

Figure 4. Conversion from one land use type to another (%) in Vilnius County from 2000 to 2020 (only the part of the land that has undergone change is shown).

Spatial changes in land use by type were most pronounced in urban and forest areas. The urban area has increased the most around Vilnius city itself, reflecting suburban development. Conversely, the urban area decreased mainly in the elderships further away from Vilnius. Forest cover has increased in most of the elderships, except in the most densely populated ones of Vilnius city. The area of agricultural land decreased in all elderships, mostly replaced by forests. No spatial trends in the area of wetlands and open water bodies were identified.

3.2. Ecosystem Service Change

The average value of food supply was 0.16 tons per ha in 2000 and 0.24 ton per ha in 2020 (Figure 5). The total food supply in Vilnius County increased from 0.18 million tons in 2000 to 0.27 million tons in 2020. The largest surplus of food supply was in the elderships located in the northwestern part of the County. Vilnius city and suburbs had the lowest values of food supply (Figure 5).

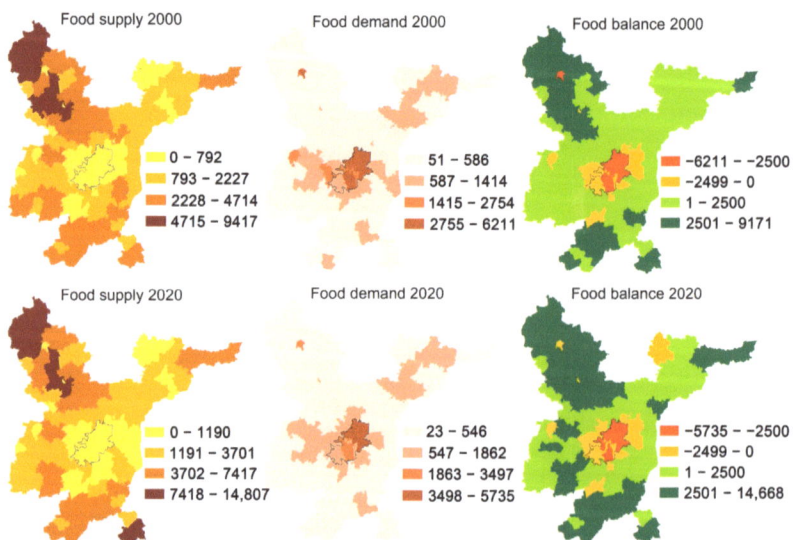

Figure 5. Spatial distribution of food provision service (tones per year)—supply (**left**), food demand (**middle**) and mismatch (**right**) in Vilnius County in 2000 and 2020.

Food service demand showed a decrease of 18.2% from 0.11 million tons in 2000 to 0.09 million tons in 2020. The average value of food demand decreased from 1.15 to

0.92 tons per ha during this period. In 2000, one third of the elderships (32 out of 105) could not meet their food demand and the proportion of elderships with food demand in 2020 increased to 32.4%. During this period, food demand decreased mainly in 94% of the territory. Most of the elderships unable to meet food needs were in the city and the surrounding peri-urban area, where a negative food supply balance was identified (Figure 5). In the elderships further away from the city, most rural districts were able to meet the food demand. The largest surpluses of food stocks were found in the communes located in the northwestern part of the county.

The average value for carbon storage per ha increased from 12.16 tons per ha in 2000 to 13.05 tons per ha in 2020 with a rate of about 0.04 tons per ha every year. The total service supply increased from 13.56 million tons in 2000 to 14.56 million tons per ha in 2020 in the county. The carbon sequestration has increased in most of the study areas (89%). The highest carbon sequestration values were in the most heavily forested elderships farther away from the city center (Figure 6). The city center tended to show the lowest values of carbon sequestration.

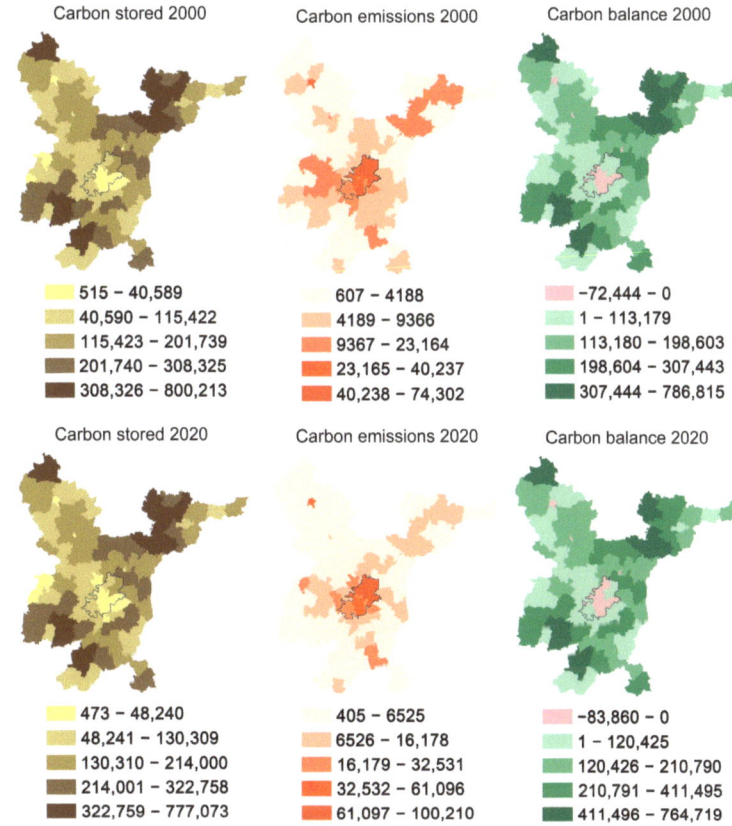

Figure 6. Carbon sequestration (**left**), carbon emissions (**middle**) and mismatch (**right**) in Vilnius County in 2000 and 2020 (tones per year).

Between 2000 and 2020, there was a slight increase in carbon emissions—from 116,546 in 2000 to 123,380 in 2020. Carbon emissions increased, as did sequestration in most of the study areas (70%). Carbon sequestration demand decreased in 2000, 21 out of 105 elderships could not meet their carbon sequestration demand, and the number of such elderships has increased by only one by 2020 (Figure 6). All these elderships were located in urban areas, in the city center and smaller towns across Vilnius County.

Outdoor recreation supply values remained stable over time, with an average value of approximately 0.73 million ha. Recreation supply has increased in half of the study area and decreased in the other half. In the city center area, outdoor recreation had the lowest values (Figure 7). In other parts of the area, medium and high recreational values were unevenly distributed, with the highest values in the areas furthest away from the city center, where the largest areas of forest were located (Figure 7).

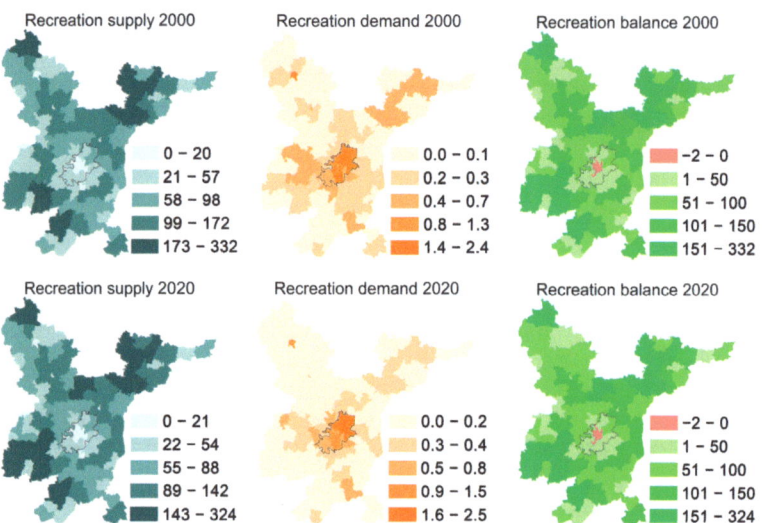

Figure 7. Outdoor recreation service (km^2): supply (**left**), demand (**middle**), and balance (**right**) in Vilnius County in 2000 and 2020.

The demand for outdoor recreation showed a decrease between 2000 and 2020 from 4221 ha to 4054 ha. Over the same period, the number of elderships unable to meet the demand for outdoor recreation was stable (10%). All of these elderships were located in the city of Vilnius, mainly old inner-city neighborhoods with a high population density (Figure 7). Demand for outdoor recreation has declined over the study period in almost the whole area (90%).

3.3. ES Supply–Demand Mismatches

The spatial distribution of food provision service showed a mismatch according to the supply and demand ratio (Figure 8). Lower food supply and higher food demand in the city center led to a shortage and a negative supply–demand ratio. The food supply–demand ratio increased with the distance from the city center: elderships within 20 km had a negative balance of food supply, while more rural areas had a positive ratio (Figure 8). The average balance has increased due to an increase in food supply in 2020 compared to 2000 by 0.72 million tons.

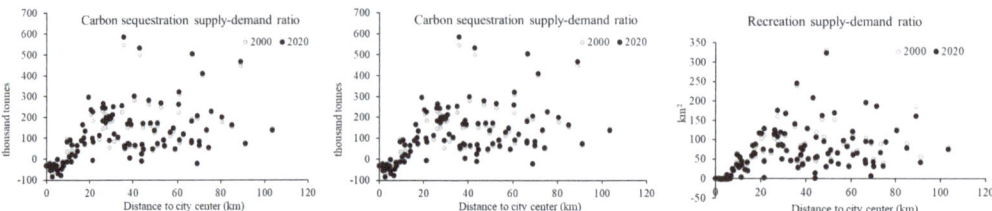

Figure 8. Mismatches between food (**left**), carbon sequestration (**middle**), and recreation (**right**) supply and demand in Vilnius by distance from the city center (km) in 2000 and 2020.

In Vilnius city and its suburbs (up to 20 km from the city center), the carbon sequestration balance was significantly lower compared to more distant areas, where a surplus of 150–200 thousand t of carbon sequestration in the supply–demand ratio was determined (Figure 8). The elderships close to the center of the city had the lowest balance between the supply–demand ratio for outdoor recreation (Figure 8). In those areas 20 km or more from the city center, the surplus of supply–demand for outdoor recreation was between 70 and 110 km^2. The balance of supply–demand for outdoor recreation has remained stable between 2000 and 2020, despite a slight decrease in demand.

Most rural areas could meet their ecosystem service needs. Conversely, many Vilnius urban and suburban elderships had at least one ecosystem service for which they could not meet the demand (Figure S1). Thus, they would have to compensate for this from other regions. Comparing the temporal changes in the ES supply–demand ratio in 2000 and 2020, there was one more eldership in 2020 where there was a mismatch between supply and demand.

3.4. Influence of Land Use Change on ESDR and CESDR

The urban land ratio had a negative influence on the provision of ecosystems services over the study period—the highest values of ESDR were under the lowest area of urban fabric. The highest decrease in service provision with the increase in urban area was characteristic for food supply. The forest land ratio had a positive influence on recreation service and carbon sequestration provision, with a minimum effect on the food supply service (Figure S2). The increasing urban land ratio had a negative impact on the CESDR ($p < 0.001$) and explained about 60% of the variability in 2000 and 2020. In contrast, the proportion of forest land had a positive impact on the CESDR and explained 40–42% of the variability. The proportion of variance in CESDR explained by the urban and forest land ratio was not changed during the study period (Figure 9).

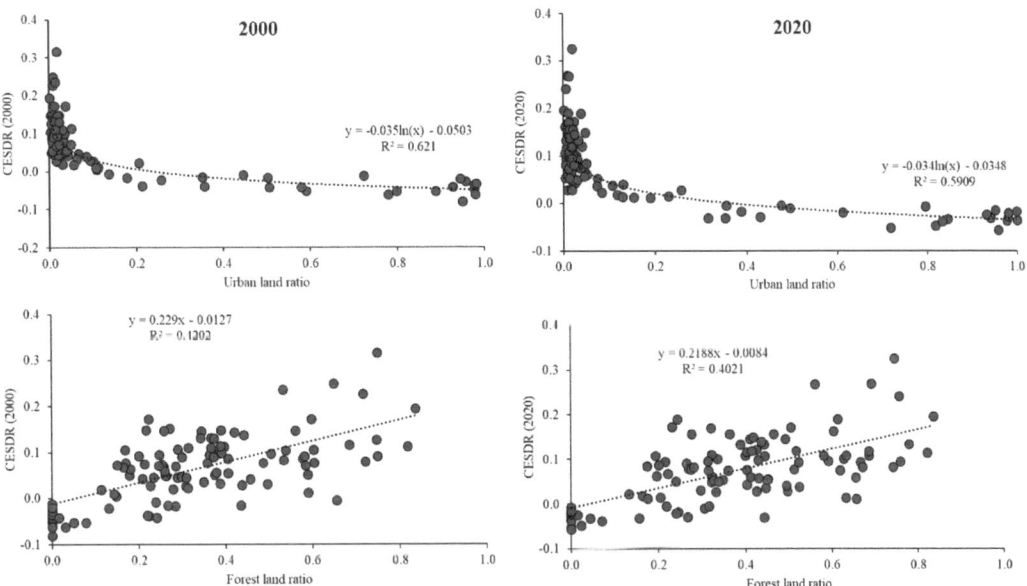

Figure 9. Influence of urban land (**upper**) and forest land (**bottom**) ratios on comprehensive supply-demand ratio (CESDR) in Vilnius in 2000 and 2020.

3.5. Relationship between ESs and Urbanization Indicators

Population was significantly positively related to the increase in urbanized areas (Figure 10). With the increase in distance from the city center, the population density and urban land ratio increased. The correlation analysis showed significant relationships between ESDR and urbanization indicators. The provision of all ESs increased with the distance from the city center with the most significant effect on food provision. Other urbanization indicators—population density and urban land area—showed a negative relationship with carbon sequestration, food provision, and recreational activity. The negative relationship between urban and natural capital increased from 2000 to 2020. Also, the relationship of all ecological indicators strengthened with the distance from the city center during this period (Figure 10).

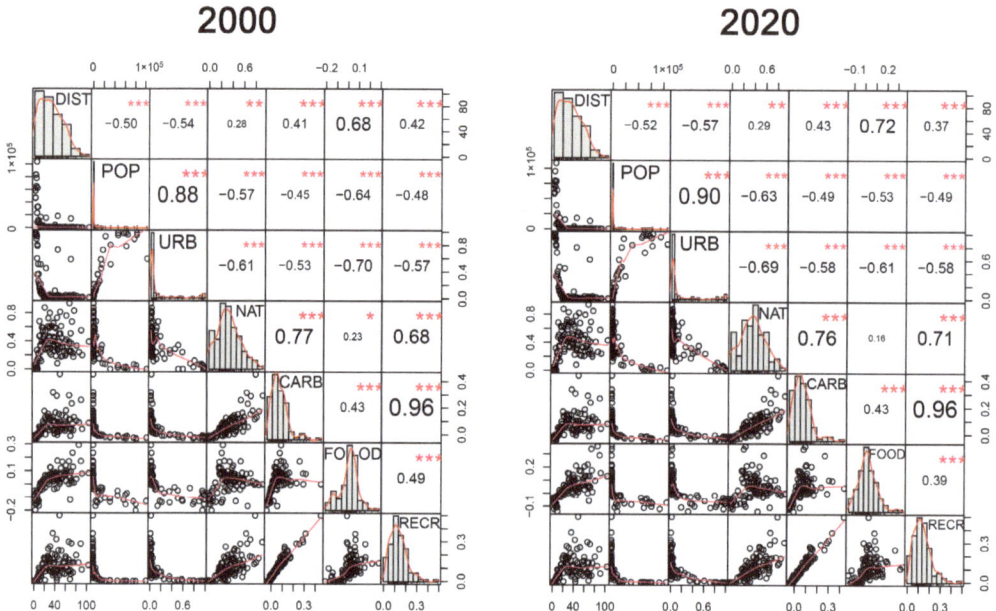

Figure 10. Relationship and fitting curves between ESs (CARB—carbon sequestration, FOOD—food provision, RECR—recreation) and urbanization indicators (DIST—distance to the city center, POP—population density, URB—urban land ratio, NAT—natural capital ratio (including forest, natural grassland, wetland, water). *—$p < 0.05$, **—$p < 0.01$, ***—$p < 0.001$.

4. Discussion

The contribution of the supply of ecosystem services to the public demand was determined in the case of Vilnius County, which is associated with changes in land use related to urbanization during the two decades. The main land use changes were related to the conversion of agricultural land to forest and urban areas. Even though the expansion of the urbanized area was relatively small, the population density increased significantly in the city center. Regarding ecosystem services, the results of the study showed that Vilnius County meets the needs of recreational services, but there are discrepancies between C sequestration and food supply services.

The highest changes in service provision were characteristic for food supply. The food supply has increased over the past along with significant land use changes, such as the conversion of agriculture land to forest and expanded urban areas. Previous studies have shown that the loss of agricultural land dying to urban sprawl did not have a significant effect on overall food production [24,25]. The total increase in food supply despite the reduction in agricultural land indicated that food supply is being ensured using a lower quantity of arable land and modern tools and technology [26]. This is particularly important for achieving greater agricultural efficiency, especially in those areas where the expansion of cultivated land will not be possible [27] and especially when predicting an increase in food demand in the future [28]. On the other hand, the decreases in population led to a decrease in the food demand but did not allow for balancing the relationship between food supply and demand in Vilnius County. Despite the decrease in food demand and increase in supply during the study period, the negative supply and demand balance remained. The provision of food was negatively influenced by the ratio of urban land—the ESDR values of the ecological supply–demand ratio were negative at the largest area of the urban fabric. This finding was consistent with the results showing food availability increasing with distance from the city center [29] and confirmed that urbanization has a significant and positive impact on food insecurity [30]. This could be explained by the loss of potential yields due to the conversion of productive land into unproductive land under the urbanization process [31]. This is also reflected in the general trend, which showed that during the period of 2005–2016, the area of agricultural land in Lithuania decreased [32].

The observed increase in C sequestration during the study period was also confirmed by the results of a study conducted in the Beijing metropolitan area that urbanization has a positive effect on C sequestration capacity [33]. However, it is generally argued that urbanization negatively affects the provision of this ecosystem service [34] by reducing the carbon sequestration capacity of vegetation due to soil compaction, low microbial activity [35], and low organic matter input [36]. As trees play an important role in carbon storage and sequestration—acting as CO_2 absorbers, fixing carbon during photosynthesis, and storing excess carbon as biomass—changes in forest cover are primarily responsible for providing this service. Although urban trees have been suggested to make a small contribution to C sequestration, offsetting the annual CO_2 emissions of cities [7,37,38], the reduction in forest area in the urban center found in our study further reduced this contribution. This was confirmed by the positive effect of forest land ratio on the provision of carbon sequestration. Agricultural land is traditionally considered a source of CO_2 due to disturbance and fertilizer use [39], but in our study, land conversion from agricultural to urbanized areas did not reduce CO_2 emissions. In general, the mismatch between supply and demand for this service increased with urbanization, indicating the loss of carbon storage in the study area during the last two decades.

Both supply and demand for recreational services have changed little during the study period. However, the data showed that the city of Vilnius was characterized by a high provision of recreational service, which confirmed that the accessibility of the population covered by the city's recreational green spaces was high in Vilnius [10]. The analysis of spatial data carried out between 1990 and 2012 showed that the relatively small decrease in greenery over time (0.53%) was determined in Vilnius, which indicates it as one of the greenest European cities with a sufficiently high recreation potential [40]. Research shows that many premature deaths in cities could be prevented by increasing exposure to green spaces, while contributing to sustainable, livable, and healthy spaces [41].

The increasing proportion of urban land has had a total negative effect on the potential of ecosystems to provide services, as shown by the CESDR. In order to ensure the provision of ecosystem services, the urban land ratio needs to be no more than a third of the area. This was typical for all analyzed ecosystem services, which increased further away from the city, while population density and urbanized area decreased [29]. Even a small increase in forest area significantly increases the provision of ecosystem services. Both indicators of urbanization, the proportion of urbanized land and population density, were negatively related to the provision of ecosystem services. The research results were confirmed by other studies, where land use changes caused by urbanization worsened the potential of ecosystem services [42].

According to the results of the quantitative assessment, there was a general lack of supply in the Vilnius central urban area. Such central urban areas are known as the "cold spot" of ES supply zones where high building densities alter the ecological space, while high population density and a high intensity of human activities reduce the potential for the supply of ES [18]. In general, declining population in the study region was not accompanied by a decrease in urbanized land area. This transformation process significantly changed the

LULC around the city of Vilnius, expanding the urban fabric at the expense of agricultural land, and revealed several consequences. Firstly, these changes in land use have not led to a drastic reduction in the supply of ecosystem services. Secondly, the distribution of the population in the suburbs did not create a high demand for ecosystem services for which the supply of ecosystem services would not be ensured. Finally, it was not land use change but population growth in the city center itself that showed the greatest supply/demand mismatch, indicating unsustainable urban development. The demand and supply of only three ecosystem services examined showed some discrepancies, so a larger number of the examined services could be included and useful in the future. A more detailed analysis of the situation would allow a decision to be made on how to change the landscape in order to achieve a sustainable balance between resource supply and use. Understanding the relationship between changes in ecosystems and the services they provide is important for sustainable urban planning and development, environmental protection, and decision making. To better understand changes in ecosystem service mismatches and flows along the urban–rural gradient, further research is needed to assess more spatial and temporal patterns of changes in ecosystem service supply and demand.

5. Conclusions

In this study, analyzing land use changes related to urbanization in the case of Vilnius County, the mismatch between the supply of ecosystem services and the public demand was evaluated. The results of this study demonstrated that the association between land use changes and the provision of ESs may be positive or negative on a local scale. Looking into land use changes—the conversion of agricultural land to forestland and urban areas—resulted in an increase in food supply and C emissions. Even the relatively less intensively urbanized studied area compared to megacities showed a mismatch between the supply and demand of ESs. This spatial mismatch, which indicates the EU's "cold spot" in the city center, would accelerate the ecological degradation of ecosystems. Examining the spatial and temporal relationships between LULC and ESs provides a clear rationale for the need to ensure sustainable land use and decision making in important policy priority areas. A clear assessment of whether the provision of ecosystem services meets their needs at different levels of urbanization will help to decide on effective land management solutions.

Supplementary Materials: The following supporting information can be downloaded at: https://www.mdpi.com/article/10.3390/land13040454/s1, Figure S1: Number of elderships with mismatched ES in Vilnius 2000 and 2020; Figure S2: Influence of urban and forest land ratios on ecological supply–demand ratio (ESDR) in Vilnius in 2000 and 2020.

Author Contributions: G.S.: conceptualization, data curation, methodology, writing—review and editing, and supervision. G.D.: investigation, methodology, formal analysis, writing—original draft, and visualization. All authors have read and agreed to the published version of the manuscript.

Funding: This research received no external funding.

Data Availability Statement: Datasets are available from the authors upon request. The data are not publicly available due to intellectual property/confidentiality issues.

Conflicts of Interest: The authors declare no conflicts of interest.

References

1. Gao, J.; O'Neill, B.C. Mapping Global Urban Land for the 21st Century with Data-Driven Simulations and Shared Socioeconomic Pathways. *Nat. Commun.* **2020**, *11*, 2302. [CrossRef] [PubMed]
2. Chakraborty, S.; Maity, I.; Dadashpoor, H.; Novotný, J.; Banerji, S. Building in or out? Examining Urban Expansion Patterns and Land Use Efficiency across the Global Sample of 466 Cities with Million+ Inhabitants. *Habitat Int.* **2022**, *120*, 102503. [CrossRef]
3. González-García, A.; Palomo, I.; González, J.A.; López, C.A.; Montes, C. Quantifying Spatial Supply-Demand Mismatches in Ecosystem Services Provides Insights for Land-Use Planning. *Land Use Policy* **2020**, *94*, 104493. [CrossRef]
4. Cumming, G.S.; Buerkert, A.; Hoffmann, E.M.; Schlecht, E.; Von Cramon-Taubadel, S.; Tscharntke, T. Implications of Agricultural Transitions and Urbanization for Ecosystem Services. *Nature* **2014**, *515*, 50–57. [CrossRef] [PubMed]

5. Ala-Hulkko, T.; Kotavaara, O.; Alahuhta, J.; Hjort, J. Mapping Supply and Demand of a Provisioning Ecosystem Service across Europe. *Ecol. Indic.* **2019**, *103*, 520–529. [CrossRef]
6. Baró, F.; Haase, D.; Gómez-Baggethun, E.; Frantzeskaki, N. Mismatches between Ecosystem Services Supply and Demand in Urban Areas: A Quantitative Assessment in Five European Cities. *Ecol. Indic.* **2015**, *55*, 146–158. [CrossRef]
7. Chen, J.; Jiang, B.; Bai, Y.; Xu, X.; Alatalo, J.M.; Bai, Y. Quantifying Ecosystem Services Supply and Demand Shortfalls and Mismatches for Management Optimisation. *Sci. Total Environ.* **2019**, *650*, 1426–1439. [CrossRef] [PubMed]
8. Larondelle, N.; Lauf, S. Balancing Demand and Supply of Multiple Urban Ecosystem Services on Different Spatial Scales. *Ecosyst. Serv.* **2016**, *22*, 18–31. [CrossRef]
9. Misiune, I.; Julian, J.P.; Veteikis, D. Pull and Push Factors for Use of Urban Green Spaces and Priorities for Their Ecosystem Services: Case Study of Vilnius, Lithuania. *Urban For. Urban Green.* **2021**, *58*, 126899. [CrossRef]
10. Pinto, L.V.; Ferreira, C.S.S.; Inácio, M.; Pereira, P. Urban Green Spaces Accessibility in Two European Cities: Vilnius (Lithuania) and Coimbra (Portugal). *Geogr. Sustain.* **2022**, *3*, 74–84. [CrossRef]
11. Peng, J.; Tian, L.; Liu, Y.; Zhao, M.; Hu, Y.; Wu, J. Ecosystem Services Response to Urbanization in Metropolitan Areas: Thresholds Identification. *Sci. Total Environ.* **2017**, *607–608*, 706–714. [CrossRef] [PubMed]
12. Ren, Q.; Liu, D.; Liu, Y. Spatio-Temporal Variation of Ecosystem Services and the Response to Urbanization: Evidence Based on Shandong Province of China. *Ecol. Indic.* **2023**, *151*, 110333. [CrossRef]
13. Sharma, S.; Nahid, S.; Sharma, M.; Sannigrahi, S.; Anees, M.M.; Sharma, R.; Shekhar, R.; Basu, A.S.; Pilla, F.; Basu, B.; et al. A Long-Term and Comprehensive Assessment of Urbanization-Induced Impacts on Ecosystem Services in the Capital City of India. *City Environ. Interact.* **2020**, *7*, 100047. [CrossRef]
14. Leitão, I.A.; Ferreira, C.S.S.; Ferreira, A.J.D. Assessing Long-Term Changes in Potential Ecosystem Services of a Peri-Urbanizing Mediterranean Catchment. *Sci. Total Environ.* **2019**, *660*, 993–1003. [CrossRef] [PubMed]
15. Wang, S.; Hu, M.; Wang, Y.; Xia, B. Dynamics of Ecosystem Services in Response to Urbanization across Temporal and Spatial Scales in a Mega Metropolitan Area. *Sustain. Cities Soc.* **2022**, *77*, 103561. [CrossRef]
16. Das, M.; Inácio, M.; Das, A.; Barcelo, D.; Pereira, P. Mapping and Assessment of Ecosystem Health in the Vilnius Functional Zone (Lithuania). *Sci. Total Environ.* **2024**, *912*, 168891. [CrossRef] [PubMed]
17. Chen, W.; Chi, G.; Li, J. The Spatial Association of Ecosystem Services with Land Use and Land Cover Change at the County Level in China, 1995–2015. *Sci. Total Environ.* **2019**, *669*, 459–470. [CrossRef] [PubMed]
18. Shi, Y.; Shi, D.; Zhou, L.; Fang, R. Identification of Ecosystem Services Supply and Demand Areas and Simulation of Ecosystem Service Flows in Shanghai. *Ecol. Indic.* **2020**, *115*, 106418. [CrossRef]
19. Ubarevičienė, R.; van Ham, M. Population Decline in Lithuania: Who Lives in Declining Regions and Who Leaves? *Reg. Stud. Reg. Sci.* **2017**, *4*, 57–79. [CrossRef]
20. Lazauskaitė, D.; Griškevičiūtė-Gečienė, A.; Šarkienė, E.; Zinkevičienė, V. Quality Analysis of Vilnius City Suburban Spatial Development. In Proceedings of the 9th International Conference "ENVIRONMENTAL ENGINEERING", Vilnius, Lithuania, 22–23 May 2014. [CrossRef]
21. Ubarevičiene, R.; Burneika, D. Fast and Uncoordinated Suburbanization of Vilnius in the Context of Depopulation in Lithuania. *Environ. Socio-Econ. Stud.* **2020**, *8*, 44–56. [CrossRef]
22. Copernicus CORINE Land Cover. Available online: https://land.copernicus.eu (accessed on 3 April 2023).
23. Russo, A.; Cirella, G. Modern Compact Cities: How Much Greenery Do We Need? *Int. J. Environ. Res. Public Health* **2018**, *15*, 2180. [CrossRef] [PubMed]
24. Kroll, F.; Müller, F.; Haase, D.; Fohrer, N. Rural-Urban Gradient Analysis of Ecosystem Services Supply and Demand Dynamics. *Land Use Policy* **2012**, *29*, 521–535. [CrossRef]
25. Koch, J.; Wimmer, F.; Schaldach, R. Analyzing the Relationship between Urbanization, Food Supply and Demand, and Irrigation Requirements in Jordan. *Sci. Total Environ.* **2018**, *636*, 1500–1509. [CrossRef] [PubMed]
26. Wang, J.; Zhou, W.; Pickett, S.T.A.; Yu, W.; Li, W. A Multiscale Analysis of Urbanization Effects on Ecosystem Services Supply in an Urban Megaregion. *Sci. Total Environ.* **2019**, *662*, 824–833. [CrossRef] [PubMed]
27. Lambin, E.F.; Meyfroidt, P. Global Land Use Change, Economic Globalization, and the Looming Land Scarcity. *Proc. Natl. Acad. Sci. USA* **2011**, *108*, 3465–3472. [CrossRef] [PubMed]
28. Fróna, D.; Szenderák, J.; Harangi-Rákos, M. The Challenge of Feeding the World. *Sustainability* **2019**, *11*, 5816. [CrossRef]
29. Hara, Y.; Tsuchiya, K.; Matsuda, H.; Yamamoto, Y.; Sampei, Y. Quantitative Assessment of the Japanese "Local Production for Local Consumption" Movement: A Case Study of Growth of Vegetables in the Osaka City Region. *Sustain. Sci.* **2013**, *8*, 515–527. [CrossRef]
30. Kousar, S.; Ahmed, F.; Pervaiz, A.; Bojnec, Š. Food Insecurity, Population Growth, Urbanization and Water Availability: The Role of Government Stability. *Sustainability* **2021**, *13*, 12336. [CrossRef]
31. Chen, A.; Partridge, M.D. When Are Cities Engines of Growth in China? Spread and Backwash Effects across the Urban Hierarchy. *Reg. Stud.* **2013**, *47*, 1313–1331. [CrossRef]
32. Ambros, P.; Granvik, M. Trends in Agricultural Land in EU Countries of the Baltic Sea Region from the Perspective of Resilience and Food Security. *Sustainability* **2020**, *12*, 5851. [CrossRef]
33. Liu, R.; Wang, M.; Chen, W. The Influence of Urbanization on Organic Carbon Sequestration and Cycling in Soils of Beijing. *Landsc. Urban Plan.* **2018**, *169*, 241–249. [CrossRef]

34. Zhang, Y.; Liu, Y.; Zhang, Y.; Liu, Y.; Zhang, G.; Chen, Y. On the Spatial Relationship between Ecosystem Services and Urbanization: A Case Study in Wuhan, China. *Sci. Total Environ.* **2018**, *637–638*, 780–790. [CrossRef] [PubMed]
35. Zhang, F.; Zhong, J.; Zhao, Y.; Cai, C.; Liu, W.; Wang, Q.; Wang, W.; Wang, H.; Jiang, X.; Yuan, R. Urbanization-Induced Soil Organic Carbon Loss and Microbial-Enzymatic Drivers: Insights from Aggregate Size Classes in Nanchang City, China. *Front. Microbiol.* **2024**, *15*, 1367725. [CrossRef] [PubMed]
36. Yang, J.-L.; Zhang, G.-L. Formation, Characteristics and Eco-Environmental Implications of Urban Soils—A Review. *Soil Sci. Plant Nutr.* **2015**, *61*, 30–46. [CrossRef]
37. Nowak, D.J.; Crane, D.E. Carbon Storage and Sequestration by Urban Trees in the USA. *Environ. Pollut.* **2002**, *116*, 381–389. [CrossRef] [PubMed]
38. Tang, Y.; Chen, A.; Zhao, S. Carbon Storage and Sequestration of Urban Street Trees in Beijing, China. *Front. Ecol. Evol.* **2016**, *4*, 53. [CrossRef]
39. Lal, R. Carbon Management in Agricultural Soils. *Mitig. Adapt. Strateg. Glob. Chang.* **2007**, *12*, 303–322. [CrossRef]
40. Kaveckis, G. Greenest Capital of the Baltic States—A Spatial Comparison of Greenery. *Balt. J. Real Estate Econ. Constr. Manag.* **2017**, *5*, 160–176. [CrossRef]
41. Barboza, E.P.; Cirach, M.; Khomenko, S.; Iungman, T.; Mueller, N.; Barrera-Gómez, J.; Rojas-Rueda, D.; Kondo, M.; Nieuwenhuijsen, M. Green Space and Mortality in European Cities: A Health Impact Assessment Study. *Lancet Planet. Health* **2021**, *5*, e718–e730. [CrossRef] [PubMed]
42. Xiao, R.; Lin, M.; Fei, X.; Li, Y.; Zhang, Z.; Meng, Q. Exploring the Interactive Coercing Relationship between Urbanization and Ecosystem Service Value in the Shanghai–Hangzhou Bay Metropolitan Region. *J. Clean. Prod.* **2020**, *253*, 119803. [CrossRef]

Disclaimer/Publisher's Note: The statements, opinions and data contained in all publications are solely those of the individual author(s) and contributor(s) and not of MDPI and/or the editor(s). MDPI and/or the editor(s) disclaim responsibility for any injury to people or property resulting from any ideas, methods, instructions or products referred to in the content.

Article

The Ecological Potential of Poplars (*Populus* L.) for City Tree Planting and Management: A Preliminary Study of Central Poland (Warsaw) and Silesia (Chorzów)

Jan Łukaszkiewicz [1], Andrzej Długoński [1], Beata Fortuna-Antoszkiewicz [1] and Jitka Fialová [2,*]

[1] Department of Landscape Architecture, Institute of Environmental Engineering, Warsaw University of Life Sciences—SGGW, ul. Nowoursynowska 159, 02-776 Warsaw, Poland; jan_lukaszkiewicz@sggw.edu.pl (J.Ł.); andrzej_dlugonski@sggw.edu.pl (A.D.); beata_fortuna_antoszkiewicz@sggw.edu.pl (B.F.-A.)

[2] Department of Landscape Management, Faculty of Forestry and Wood Technology, Mendel University in Brno, 613 00 Brno, Czech Republic

* Correspondence: jitka.fialova@mendelu.cz

Citation: Łukaszkiewicz, J.; Długoński, A.; Fortuna-Antoszkiewicz, B.; Fialová, J. The Ecological Potential of Poplars (*Populus* L.) for City Tree Planting and Management: A Preliminary Study of Central Poland (Warsaw) and Silesia (Chorzów). *Land* **2024**, *13*, 593. https://doi.org/10.3390/land13050593

Academic Editors: Alessio Russo and Shiliang Liu

Received: 25 February 2024
Revised: 7 April 2024
Accepted: 25 April 2024
Published: 29 April 2024

Copyright: © 2024 by the authors. Licensee MDPI, Basel, Switzerland. This article is an open access article distributed under the terms and conditions of the Creative Commons Attribution (CC BY) license (https:// creativecommons.org/licenses/by/ 4.0/).

Abstract: Urban environments face escalating challenges due to uncontrolled urbanization, rapid population growth, and climate changes, prompting the exploration of sustainable solutions for enhancing urban green spaces (UGSs). For this reason, poplars (*Populus* L.), due to their rapid growth, wide range adaptability to environmental conditions and versatility of use, have emerged as very promising. This comprehensive review synthesizes current knowledge regarding poplar's application in urban landscapes, emphasizing its multifaceted contributions and benefits. However, challenges arise from the variable lifespans of different poplar cultivars, necessitating strategic management approaches. Selecting cultivars based on growth rates, root system characteristics, and adaptability to urban conditions is pivotal. Adaptive replanting strategies, incorporating species with varying lifespans, offer solutions to maintain continual greenery in urban landscapes. Collaborative efforts between researchers, urban planners, and policymakers are essential for devising comprehensive strategies that maximize benefits while addressing challenges associated with their variable lifespans. In conclusion, harnessing poplar's potential in urban greenery initiatives requires a balanced approach that capitalizes on their benefits while mitigating challenges. Further research and adaptive strategies are crucial for sustained and effective utilization to create resilient and vibrant urban landscapes.

Keywords: urban green spaces; urbanized landscape; urban environments; climate changes; aesthetical values of trees; *Populus* ×*berolinensis*

1. Introduction

Trees play a crucial role in fostering sustainable urban ecosystems [1–13]. In urbanized and open landscapes, trees can provide a diverse range of ecosystem services—among others, the mitigation of environmental degradation and the enhancement of biodiversity. The impact of high greenery on cultural services is also invaluable [14–21]. Planting and maintaining trees in urban areas is widely recognized as a form of environmental biotechnology (phytoremediation)—the simplest and most direct means to reduce air, water, and soil pollution and increase carbon sequestration due to the vast biologically active surface areas that trees can produce [6,22–31].

Unfortunately, on average, in cities (downtowns), trees appear to have gradually decreasing lifespans or health and safety quality. Because of anthropopression, large, mature, and old trees are declining, and a percentage of new-planted trees do not compensate for the benefits of the old trees. In this way, very often, one large urban tree provides benefits equivalent to a few dozen young, new plantings [6,32–36]. At the same time, the need arises to obtain the effect of planting healthy and young trees as quickly as possible because only such specimens can provide the desired ecosystem services to face challenges like global climate changes [37–40]. Cities are mostly affected by poor soil properties and drought risks and are full of degraded areas after earlier periods of development. In this way, selected groups of trees available for heap difficult conditions must be considered in the future. Not all areas are able to be rained on or can be considered organic ground. Thus, as we look into the past, such obligations were not able to be met due to the poor conditions of the 20th century in Eastern Europe. However, some trees were first chosen for planting in such degraded areas where new developments were planned. In Polish cities, we have a lot of examples of the use of poplars and clones. Clones are represented mostly by *Acer negundo* L. and this species is mostly used in degraded areas and does not represent aesthetic value. On the other hand, we have observed other trees that are more ecologically valuable, which are poplars.

In such a context, poplar trees (*Populus* L.) have garnered attention in previous decades for their great potential to contribute significantly to the quality and quantity of urban greenery. It is just worth mentioning that before the Second World War, and then from the 1950s up to the end of 1970s, in some European countries (e.g., Poland or the Czech Republic), the mass afforestation of urbanized, industrial, and open landscapes was carried out with a significant share of poplars [41]. Difficult urban conditions make poplars valuable for introduction into UGSs because of their wide range of critical predispositions, including the following:

- Achieving quite large dimensions connected to rapid growth and an increase in biomass (e.g., LAI);
- Having a quick impact on the local microclimate, e.g., shading, transpiration, protection against wind and noise, etc.;
- Their phytoremediation abilities like the filtration of particular matter (PM) or gas pollution from air or the absorption of heavy metals from water and soil;
- Their high adaptability to various soil and water conditions;
- Their considerable tolerance for pollution of air, water, and soil;
- Their mass production of nursery material;
- Their good planting efficiency with the minimum necessary maintenance [41–44].

In general, the advantages of poplars, making them a promising planting material for improving urbanized landscapes, are connected to their rapid growth (allowing them to obtain a large amount of plant biomass in a relatively short time), their high adaptability to various soil conditions, and the diverse environmental benefits that they can provide [42–44].

On the other hand, the use of poplars in urban areas is limited due to some inconvenient features of these plants like their relatively short lifespans; low wood resistance (which makes them affected more by storms); shallow root systems, possibly damaging infrastructure; potential production of root suckers; quite long period of leaf fall (cleaning); relatively higher susceptibility to insect and fungal diseases; or production of seeds with cottony hairs, polluting the environment in the spring (only female specimens) [41–45].

In the context of urban environments, cities can expand the realized climatic niches of various species. This issue was demonstrated in a study involving five poplar species worldwide, *Populus balsamifera*, *Populus deltoides*, *Populus nigra*, *Populus tremula*, and *Populus tremuloides*, in cities across the globe [46]. Another research effort presents a global risk assessment of over 3000 tree species, including 19 poplar species planted in 119 cities worldwide. The study revealed that 15 species are potentially at risk due to increases in temperature in 100 cities globally [47]. Consequently, countries like China have noted the impacts of climate change on three poplar species, recognizing it as an important factor to be assessed, as indicated in the Global Ecology and Convention report [48].

Regarding the undoubted advantages of poplars (*Populus* L.) for diverse urban environments worldwide and keeping in mind their obvious limitations, the problem of to solve is how to ensure their optimal integration into urban green spaces (UGSs) for maximizing their environmental benefits, mitigating potential challenges, and ensuring their sustainable use. This publication aims to synthesize and critically examine the use of poplars in urban landscapes, emphasizing their aesthetic, ecological, economic, and social significance. Our goal was also to consolidate existing knowledge on utilizing poplars in urban greenery, identify critical areas for further research, and advocate for their strategic incorporation into urban planning for a more sustainable and resilient urban future.

Here, we have formulated our research thesis about the hypothesis that we aim to prove: the rational use of poplars in 21st-century cities worldwide, including those in Europe, the Middle East, and Asia, is essential for maintaining healthy greenery and urban woodlots. Proper selection of species/cultivars and their adaptation to habitat conditions are crucial for addressing ongoing climate change and enhancing quality of life for urban populations. While poplars are often undervalued in urban settings, their properties, such as rapid growth, large size, and tolerance to anthropogenic pressures, today make them invaluable in shaping urban forests amid changing climates.

2. Methodology

2.1. Research Framework

The research framework of the present study is described below. It consisted of a few main stages presented in Figure 1. The formulation of the main goal of our research allowed us to start the first stage of work. An extensive literature search was conducted to compile examples of research concerning poplars, especially in the context of their potential for use in urban green areas. Issues identified during the literature review led us to synthesize poplar species' traits, making them unique for use in diverse urban settings. It was also a theoretical background for case studies presented in the following stages of research. The next step involved collecting the results of field observation and investigations by the authors of woodlot forms consisting of various poplar taxa in urban areas. Field studies used techniques of dendrological inventories and included, among others, taxonomic identification, the spatial forms of woodlots (the horizontal and vertical structure), the measurement of parameters of representative trees by determining their conditions and health statuses, assessments of the dates of planting and the ages of representative trees, and photographic documentation. The analysis of collected data allowed us to show, in particular, one representative cultivar of balsamic poplar (*Populus* ×*berolinensis* (K. Koch) Dippel) for its up-and-coming use features in urbanized landscapes.

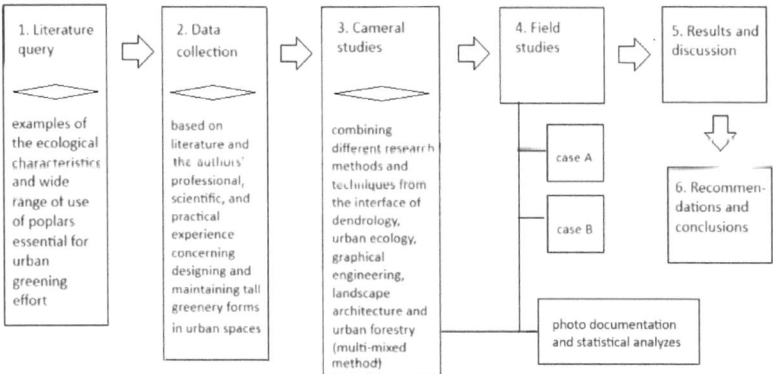

Figure 1. The research framework (the author's own elaboration).

The fourth step consisted of cameral studies of multi-mixed methods, combining different research methods and techniques from the interface of dendrology, urban ecology, graphical engineering, landscape architecture, and urban forestry.

Firstly, it consisted of a collection of archival photographs illustrating the Ziętek Promenade in Chorzów and Rakowiecka Street at the former headquarters of the Warsaw University of Life Sciences (WULS-SGGW) and the graphic processing of archive materials in GIMP and Inkscape graphical applications. The analysis of the photos was performed as follows:

Case A. Rakowiecka Street in Warsaw, Poland. Images from individual years (1925–2014) collected during the literature search were superimposed and aligned with the original 1930 photograph of the first tree planting next to the building, then scaled and aligned using perspective and grid tools in the GIMP graphics tool. The building's facade was measured using a Nikon laser rangefinder (model Forestry Pro 2, Nikon Vision Co., Ltd., Shanghai, China) for verification purposes. Then, the profile of the poplar at the entrance to the building was drawn for individual years, and the dimension was measured by reading the data from the grid. Subsequently, auxiliary illustrations were made showing the patterns of poplars in specific periods and the parameters obtained. On this basis, the ages of the trees were estimated.

Case B. The Ziętek's Promenade in Chorzów, Poland. First, three pictures illustrating the view of the even row of Berlin poplar trees were selected from the query literature (those from the 1950s, the transition from the 1950s and 1960s, and 2014). Then, the lengths and heights of characteristic elements in the field were measured with a rangefinder; lamps measured the distances between the poplars in the alley (dimensions similar to those of previous years). These dimensions were plotted based on illustrations, and then the heights and widths of the tree bark in specific periods were estimated. Average values for plant growth for each year were used, i.e., 1959, 1965, and 2014. Subsequently, auxiliary illustrations were made to show the patterns of poplars in specific periods and obtain parameters.

Secondly, we tested the relationship between the selected dendrometric parameters (e.g., height vs. age) using Statistica 13.0 software. A representative group of 10 young Berlin poplars from a newly replanted section of the Ziętek's Promenade measured in 2016, 2021, and 2023, as well as selected preserved poplars on Rakowiecka Street in Warsaw (data from 2012 and 2024), were subjected to a statistical study (Sperman's rank correlation analysis).

Finally, we discussed the results achieved and formulated concussions.

2.2. Materials and Methods

A mixed-method approach was employed in this study. The methods used can be divided into two groups: analytical and field-based methods.

The first group of methods involved a meticulous review of the literature data and its critical analysis. To address the gaps in our understanding of poplars' urban use, we diligently utilized major databases, employing a broad set of keywords indirectly related to these trees. This was carried out to explore and compile examples of the ecological characteristics and wide range of uses of poplars crucial for urban greening efforts. Various combinations of keywords, such as "urban green spaces", "green areas", "urban environments", "climate changes", "trees' values", "poplars", and "phytoremediation", were used to search online literature databases, including Scopus, ISI Web of Knowledge, EBSCO, and Google Scholar. Additionally, data collected in this research stage included the literature sources and the authors' professional, scientific, and practical experience in designing and maintaining tall greenery in urban areas.

An essential part of the fieldwork involved performing queries of archival sources (books, digital archives) to identify maps and historical photographs from locations selected for field studies (Case A and Case B). The obtained materials allowed for determining the growth rate of tree height over the past decades via comparison with other objects of known dimensions in precisely dated photographs.

In the fieldwork group, the authors conducted extensive long-term observations and detailed studies at both sites (Case A and Case B). The essential source of data comprised dendrological inventories conducted on Rakowiecka Street in Warsaw and the Ziętek's Main Promenade in Silesia Park in Chorzów (materials in the authors' collections). Tree inventories were conducted in similar seasons of the year (summer–autumn period), considering parameters such as trunk circumference at a height of 1.3 m, tree height, crown width, the assessment of health condition, and habit maintenance. The research material was obtained and supplemented with annual observations of tree conditions in the spring–summer and autumn periods. The Berlin poplars selected for this study grow linearly along street tree lanes and park promenades. Hence, tree applications in Cases A and B are comparable. Photographic documentation serves as supplementary material to field measurements. It is a valuable form of research documentation, allowing for the presentation of examples illustrating the research problem and the locations and objects of direct field studies.

The data gathered through both analytical and field-based methods provide robust material for statistical analysis. In this study, we employed correlation analysis using the *t*-test (after conducting a test of normal distribution compliance using the Shapiro–Wilk test). The data analysis was performed using the reliable Statistica 13.0 software.

3. Results and Discussion

Our review of scientific databases has shown that there is only very limited research devoted to the direct use of poplar cultivars in woodlots located in UGSs. But, indirectly, some of these studies also indicate the practical potential of poplars for use in anthropogenic environments. The conducted review of the literature revealed that poplars offer significant potential in current research for providing ecosystem services and mitigating pollutants in urban areas, recognizing for their role in improving air quality and acting as natural barriers for contaminant retention [41,49–53]. In cities, poplars are valuable trees for the following tasks:

- Obtaining quick visual results in urban green spaces and cityscapes;
- Reinforcing the environment and shaping a favorable microclimate in a relatively short time period;
- Covering and masking buildings and unattractive objects and views, e.g., warehouses, landfills and heaps, etc.

Integrating poplars into urban greenery initiatives presents challenges and opportunities that must be carefully navigated, including selecting species considering their environmental impacts [54], managing potential trace metal contamination [55], ensuring long-term maintenance, addressing community concerns, and recognizing ecological and psychological benefits for sustainable urban environments.

3.1. The Contribution of Poplars to the Urban Environment

As mentioned, urban landscapes in the 21st century suffer from deforestation, which stems from numerous factors. In context of rapid global climate changes (droughts, hurricanes, floods, etc.) and the need to develop sustainable urban environments, this phenomenon could be mitigated through comprehensive reforestation initiatives and the development of urban green spaces [56–60] connected to effective urban planning [58,59]. The expansion of urban areas and changes in land cover pose threats to urban vegetation, emphasizing the importance of sustainable management [61]. In this case, strategic tree selection is one of the necessary comprehensive solutions that provides ecosystem services such as carbon sequestration, biodiversity support, and air and water pollution mitigation. Poplars could stand out in this field due to their suitability for resolving the contemporary challenges facing urban and industrial environments [42–44] (Figure 2).

It has been proved that poplar trees can efficiently remove airborne particulate matter (PM) and associated metals through phytoremediation, contributing to urban air quality enhancement, while their adaptation to adverse environments and fast growth make them an interesting alternative for urban landscaping [52,53]. Cultivars of poplars—if used properly (carefully selected, e.g., in relation to spatial and site conditions)—could effectively improve air quality by capturing PMs, thereby reducing air pollution levels [62].

Figure 2. (**A**) Original group of *Populus* ×*canadensis* Moench (*P.* ×*euroamericana* Guiner) in Pole Mokotowskie Park, 2019, Warsaw, Poland, and (**B**) a new group of *Populus* ×*canadensis* 'Koster' in 2020 (author: J. Łukaszkiewicz). (**C**) *Populus simonii* Carrière 'Fastigiata' as urban street trees along a pedestrian route with sheltering functions (author: J. Łukaszkiewicz, 2018). (**D**) *Populus* ×*canadensis* Moench 'Marilandica' planted in by-water shelterbelts beside Żerański Canal, Warsaw, Poland (author: J. Łukaszkiewicz, 2016) and *Populus* ×*canadensis* Moench 'Marilandica' planted in by-water shelterbelts beside Żerański Canal, Warsaw, Poland (author: J. Łukaszkiewicz, 2016). (**E**) *Populus* ×*canadensis* Moench (*P.* ×*euroamericana* Guiner) as a high greenery screen for multi-story residential buildings with sheltering functions against the traffic on Sobieskiego Street, Warsaw, Poland (author: J. Łukaszkiewicz, 06.2019). (**F**) *Populus nigra* L. 'Italica' planted repeatedly in regular groups of four trees each, creating a green wall along the interior borders of the "Field of Mars", one of the most important interior areas of Silesia Park in Chorzów, Poland (author: B. Fortuna-Antoszkiewicz, 2014).

Poplars could also be effective at carbon sequestration by accumulating trace elements from polluted urban soils, serving as bioindicators for urban environmental pollution assessment [63]. Furthermore, poplar trees can modify soil microbial communities, enhance soil stability, and provide valuable habitats for ground beetles and entomofauna, enriching urban biodiversity [64].

Because of their fast growth, poplars could be pivotal for counteracting the urban heat island effect and improving urban microclimates. Urban tree cover supplemented with plantings of different poplar taxa could more quickly mitigate the urban heat island effect. Poplar trees enhance the urban thermal environment through transpiration, stimulating urban "cold islands" [65–70].

Of course, in the case of poplars in cities, their cultivation and maintenance is necessary. Considering diverse research findings to ensure the optimal growth and sustainability of poplars in urban settings while minimizing risks, effective maintenance practices such as pruning, soil amendment, and irrigation are crucial [71–78]. Challenges such as insect pests and heavy metal accumulation in poplar trees require the careful management and selection of clones, especially when monoculture plantations are planned to be introduced [63,79–82].

3.2. The Fast Increments of Poplars

In the past few decades, a lot of practical knowledge about the use of poplars in plantations or tree stands has been accumulated (e.g., [45,83]). For instance, in Central and Eastern Europe, first before the Second World War and then in a period from the 1950s to the 1970s, the mass afforestation of urban, industrial, and open landscapes was carried out, motivated by economic purposes, counteracting environmental degradation and improving landscape values (e.g., technical and shielding functions, phytoremediation, windbreaks, anti-snow, anti-erosion and others). In Poland, for instance, it is estimated that that at that time, the share of various poplars' taxa in shelterbelts and woodlots reached up to $\frac{1}{4}$ (25%) of total trees species [41,50,84–86]. The intended objective was to achieve the desired effect in the possible shortest possible time.

Traditionally, poplars have been considered to be one of fastest-growing trees in Europe and Asia (similarly to some willow species), e.g., *P. alba* L. and *P. nigra* L. 'Italica' increase in height by ± 1.7 m/year (juvenile life phase), reaching a maximum rate of shoot elongation of ca. 17–25 cm/week (referenced to climate zone 6) [41,45,87,88].

The research from databases indicates that the use of poplar species in urban greenery initiatives, including *P. tomentosa*, *P.s alba* 'Berolinensis', and *P. nigra*, is driven by their rapid growth, strong adaptability, and tolerance to environmental stressors [53,89,90]. These species are favored for their ability to thrive in diverse urban settings, offering benefits such as flood tolerance and resilience to adverse conditions [53,91].

Populus tomentosa exhibits rapid growth and ecological adaptability, making it ideal for establishing greenery in urban areas [89]. Similarly, the hybrid triploid *P. alba* 'Berolinensis' demonstrates fast growth and high stress tolerance, making it suitable for urban forestry [53]. *P. nigra*'s flood tolerance further enhances its suitability for use in urban settings, particularly in flood-prone areas [91]. However, its susceptibility to diseases like leaf rust poses challenges [92]. Therefore, to ensure long-term success, general careful disease management is crucial when selecting poplar cultivars for urban greenery projects.

3.3. The Short Lifespan of Poplars

One of the controversies that has arisen in relation to poplars is the very short lifespan of these trees. Short-lived (a few dozen years) status is mainly characterized by taxa that are hybrids of botanical species or their varieties, including cultivated varieties or hybrids (cultivars are marked as 'cv' and hybrids are marked with '×' preceding the second part of systematic name), e.g., *P. nigra* L. 'Italica', *P. simonii* Carrière 'Fastigiata'), or *P. ×canadensis* Moench (*P. ×euroamericana* Guiner) [41,87]. Species that are definitely more durable, achieving medium- and long-lived status (life expectancy of ± 150 (200) years in natural conditions), are the following:

- Typical botanical species, e.g., white poplar (*P. alba* L.) and black poplar (*P. nigra* L.);
- The intra- and inter-sectional hybrids of the balsam poplar *TACAMAHACA* and some cultivated varieties, e.g., *P. simonii* Carrière and *P. maksymowiczii* Henry, as well as balsam poplar hybrids, e.g.,: *P. ×berolinensis* Dippel, and hybrids like 'NE 49' and 'NE 42' (Table 1).

Table 1. Selected examples of the balsam poplars section (*TACAMAHACA*) species and hybrids used, among others, in the afforestation of urban landscapes (authors' elaboration based on [84–86,93,94]).

Balsam Poplar Section (*TACAMAHACA*)	
Subsection	Selected Representative
American balsam poplars	*P. balsamifera* (syn. *P. tacamahaca*) *P. trichocarpa*
Asian balsam poplars	*P. simonii* *P. maximowiczii* *P. laurifolia*
Intra- and inter-sectional hybrids of balsam poplars	Europe: *P.* ×*berolinensis* (K. Koch) Dippel 'Berlin'—Berlin poplar (male hybrid) *P.* ×*berolinensis* (K. Koch) Dippel 'Petrowskyana'—Berlin poplar—so-called "Tsar's"; *P.* ×*berolinensis* 'Razumovskyana' (both are female hybrids s) USA (in the 1920s by E.J. Schreiner and A.B. Stout—the so-called Schreinerian hybrids): Populus 'NE 49' or 'Hybrid 194' (*P.* x 'Hybrid 194') (male hybrid) Populus 'NE 42' or 'Hybrid 275' (*P.* x 'Hybrid 275') (male hybrid) Populus 'NE 44' or 'Hybrid 277' (*P.* x 'Hybrid 277') (male hybrid) Populus 'Androscoggin' (male hybrid), 'Geneva', 'Oxford' (female hybrids)

3.4. The Use of Berlin Poplar in the Afforestation of Urban Landscapes—Field Research

Regarding the literature on the use of poplars in urban conditions [84–86,93,94], our field research (Section 2) focuses on one selected cultivar—the Berlin poplar (Section 3.3). Our decision to study this particular cultivar was dictated by its many interesting characteristics, making it, in the past, a good choice for planting in cities. Therefore, by examining selected locations in Poland, we wanted to determine how the Berlin poplar performed in urban conditions. This cultivar was bred around 1870 in the botanical garden in Berlin as the male form of *Populus* ×*berolinensis* (K. Koch) Dippel 'Berlin'. Then, at the end of the 19th century, at the Petrovsko-Razum Agricultural Academy near Moscow, female forms were bred: *P.* ×*berolinensis* (K. Koch) Dippel 'Petrowskyana' (*P.* ×*petrowskyana* (Regel) C. K. Schneid.—the so-called Tsar's poplar) and *P.* ×*berolinensis* (K. Koch) Dippel 'Razumovskyana'. It tolerates dry urban environments and dry soil very well; it has also been successfully tested on a gravel–sand base with an inaccessible groundwater level. It grows well even in sloping localities. Conversely, *P.* ×*berolinensis* (K. Koch) Dippel is susceptible to cancer, as it is planted in locations with high groundwater levels, prone to flooding or with high air moisture levels. This cultivar is suitable for urban areas due to its narrow oval crown, which will keep its shape even when old. It has a straight, continuous trunk. That is why it is especially recommended for planting in lines or rows, e.g., along streets or avenues. Berlin poplar wood is stronger than that of other poplar species and cultivars. That is why these trees growing in the alleys do not suffer much from fractures in the canopy. Despite the inhibiting factors of the urban environment, this poplar can grow very fast, reaching up to 30 m in height and achieving a trunk diameter of 1.0 m at breast height (DBH). Leaves can also accumulate PM from the air, so Berlin poplars also possess phytoremediation abilities.

Berlin poplar's lifespan varies, though in good health, it can easily reach over 60 years and even more (as indicates our observations). However, exchanging overmatured trees over 40 years old for younger specimens is sometimes advisable in urban areas. Such a policy of exchanging older generations of trees is similar to how urban plantations are maintained. A distinctive feature of the female forms of the Berlin poplar is the production of seed down, which is abundantly secreted by the trees in May. In the context of these trees' use, especially in streets, avenues, and squares, this is a very undesirable factor, which may require abandoning the use of female forms in urban areas. Attention should also be paid to the shallow root systems of these trees, which may cause damage to paved surfaces, and as a consequence, they should be planted in locations that allow sufficient space for rooting (e.g., wide grassy sections along streets, etc.). During past decades, Berlin poplars were used in urban locations in Central European countries, like the Czech Republic or Poland.

As, compared to many other species and cultivars, Berlin poplars are much better suited for planting in cities, our research was focused on two representative locations in Poland where such trees have been used in the context of the urban environment. The first research area is Rakowiecka Street in Warsaw, and the second is the great Ziętek Promenade in Silesia Park in Chorzów.

Case A. The research area of Rakowiecka Street in Warsaw, Poland.

Rakowiecka Street in Warsaw began to be intensively built in the second half of the 19th century, and this process culminated in the 1950s (Figure 3). The characteristic number 8 building of the WULS-SGGW's (Warsaw University of Life Sciences) headquarters was partially put into service in 1929. In 1930, Rakowiecka Street was upgraded, with road infrastructure with green belts planted with Tsar's Berlin poplars female hybrids (*Populus* ×*berolinensis* (K. Koch) Dippel 'Petrowskyana'—*P.* ×*petrowskyana* (Regel) C. K. Schneid.); the young trees were 3.5 m high, with spacing of 8.0–10.0 m in each row [95].

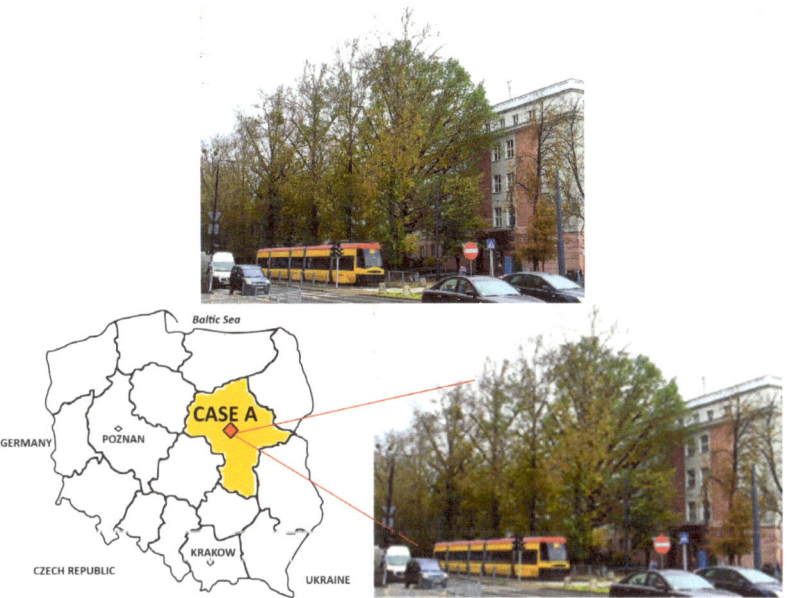

Figure 3. The magnificent streetside row of *Populus* ×*berolinensis* (K. Koch) Dippel 'Petrowskyana' along Rakowiecka Street, Warsaw, Poland (Case A)—the adjusted size of the trees relative to the street scale is visible (author: J. Łukaszkiewicz, September 2017).

The street buildings and rows of poplars survived in general the destruction of World War II. In the 1940s, the trees were large enough to provide shade for the WULS-SGGW building and protection against noise and pollution from the street [95] (Figure 4). However, in the late 1970s, specific problems were already noticed related to the collision of rapidly growing trees with the infrastructure of the buildings and the traction network [96]. In addition, another problem was the shallow root system of the poplars that lifted the paving slabs. Trees were planted too shallowly, and permeable surfaces were not used. One of the significant inconveniences of lining the street with the female form of Berlin poplar (*Populus* ×*berolinensis* (K. Koch) Dippel 'Petrowskyana') was the annual abundant release of seed down (cotton-like), which disturbed people and caused much littering of the street and apartments in nearby houses. In the following years, Berlin poplars were gradually felled during infrastructure modernization projects or due to weather anomalies. In 2024, only a few trees remain from the avenue that existed in past decades, which are the subject of our further measurement and analysis.

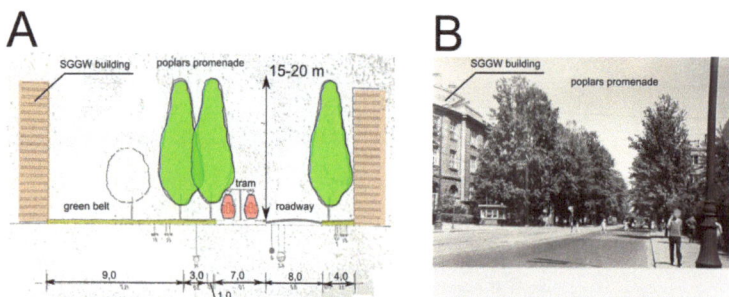

Figure 4. The cross section of Rakowiecka Street in Warsaw, in an eastwards direction (**A**), lined with two-sided rows of *Populus ×berolinensis* (K. Koch) Dippel 'Petrowskyana' (1973), and a view of Rakowiecka Street near the WULS-SGGW's headquarters building (**B**) in 1973 [95].

We analyzed historical iconographic materials [96,97] and performed field measurements, i.e., measurements of trees, the street, and the facade of the WULS-SGGW building. The results are presented in Figures 5–8 and Table 2. Figure 8a, using evidence from 1925, shows the measured heights of the building's floors during field tests. Based on the analysis of the remaining illustrations (Figure 5B–H), it can be concluded that the trees planted in the 1920s and 1930s were homogeneous plant material with equal parameters (height of approximately 3.5 m). The trees reached the windows of the high ground floor of the WULS-SGGW building (Figure 5B). In 1937, it can be seen that the trees reached the windows of the first floor of the WULS-SGGW building, reaching a height of approximately 8.5 m (Figure 5C). In 1939, the trees reached the windows of the second floor of the WULS-SGGW building, reaching a height of approximately 11.0 m (Figure 8D). In the illustration from 1947, the trees reach the roof of the SGGW building, reaching a height of approximately 20 m (Figure 8E). In 1955, the trees obscured the SGGW building and reached a height of approximately 21 m (Figure 5F). In 1975, the trees were already mature and reached a height of approximately 22–25 m (Figure 5G). At the beginning of the 21st century, the poplars remaining from the original forest cover had already reached their maximum size; in 2012, their height was 24.0 m, while in 2017, it was 28.0 m, and in 2024, it was 30.5 m. Due to collisions with buildings and technical infrastructure, the trees were cut, and their shapes were deformed. The illustration from 2017 shows partial losses in the street trees, and the trees in the planted avenue were becoming old and falling out (Figure 5H).

Field measurements showed that *Populus ×berolinensis* (K. Koch) Dippel 'Petrowskyana' trees planted in 1920s and 1930s along Rakowiecka Street in Warsaw (section Boboli Street—Niepodległości Av.) achieved, in 2013, an average trunk girth of 220–235 cm (measured at 1.3 m—breast's high), and in 2024, they achieved an average girth of 241.3 cm (Table 2). Tree age parameters were calculated and correlated with illustrations in individual years and heights (Figure 6). On this basis, it was estimated that in 1930, the trees were approximately 3 years old; in 1937, they were approximately 8–10 years old; in 1937, they were approximately 12 years old; in 1947, they were approximately 20 years old; in 1955, they were approximately 28 years old; in 1975, they were approximately 48 years old; and in 2017, they were already approximately 90–95 years old. Now, in 2024, they are approx. 97–102 years old. The figure shows averaged values or those rounded to the upper or lower values for better data visualization.

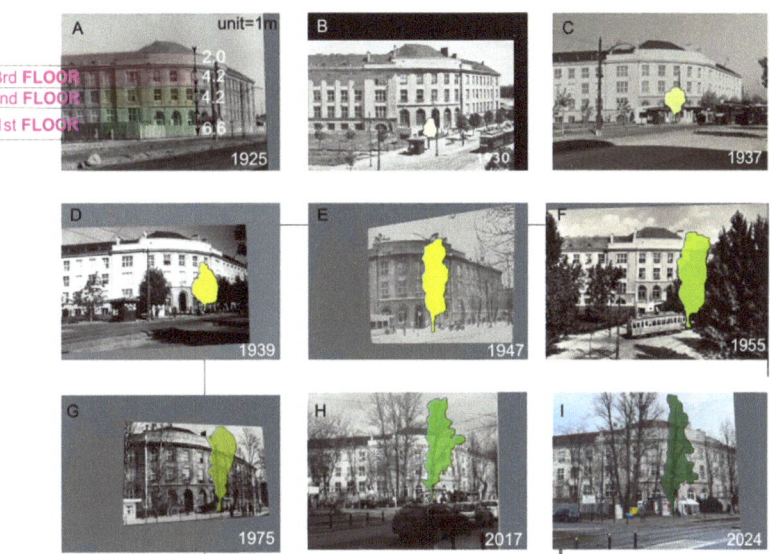

Figure 5. The sequence of illustrations (**A–I**) of Rakowiecka Street in Warsaw showing the growth of *Populus ×berolinensis* Dippel 'Petrowskyana' (1925–2024) near the WULS-SGGW's former headquarters building [authors' own elaboration based on historical photos of Rakowiecka Street in Warsaw (photos from [97] and own photo: Łukaszkiewicz, 23 February 2024)].

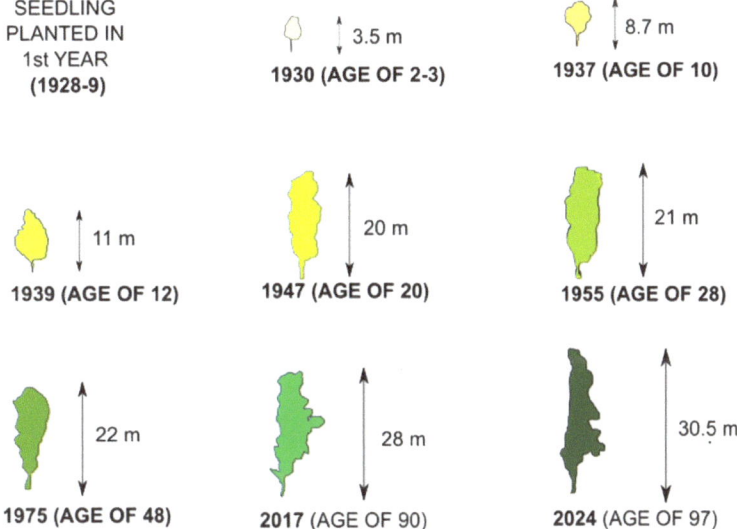

Figure 6. The schematic habits and estimated heights of *Populus ×berolinensis* (K. Koch) Dippel 'Petrowskyana' trees along Rakowiecka Street in Warsaw in the chosen years of the period 1930–2024, located near the WULS-SGGW's former headquarters building. The authors' elaboration was based on the pictures in Figure 8 (photos A-I [97] and their own photo: Łukaszkiewicz, 23 February 2024).

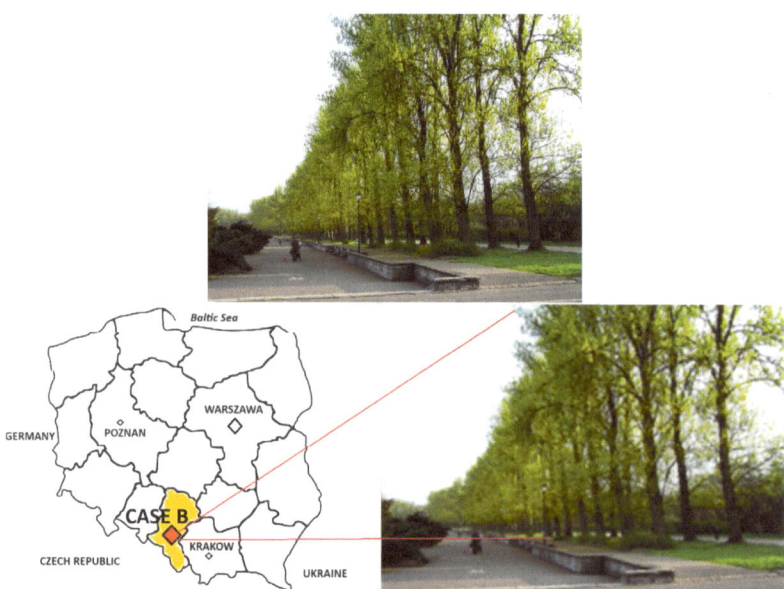

Figure 7. The form of a high, even wall of male hybrid poplars: *Populus* ×*berolinensis* (K. Koch) Dippel 'Berlin' in Silesia Park's Ziętek Promenade, Chorzów, Poland (Case B) (B. Fortuna-Antoszkiewicz, May 2014).

Table 2. Descriptive statistics of the parameters of the studied Berlin poplars near the SGGW building on Rakowiecka Street in Warsaw in 2012 and 2024 (n—number of selected trees, *t*-test; *p*—statistical significance; SD—standard deviation).

Parameter	Year	Age [Years]	Descriptive Statistics			*t* Test
			Mean (+/−SD)	Me	Min–Max	
height [m]	2012 [n = 16]	85	23.87 (1.86)	224.00	20.0–27.0	$p = 0.0175$
	2024 [n = 12]	97	25.5 (1.40)	25.60	23.5–27.5	$t = -2.536$
circumference [cm] at 1.3 breast height	2012 [n = 16]	85	225.87 (50.98)	224.00	116.00–321.00	$p = 0.4341$
	2024 [n = 12]	97	241.33 (50.93)	244.50	146.00–339.00	$t = -0.794$

Table 2 shows the descriptive statistics of average measurements of tree height and trunk circumference at a height of 1.3 m and compares the significance of the differences in the average values of these measurements. In order to compare the significance of differences between the average values of tree height and trunk circumference at a height of 1.3 m, Student's *t*-test was used (after performing a test of compliance with the normal distribution using the Shapiro–Wilk test). In the case of the data from 2024, the number of trees (n = 12) was lower than in 2012 (n = 16) because some trees fell due to weather anomalies, poor health, or infrastructure modernization. Based on the results obtained, it can be concluded that in the tested sample, both the trunk circumferences and the heights of the trees increased over the 12 years, but statistically significant growth was only observed in the case of height.

We also checked whether the increase in tree height and circumference depended on the spacing between the examined trees; Spearman's rank correlation analysis was used. The average spacing between the examined trees was 10.3 m, the smallest distance between them was 8.0 m, and the largest distance was 16.0 m (SD = 2.25 m). A negative relationship was observed in the case of height, but despite the coefficient value r = 0.64 indicating a relatively significant relationship, it was statistically insignificant (r = −0.64; *t* = −2.21;

$p = 0.0627$). In the case of trunk circumference, a similar direction of relationship was observed, which was also statistically insignificant, but the strength of this relationship was much weaker ($r = -0.14; t = -0.36; p = 0.7256$). This direction means that as the distance between trees decreases, the value of the analyzed parameters increases.

Case B. The research area of Ziętek Promenade in Silesia Park in Chorzów, Poland.

Silesia Park (The Voivodship Park of Culture and Recreation in Chorzów, Poland), with a total area of ca. 600 ha, was established on land of poor quality, partially degraded by mining and metallurgy industries. The main objective was to improve the quality of life for residents of Silesia by creating the bulk enclave of greenery combined with a versatile program for active recreation in this partially degraded area. After many years, Silesia Park has become an example of successful restoration and naturalization of the anthropogenic landscape [98,99].

The Ziętek Promenade in Silesia Park is the main, wide walkway connecting the most attractive park program elements, and it is the backbone of the park's composition. Regular tree arrangements, such as multi-row bosquets in a checkerboard pattern, are the leading theme of the spatial composition. The original design, including vegetation, is preserved and visible. The main feature of that park section is a magnificent double row of male hybrid Berlin poplars (*Populus* ×*berolinensis* (K. Koch) Dippel 'Berlin'). It runs along the banks of the Park's great pond, constituting a uniform "living wall", which ideally fits the scale of the vast park's interior. In 2016, at the end of the promenade near Al. Główna, a section of an old poplar row was exchanged for plantings of new trees of the same specimen [6,98–100].

Figure 8. The analyzed sequence of historical photographs of Silesia Park in Chorzów, Poland, showing the double row of Berlin poplars (*Populus* ×*berolinensis* (K. Koch) Dippel 'Berlin') in the Park in Chorzów (the authors' elaboration): (**A**) a view of the promenade from the late 1950s [101], (**B**) a view of the promenade from the 1960s or 1970s [102]; (**C**) a view of the avenue of adult poplars in 2014 (photo by J. Łukaszkiewicz, 2014) [98].

In the first step, we analyzed iconographic materials and performed field measurements. The results are presented in Figures 8 and 9 and Tables 3 and 4. Based on the data obtained, it can be concluded that the trees planted in the 1950s on the Park's promenade reached heights of 4.5 m at the end of the 1950s (the estimated year was 1959), which were equal to the height of the park lamp (Figure 8A). On this basis, the remaining parameters were estimated: widths equal to 2.0 m and spacing equal to 6.0 m. Next, the average height of poplars in the 1960s and 1970s was estimated (the estimated year was 1965), which was 9.0 m, and the width of the crown was 6.0 m because the trees touched each other with their crowns, building a homogeneous wall-like row of trees (Figure 8B,C). In turn, the analysis of photos and our own field tree inventory from 2014 shows that their parameters were, on average, as follows: height—approx. 24.0 m; crown width—ca. 8.0 m; mean trunk girth at 1.3 m—ca. ± 197 cm [98].

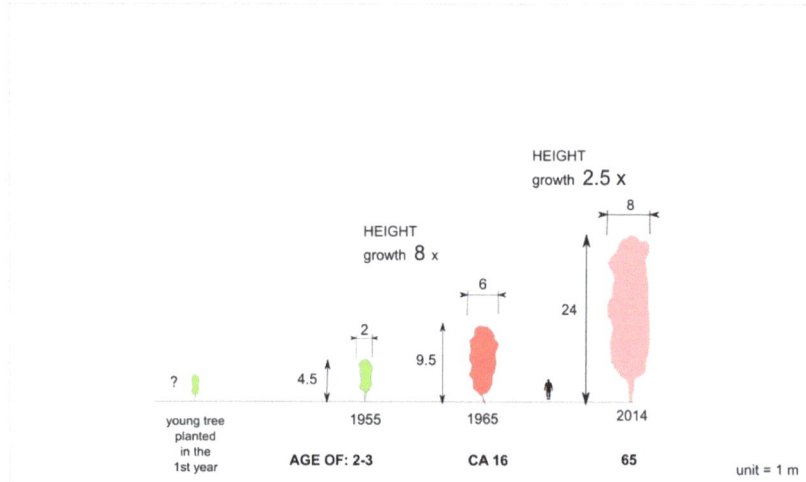

Figure 9. The increase in the height and spacing of the crowns of historic Berlin poplars (*Populus* ×*berolinensis* (K. Koch) Dippel 'Berlin') in Silesia Park in Chorzów in the years 1955–2014 (authors' elaboration based on [98,101,102]).

Table 3. Case B. Descriptive statistics of the parameters of young studied Berlin poplars from the Ziętek's Promenade in Silesia Park in Chorzów in 2016, 2021, and 2023 (n—number of selected trees; SD—standard deviation).

Parameter	Year	Age [Years]	Descriptive Statistics		
			Mean (+/−SD)	Me	Min–Max
height [m]	2016 [n = 10]	3	4.96 (0.10)	5.0	4.8–5.1
	2021 [n = 10]	8	8.40 (0.51)	8.5	7.6–9.5
	2023 [n = 10]	10	13.90 (0.42)	14.0	13.5–14.5
circumference [cm] at 1.3 breast height	2016 [n = 10]	3	12.90 (0.87)	13.0	12.0–14.0
	2021 [n = 10]	8	63.65 (6.11)	62.0	57.0–76.0
	2023 [n = 10]	10	79.80 (5.12)	82.0	74.0–86.0
crown average diameter [m]	2016 [n = 10]	3	2.17 (0.26)	2.0	2.0–2.5
	2021 [n = 10]	8	4.50 (0.33)	4.5	4.0–5.0
	2023 [n = 10]	10	7.90 (0.26)	8.0	7.5–8.2

Table 4. Case B. Descriptive statistics of the parameters of the examined poplars on the Ziętek's Promenade in Chorzów Park in 2016, 2021, and 2023 (*t*-test; *p*—statistical significance; SD—standard deviation).

Age	Parameter	Growth/1 Year	Correlation		
			R(X.Y)	t	p
2–8	height [m]	0.69	0.9740	16.1	<0.0001
	circumference 1.3 [cm]	10.15	0.9831	20.1	<0.0001
	crown diameter [m]	0.66	0.9689	14.6	<0.0001
8–10	height [m]	0.83	0.9853	20.8	<0.0001
	circumference 1.3 [cm]	5.38	0.8146	5.1	<0.0002
	crown diameter [m]	1.33	0.9837	19.8	<0.0001

On this basis, the ages of the trees were assessed approximately: ±3 years in 1959, ±10 years in 1965, and ±65 years in 2014 (Figure 9). Knowing the poplars' average height for individual years, it can be concluded that these trees, with favorable conditions for development (city park space), achieved their most significant growth of 8.5 times in approximately 6–10 years (between 1959 and 1965) and achieved relatively slower growth during the mature phase of life—over the next ca. fifty years (between 1965 and 2014), growth was equal to only 2.5 times. For better data visualization, the following figure uses average values or extreme ranges (Figure 9).

Due to the ageing of the original poplars and for safety reasons, the gradual replacement of trees on the Ziętek's Promenade began in 2016 (Figure 10). Following the original project assumptions, new Berlin poplars of the same variety were planted in place of the old trees. In the first section of the poplar row, replaced on April 2016, 37 new young poplars were planted (planting material with a girth of 12–14 cm).

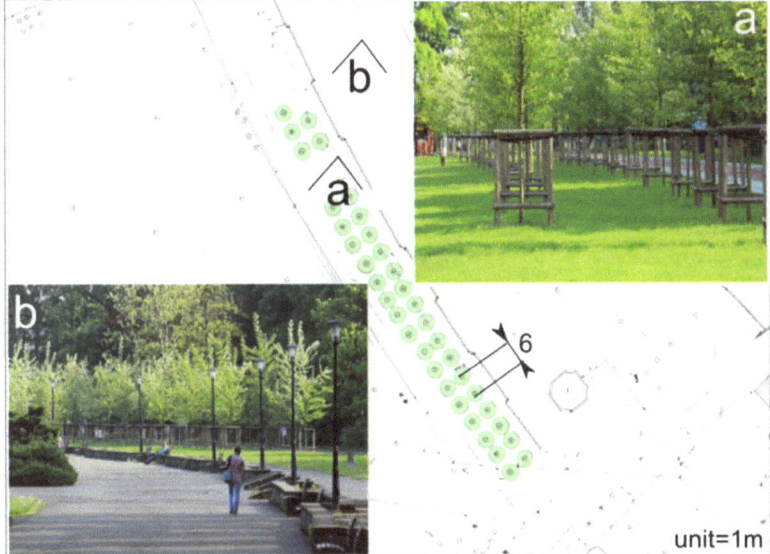

Figure 10. The perfect condition of young (**a**) Berlin poplar trees ((*Populus* ×*berolinensis* (K. Koch) Dippel 'Berlin') planted in exchange for one section of the Ziętek's Promenade in Chorzów in 2016, together with a general view (**b**) of the promenade's main walkway (author: J. Łukaszkiewicz 2018).

Our next step was to value descriptive statistics, estimated for measurements collected from ten newly planted Berlin poplars (*Populus* ×*berolinensis* (K. Koch) Dippel B'erlin') in 2016, 2021, and 2023. Based on the results obtained, it should be concluded that with age, there is a systematic increase in all the parameters examined (height, trunk circumference, crown width).

Next, the average annual increase in the studied parameters and the strength of the relationship between age and the average growth rate of the studied trees was estimated. Sperman's rank correlation analysis was used to estimate the strength of the relationship, and the growth rate was estimated by using the obtained average values of the studied characteristics and dividing them by the appropriate number of years. Table 3 summarizes the results of these analyses. They show that between the fifth and eighth years of life of Berlin poplars, the trunk circumference grows most intensively, while in the following years, trunk growth slows down in favor of crown growth (Table 4).

The conducted field observations of older *Populus* ×*berolinensis* specimens from Warsaw's Rakowiecka Street (Case A) and the Ziętek's Promenade in the Silesia Park in Chorzow (Case B) reveal their resilience in urban settings, with some specimens surviving for up to 100 years. While trees planted in the 1930s on Rakowiecka Street have largely perished, Berlin poplars in Chorzów have shown durability over several decades [103]. However, urban conditions in Warsaw have led to premature fall-outs, with trees succumbing as early as age 40 [103].

Statistical analyses based on parameters like tree height, trunk circumference, and crown growth indicate significant changes in the early years of a tree's life, with a clear correlation between tree age and growth parameters observed in the Ziętek's Promenade in the Silesia Park in Chorzow. Older trees in Warsaw also showed increases in height and trunk circumference, albeit they were statistically significant only for the height parameter. Growth rates decrease notably after the eighth year of a tree's life, with advanced-age trees exhibiting slower growth [103].

Despite limitations in this study's scope, it suggests a rapid increase in tree size in the initial years of growth, allowing for replacing older trees prone to felling or decay. Considering the lifespan of *Populus* ×*berolinensis*, young trees could replace older specimens within 37 years for height growth or 24 years for trunk circumference. However, sustained care can enable these trees to thrive even in poor urban conditions, as seen on Rakowiecka Street. Although trunk circumference measurements have proven valuable for determining tree growth, older trees may also require height measurements to obtain more reliable growth results. It is crucial to conduct precise measurements, especially when assessing tree growth in urban environments.

The selected examples demonstrate the potential applications of poplar trees, such as those in Chorzów's Silesia Park, which fulfil their functions effectively. Careful selection of tree species adapted to local conditions is essential to avoid previous mistakes in urban forestry. Regularly monitoring and using ecological surfaces are recommended to mitigate infrastructure damage caused by rapid tree growth.

While rapid growth may lead to visually appealing urban greenery, caution is advised in selecting tree species and managing their lifespans. Replacing older trees with younger specimens ensures the continuity of urban greenery, while slower-growing species like lime or maple contribute to the diversity and longevity of urban forests.

Flexibility in managing old trees is necessary for overmatured poplars, as rigid approaches may lead to unnecessary removals. The introduction of tree replacement patterns, often used in many Western countries, where similar species are used to preserve the unique character of urban green spaces, should be promoted.

3.5. Poplar in Cities as Suitable but Under-Rated Trees in Urban Greenery

Our review of scientific databases has shown that very little existing research is devoted to using poplar cultivars in woodlots in urban green spaces (UGSs). This is, among other things, due to some long-lasting myths that these trees are unsuitable for keeping in cities, e.g., because they are not aesthetically valuable and are very short-lived. In fact, concerning global climatic changes, poplars can offer various ecological benefits in urban or industrial areas, such as improving air quality and carbon sequestration due to their fast growth, as well as achieving big sizes and having high adaptability to unarboristic site conditions [104–107].

The presented cases of *Populus* ×*berolinensis* indicate that using such trees is reluctantly recommended nowadays in the context of providing sufficient support for mitigating urban heat islands and improving the local microclimates of streets, avenues, parks, etc. [103]. Integrating poplars into urban landscapes also contributes to their sustainability [108,109], and their fast growth and adaptability make them suitable for urban areas [100,110–112]. However, their variable lifespans pose challenges for sustained urban greenery [113,114]. In such a case, the careful selection of cultivars and management practices is crucial [115,116]. Also, collaborative efforts are needed for effective urban greenery planning and devising comprehensive strategies that maximize the benefits provided by poplars while addressing challenges associated with their diverse lifespans [115].

Considering that street trees have an average life expectancy of 13–20 years (e.g., [117]), the so-called short lifespans of some taxons/cultivars of poplar are not a factor that neglects their efficient usage in urban areas. The variable lifespans of poplars might necessitate more frequent replanting cycles (the urban plantation-like system) after a few decades when they mature [100]. Their quick growth and establishment align with the urgency of creating green spaces and forests in urban areas, addressing immediate environmental concerns, and fostering a more sustainable and livable urban landscape [118–120].

3.6. Benefits of Poplars against Other Species for Cities

Notably, the deterioration in the visual condition of trees in Poland during the 1990s and the 20th century, more generally, was attributed partially to inadequate management systems. Managers often struggled to keep pace with emerging trends, and legal frameworks sometimes needed clarification. In Poland, decisions regarding tree planting in urban areas are made by local authorities, which may require more representation from urban forestry specialists like landscape architects or foresters.

Furthermore, poplar trees are not a recent addition to urban landscapes. They were successfully introduced to cities in the 20th century. Still, they were eventually supplanted by other species, such as maples (primarily *Acer* sp.) and linden trees (*Tilia* sp.), due to their susceptibility to fragility and poor sanitation. However, relying solely on these alternative species may not be sustainable, as they struggle to adapt to climatic changes and temperature fluctuations, as evidenced by recent urban ecology research. Therefore, there is a need to revert to proven varieties of poplar trees, capable of rapid growth and surviving up to 100 years in challenging urban conditions, as exemplified by preserved specimens in Warsaw.

Despite evidence supporting the need for replacement after 60 years, this practice is not widely adopted in Poland due to a misguided trend termed the "civilization of the environment". This involves a pseudo-protection of trees, particularly by older capital residents, perpetuating outdated practices. Indeed, we also observe that society's involvement in these issues has become more relevant. However, its opinion should not undermine the opinion of forestry specialists. Therefore, educating society in this area is essential, as it has been a good practice for many years, especially in Western European countries. Thus, this education should also be implemented in Poland [121], and a return to some old varieties and best practices, including maintenance procedures and monitoring, is essential to ensure the sustainability of urban greenery.

Our research underscores the significance of poplars as a valuable but often overlooked species in urban environments. While modern studies recommend only a few species for urban planting, monocultures pose risks to city landscapes. Furthermore, as emphasized in our revised paper, it is crucial to reintroduce species historically planted in significant locations. However, these arguments require further elaboration and more substantial justification.

Our study also highlights the water sensitivity of poplar trees and emphasizes the need for improved urban green management practices. Despite the challenges posed by the diverse global distribution of poplar species, addressing common environmental issues can facilitate future research on a worldwide scale.

In summary, while poplar trees were once favored for planting in the 20th century, they are rarely selected in contemporary urban forestry due to past mismanagement and misconceptions. However, their exclusion based on past mistakes must be addressed, and efforts should be made to incorporate them into future urban planting initiatives.

4. Conclusions

The following conclusions underscore the intricate balance between the advantages and challenges of integrating poplar trees into urban environments, underscoring the necessity for comprehensive and adaptable strategies in urban greening efforts:

- Berlin poplars serve, in the presented research, as exemplars of trees that have been successfully utilized in parks and urban street settings. They exhibit rapid growth even under adverse urban conditions, such as drought, limited rooting space, and pollution exposure.
- While poplar trees can demonstrate acceptable longevity in urban conditions, their survival rates mayvary between park and street environments, e.g., urban stressors have caused the premature decline of *Populus ×berolinensis* on Rakowiecka street in Warsaw (Poland).
- The literature and empirical evidence suggest that poplar trees' alleged short lifespans are less pronounced in downtowns, where trees struggle to reach maturity. It prompts a call to recognize various poplar cultivars as resilient tree species capable of enhancing urban biodiversity.
- Proper management practices, including root system care and selective pruning, are crucial for effectively utilizing poplar trees in urban green spaces, such as street tree lanes and park alleys. The selection of poplar cultivars tailored to specific urban functions is paramount for achieving the desired outcomes. Careful consideration should be given to avoiding female cultivars that produce abundant seed down.
- Given their rapid growth, adaptability, and positive environmental impact, poplar trees can be readily replaced in urban spaces through cyclic renewal practices akin to urban plantation schemes. A notable example is the reforestation of the main promenade in Silesia Park in Chorzów (Poland), where young poplar saplings are estimated to completely replace large old trees over 25 m in height within approximately 20 years. This highlights their significant role in urban landscapes from both ecological and aesthetic perspectives.
- Our studies on the growth of *Populus ×berolinensis* (Cases A and B) are of a pilot nature. However, the obtained results encourage continued research into the wide use of poplars in cities in the context of challenges related to climate change and maintaining tree cover in urban forests and urban green spaces in the future.

Author Contributions: Conceptualization, J.Ł., B.F.-A. and J.F.; methodology, J.Ł. and J.F.; software, J.Ł. and A.D.; validation, J.Ł. and J.F.; formal analysis, J.Ł.; investigation, J.Ł., B.F.-A. and J.F.; resources, B.F.-A.; data curation, J.F. and A.D.; writing—original draft preparation, J.Ł.; writing—review and editing, J.F.; visualization, J.Ł. and A.D.; supervision, J.F. All authors have read and agreed to the published version of the manuscript.

Funding: This research received no external funding.

Data Availability Statement: No new data were created.

Acknowledgments: We thank Justyna Marchewka from the Department of Human Biology, Institute of Biological Sciences, Faculty of Biology and Environmental Sciences, and Cardinal Stefan Wyszynski University in Warsaw for her help with the statistical analysis. The authors would like to thank the reviewers for all the useful and helpful comments on our manuscript. We would also like to thank our colleagues Łukasz Drozd and Piotr Wiśniewski for helping us with field measurements in the Silesia Park.

Conflicts of Interest: The authors declare no conflicts of interest.

References

1. Fischer, L.K.; Neuenkamp, L.; Lampinen, J.; Tuomi, M.; Alday, J.G.; Bucharová, A.; Cancellieri, L.; Casado-Arzuaga, I.; Čeplová, N.; Cerveró, L.; et al. Public attitudes toward biodiversity-friendly greenspace management in Europe. *Conserv. Lett.* **2020**, *13*, e12718. [CrossRef]
2. Dade, M.C.; Mitchell, M.G.E.; Brown, G.; Rhodes, J.R. The effects of urban greenspace characteristics and socio-demographics vary among cultural ecosystem services. *Urban For. Urban Green.* **2020**, *49*, 126641. [CrossRef]
3. WHO. *Urban Green Spaces: A Brief for Action*; World Health Organization, Regional Office for Europe: Bonn, Germany, 2017. Available online: https://www.euro.who.int/en/health-topics/environment-and-health/urban-health/publications/2017/urban-green-spaces-a-brief-for-action-2017 (accessed on 1 October 2021).

4. Forman, R.T.T. Values of large-versus-small urban greenspaces and their arregement. In *Why Cities Need Large Parks*; Murray, R., Ed.; Large Parks in Large Cities; Routledge: London, UK; New York, NY, USA, 2021; pp. 18–43.
5. Matos, P.; Vieira, J.; Rocha, B.; Branquinho, C.; Pinho, P. Modeling the provision of air-quality regulation ecosystem service provided by urban green spaces using lichens as ecological indicators. *Sci. Total Environ.* **2019**, *665*, 521–530. [CrossRef]
6. Borowski, J.; Fortuna-Antoszkiewicz, B.; Łukaszkiewicz, J.; Rosłon-Szeryńska, E. Conditions for the effective development and protection of the resources of urban green infrastructure. *E3S Web Conf.* **2018**, *45*, 00010. [CrossRef]
7. Graça, M.; Queirós, C.; Farinha-Marques, P.; Cunha, M. Street trees as cultural elements in the city: Understanding how perception affects ecosystem services management in Porto, Portugal. *Urban For. Urban Green.* **2018**, *30*, 194–205. [CrossRef]
8. Konijnendijk, C.C. Evidence-based guidelines for greener, healthier, more resilient neighbourhoods: Introducing the 3–30–300 rule. *J. For. Res.* **2022**, *34*, 821–830. [CrossRef]
9. Ziter, C.; Pedersen, E.J.; Kucharik, C.J.; Turner, M.G. Scale-dependent interactions between tree canopy cover and impervious surfaces reduce daytime urban heat during summer. *Proc. Natl. Acad. Sci. USA* **2019**, *116*, 7575–7580. [CrossRef] [PubMed]
10. Yang, J.; Chang, Y.; Yan, P. Ranking the suitability of common urban tree species for controlling PM2.5 pollution. *Atmos. Pollut. Res.* **2015**, *6*, 267–277. [CrossRef]
11. Yang, Q.; Wang, H.; Wang, J.; Lu, M.; Liu, C.; Xia, X.; Yin, W.; Guo, H. PM2.5-bound SO_4^{2-} absorption and assimilation of poplar and its physiological responses to PM2.5 pollution. *Environ. Exp. Bot.* **2018**, *153*, 311–319. [CrossRef]
12. Kis, B.; Avram, S.; Pavel, I.Z.; Lombrea, A.; Buda, V.; Dehelean, C.; Şoica, C.; Yerer, M.B.; Bojin, F.; Folescu, R.; et al. Recent Advances Regarding the Phytochemical and Therapeutic Uses of *Populus nigra* L. Buds. *Plants* **2020**, *9*, 1464. [CrossRef]
13. Madejón, P.; Ciadamidaro, L.; Marañón, T.; Murillo, J.M. Long-Term biomonitoring of soil contamination using poplar trees: Accumulation of trace elements in leaves and fruits. *Int. J. Phytoremediat.* **2013**, *15*, 602–614. [CrossRef] [PubMed]
14. Roy, S.; Byrne, J.; Pickering, C. A systematic quantitative review of urban tree benefits, costs, and assessment methods across cities in different climatic zones. *Urban For. Urban Green.* **2012**, *11*, 351–363. [CrossRef]
15. Bamwesigye, D.; Fialova, J.; Kupec, P.; Yeboah, E.; Łukaszkiewicz, J.; Fortuna-Antoszkiewicz, B.; Botwina, J. Urban Forest Recreation and Its Possible Role throughout the COVID-19 Pandemic. *Forests* **2023**, *14*, 1254. [CrossRef]
16. Bose, A.; Moser, B.; Rigling, A.; Lehmann, M.; Milcu, A.; Peter, M.; Rellstab, C.; Wohlgemuth, T.; Gessler, A. Memory of environmental conditions across generations affects the acclimation potential of scots pine. *Plant Cell Environ.* **2020**, *43*, 1288–1299. [CrossRef] [PubMed]
17. Pretzsch, H.; Biber, P.; Uhl, E.; Dahlhausen, J.; Schütze, G.; Perkins, D.; Rötzer, T.; Caldentey, J.; Koike, T.; van Con, T.; et al. Climate change accelerates growth of urban trees in metropolises worldwide. *Sci. Rep.* **2017**, *7*, 15403. [CrossRef] [PubMed]
18. Crous-Duran, J.; Graves, A.; Jalón, S.; Kay, S.; Tomé, M.; Burgess, P.; Giannitsopoulos, M.; Palma, J. Quantifying regulating ecosystem services with increased tree densities on european farmland. *Sustainability* **2020**, *12*, 6676. [CrossRef]
19. Zhang, M.; Yuan, N.; Li, M. Evaluation of the growth, adaption, and ecosystem services of two potentially-introduced urban tree species in guangzhou under drought stress. *Sci. Rep.* **2023**, *13*, 3563. [CrossRef] [PubMed]
20. Vaz, A.; Castro-Díez, P.; Godoy, Ó.; Alonso, Á.; Vilà, M.; Saldaña, A.; Marchante, H.; Bayón, Á.; Silva, J.S.; Vicente, J.R.; et al. An indicator-based approach to analyse the effects of non-native tree species on multiple cultural ecosystem services. *Ecol. Indic.* **2018**, *85*, 48–56. [CrossRef]
21. Lima, I.; Scariot, A.; Giroldo, A. Impacts of the implementation of silvopastoral systems on biodiversity of native plants in a traditional community in the brazilian savanna. *Agrofor. Syst.* **2016**, *91*, 1069–1078. [CrossRef]
22. Murray, R. Why cities need large parks. In *Why Cities Need Large Parks. Large Parks in Large Cities*; Murray, R., Ed.; Routledge: London, UK; New York, NY, USA, 2021; pp. 11–17.
23. Bell, J.N.B.; Treshow, M. *Air Pollution and Plant Life. [Zanieczyszczenie Powietrza a Życie Roślin]*; WNT: Warszawa, Poland, 2016.
24. Lo, A.; Byrne, J.; Jim, C.Y. How climate change perception is reshaping attitudes towards the functional benefits of urban trees and green space: Lessons from Hong Kong. *Urban For. Urban Green.* **2017**, *23*, 74–83. [CrossRef]
25. Gawroński, S.W. Phytoremediation Role of Plants in Urbanized Areas. In *Trees and Shrubs in Environmental Reclamation, Mat. IX Zjazdu Polskiego Towarzystwa Dendrologicznego*; Nowak, G., Kubus, M., Sobisz, Z., Eds.; Konferencja Naukowa, Wirty-Ustka, 19–22 września 2018 r.; Polish Dendrology Society: Szczecin, Poland, 2018; pp. 19–27.
26. Łukaszkiewicz, J. *Zadrzewienia w Krajobrazie Miasta: Wybrane Aspekty Kształtowania Struktury i Funkcji [Trees in the City Landscape: Selected Aspects of Shaping the Structure and Function]*; SGGW: Warsaw, Poland, 2019.
27. Łukaszkiewicz, J.; Fortuna-Antoszkiewicz, B. The influence of woodlots on the photoclimate of green areas and the quality of recreation. In *Public Recreation and Landscape Protection–with Environment Hand in Hand*; Fialová, W.J., Ed.; Mendel University: Brno, Czech Republic, 2022; Volume 13, pp. 385–389. [CrossRef]
28. Popek, R.; Gawrońska, H.; Gawronski, S.W. The level of particulate matter on foliage depends on the distance from the source of emission. *Int. J. Phytoremediat.* **2015**, *17*, 1262–1268. [CrossRef] [PubMed]
29. Sgrigna, G.; Sæbø, A.; Gawroński, S.; Popek, R.; Calfapietra, C. Particulate Matter deposition on Quercus ilex leaves in an industrial city of central Italy. *Environ. Pollut.* **2015**, *197*, 187–194. [CrossRef] [PubMed]
30. Chambers-Ostler, A.; Walker, H.; Doick, K.J. The role of the private tree in bringing diversity and resilience to the urban forest. *Urban For. Urban Green.* **2024**, *91*, 127973. [CrossRef]
31. Gillner, S.; Vogt, J.; Tharang, A.; Dettmann, S.; Roloff, A. Role of street trees in mitigating effects of heat and drought at highly sealed urban sites. *Landsc. Urban Plan.* **2015**, *143*, 33–42. [CrossRef]

32. Jim, C.Y. Urban Heritage Trees: Natural-Cultural Significance Informing management and Conservation. In *Advances in 21st Century Human Settlements*; Springer: New York, NY, USA, 2017; pp. 279–305. [CrossRef]
33. Jacobsen, R.M.; Birkemoe, T.; Evju, M.; Skarpaas, O.; Sverdrup-Thygeson, A. Veteran trees in decline: Stratified national monitoring of oaks in Norway. *For. Ecol. Manag.* **2023**, *527*, 120624. [CrossRef]
34. Carmichael, C.; McDonough, M.H. The trouble with trees? Social and political dynamics of street tree-planting efforts in Detroit, Michigan, USA. *Urban For. Urban Green.* **2018**, *31*, 221–229. [CrossRef]
35. Nowak, D.J.; Aevermann, T. Tree compensation rates: Compensating for the loss of future tree values. *Urban For. Urban Green.* **2019**, *41*, 93–103. [CrossRef]
36. Riedman, E.; Roman, L.A.; Pearsall, H.; Maslin, M.; Ifill, T.; Dentice, D. Why don't people plant trees? Uncovering barriers to participation in urban tree planting initiatives. *Urban For. Urban Green.* **2022**, *73*, 127597. [CrossRef]
37. Hurley, A.; Heinrich, I. Assessing urban-heating impact on street tree growth in Berlin with open inventory and environmental data. *Urban Ecosyst.* **2023**, *27*, 359–375. [CrossRef]
38. Moffat, A. Communicating the benefits of urban trees: A critical review. *Arboric. J.* **2016**, *38*, 64–82. [CrossRef]
39. Moffat, A.; Ambrose-Oji, B.; Clarke, T.; O'Brien, L.; Doick, K.J. Public attitudes to urban trees in Great Britain in the early 2020 s. *Urban For. Urban Green.* **2024**, *91*, 128177. [CrossRef]
40. Schroeder, H.; Flannigan, J.; Coles, R. Residents' attitudes toward street trees in the UK and U.S. communities. *Arboric. Urban For.* **2006**, *32*, 236–246. [CrossRef]
41. Łukaszkiewicz, J.; Fortuna-Antoszkiewicz, B.; Rosłon-Szeryńska, E.; Wiśniewski, P. Poplars' shelterbelts and woodlots in the cultural landscape of Poland—Functions, application and maintenance. In *Public Recreation and Landscape Protection—With Sense Hand in Hand...: Conference Proceeding: 13th—15th May 2019, Křtiny (s. 281–285)*; Fialová, W.J., Ed.; Mendel University in Brno: Brno, Czech Republic, 2019; ISBN 978-80-7509-660-9. Available online: http://utok.cz/sites/default/files/data/USERS/u24/02_sbornik%20RaOP%202019.pdf (accessed on 30 December 2020).
42. Endreny, T. Strategically growing the urban forest will improve our world. *Nat. Commun.* **2018**, *9*, 1160. [CrossRef] [PubMed]
43. Livesley, S.; McPherson, E.; Calfapietra, C. The urban forest and ecosystem services: Impacts on urban water, heat, and pollution cycles at the tree, street, and city scale. *J. Environ. Qual.* **2016**, *45*, 119–124. [CrossRef] [PubMed]
44. Hiemstra, J.A.; Schoenmaker-van der Bijl, E.; Tonneijck, A.E.G.; Hoffman, M.H.A. *Trees: Relief for the City*; Plant Publicity Holland: Wageningen, The Netherlands, 2008.
45. Avşar, M.D.; Ok, T. Using poplars (*Populus* L.) in urban afforestation: KAHRAMANMARAŞ sample. *Turk. J. For.* **2010**, *11*, 127–135. Available online: https://dergipark.org.tr/tr/download/article-file/195755 (accessed on 24 April 2024).
46. Kendall, C.; Silva, S.R.; Kelly, V.J. Carbon and nitrogen isotopic compositions of particulate organic matter in four large river systems across the United States. *Hydrol. Process.* **2001**, *15*, 1301–1346. [CrossRef]
47. Esperon-Rodriguez, M.; Tjoelker, M.G.; Lenoir, J.; Baumgartner, J.B.; Beaumont, L.J.; Nipperess, D.A.; Power, S.A.; Richard, B.; Rymer, P.D.; Gallagher, R.V. Climate change increases global risk to urban forests. *Nat. Clim. Chang.* **2022**, *12*, 950–955. [CrossRef]
48. Wang, Y.; Qu, J.; Han, Y.; Du, L.; Wang, M.; Yang, Y.; Cao, G.; Tao, S.; Kong, Y. Impacts of linear transport infrastructure on terrestrial vertebrate species and conservation in China. *Glob. Ecol. Conserv.* **2022**, *38*, e02207. [CrossRef]
49. Zalesny, R.; Headlee, W. Developing woody crops for the enhancement of ecosystem services under changing climates in the north central united states. *J. For. Environ. Sci.* **2015**, *31*, 78–90. [CrossRef]
50. Levei, L.; Cadar, O.; Babalau-Fuss, V.; Kovacs, E.; Török, A.; Levei, E.; Ozunu, A. Use of black poplar leaves for the biomonitoring of air pollution in an urban agglomeration. *Plants* **2021**, *10*, 548. [CrossRef]
51. Molnár, V.; Tőzsér, D.; Szabó, S.; Tóthmérész, B.; Simon, E. Use of leaves as bioindicator to assess air pollution based on composite proxy measure (apti), dust amount and elemental concentration of metals. *Plants* **2020**, *9*, 1743. [CrossRef] [PubMed]
52. Liu, Y.; Yang, Z.; Zhu, M.; Yin, J. Role of plant leaves in removing airborne dust and associated metals on beijing roadsides. *Aerosol Air Qual. Res.* **2017**, *17*, 2566–2584. [CrossRef]
53. Chen, S.; Yu, Y.; Wang, X.; Wang, S.; Zhang, T.; Zhou, Y.; He, R.; Meng, N.; Wang, Y.; Liu, W.; et al. Chromosome-level genome assembly of a triploid poplar populus alba 'berolinensis'. *Mol. Ecol. Resour.* **2023**, *23*, 1092–1107. [CrossRef] [PubMed]
54. Fineschi, S.; Loreto, F. A survey of multiple interactions between plants and the urban environment. *Front. For. Glob. Chang.* **2020**, *3*, 30. [CrossRef]
55. Luo, X.; Yu, S.; Zhu, Y.; Li, X. Trace metal contamination in urban soils of china. *Sci. Total Environ.* **2021**, *421–422*, 17–30. [CrossRef] [PubMed]
56. Kroeger, T.; Escobedo, F.; Hernandez, J.; Varela, S.; Delphin, S.; Fisher, J.; Waldron, J. Reforestation as a novel abatement and compliance measure for ground-level ozone. *Proc. Natl. Acad. Sci. USA* **2014**, *111*, E4204–E4213. [CrossRef] [PubMed]
57. Urgilez-Clavijo, A.; Fernández, J.; Rivas-Tabares, D.; Tarquis, A. Linking deforestation patterns to soil types: A multifractal approach. *Eur. J. Soil Sci.* **2020**, *72*, 635–655. [CrossRef]
58. Cardoso, M.; Alves, H.; Costa, I.; Vieira, T. Anthropogenic actions and socioenvironmental changes in lake of juá, brazilian amazonia. *Sustainability* **2021**, *13*, 9134. [CrossRef]
59. Millward, A.; Sabir, S. Benefits of a forested urban park: What is the value of allan gardens to the city of toronto, canada? *Landsc. Urban Plan.* **2011**, *100*, 177–188. [CrossRef]

60. Pavlacky, D.; Goldizen, A.; Prentis, P.; Nicholls, J.; Lowe, A. A landscape genetics approach for quantifying the relative influence of historic and contemporary habitat heterogeneity on the genetic connectivity of a rainforest bird. *Mol. Ecol.* **2009**, *18*, 2945–2960. [CrossRef]
61. Sari, N.; Indra, T.; Kushardono, D. Urban vegetation quality assessment using vegetation index and leaf area index from spot 7 data with fuzzy logic algorithm. *Int. J. Adv. Sci. Eng. Inf. Technol.* **2022**, *12*, 738. [CrossRef]
62. Rosier, C.; Polson, S.; D'Amico, V.; Kan, J.; Trammell, T. Urbanization pressures alter tree rhizosphere microbiomes. *Sci. Rep.* **2021**, *11*, 9447. [CrossRef] [PubMed]
63. Assad, M.; Chalot, M.; Tatin-Froux, F.; Bert, V.; Parelle, J. Trace metal(oid) accumulation in edible crops and poplar cuttings grown on dredged sediment enriched soil. *J. Environ. Qual.* **2018**, *47*, 1496–1503. [CrossRef] [PubMed]
64. Piotrowska, N.; Czachorowski, S.; Stolarski, M. Ground beetles (carabidae) in the short-rotation coppice willow and poplar plants—Synergistic benefits system. *Agriculture* **2020**, *10*, 648. [CrossRef]
65. Liu, L.; Zhang, Y. Urban heat island analysis using the landsat tm data and aster data: A case study in hong kong. *Remote Sens.* **2011**, *3*, 1535–1552. [CrossRef]
66. Tang, C. A study of the urban heat island effect in guangzhou. *IOP Conf. Ser. Earth Environ. Sci.* **2022**, *1087*, 012015. [CrossRef]
67. Li, B.; Liu, Z.; Nan, Y.; Li, S.; Yang, Y. Comparative analysis of urban heat island intensities in chinese, russian, and dprk regions across the transnational urban agglomeration of the tumen river in northeast asia. *Sustainability* **2018**, *10*, 2637. [CrossRef]
68. Lee, D.; Oh, K.; Seo, J. An analysis of urban cooling island (uci) effects by water spaces applying uci indices. *Int. J. Environ. Sci. Dev.* **2016**, *7*, 810–815. [CrossRef]
69. Zhou, L.; Dickinson, R.; Tian, Y.; Fang, J.; Li, Q.; Kaufmann, R.; Tucker, C.J.; Myneni, R.B. Evidence for a significant urbanization effect on climate in china. *Proc. Natl. Acad. Sci. USA* **2004**, *101*, 9540–9544. [CrossRef]
70. Yang, Y.; Fan, S.; Ma, J.; Zheng, W.; Song, L.; Wei, C. Spatial and temporal variation of heat islands in the main urban area of zhengzhou under the two-way influence of urbanization and urban forestry. *PLoS ONE* **2022**, *17*, e0272626. [CrossRef]
71. Simiele, M.; Zio, E.; Montagnoli, A.; Terzaghi, M.; Chiatante, D.; Scippa, G.; Trupiano, D. Biochar and/or compost to enhance nursery-produced seedling performance: A potential tool for forest restoration programs. *Forests* **2022**, *13*, 550. [CrossRef]
72. Jia, D.; Li, X.; Zhang, Y.; Feng, Y.; Liu, D. Analysis on water use strategies of natural poplar in hunshandake sandy land, china. *Environ. Prog. Sustain. Energy* **2021**, *40*, e13579. [CrossRef]
73. Wang, G.; Deng, F.; Xu, W.; Chen, H.; Ruan, H. Poplar plantations in coastal china: Towards the identification of the best rotation age for optimal soil carbon sequestration. *Soil Use Manag.* **2016**, *32*, 303–310. [CrossRef]
74. Xia, J.; Zhang, S.; Li, T.; Liu, X.; Guangcan, Z. Effect of continuous cropping generations on each component biomass of poplar seedlings during different growth periods. *Sci. World J.* **2014**, *2014*, 618421. [CrossRef]
75. Scotti, R.; D'Ascoli, R.; Cáceres, M.; Bonanomi, G.; Sultana, S.; Cozzolino, L.; Scelza, R.; Zoina, A.; Rao, M.A. Combined use of compost and wood scraps to increase carbon stock and improve soil quality in intensive farming systems. *Eur. J. Soil Sci.* **2015**, *66*, 463–475. [CrossRef]
76. Wang, K.; Zhang, R.; Song, L.; Yan, T.; Na, E. Comparison of c:n:p stoichiometry in the plant–litter–soil system between poplar and elm plantations in the horqin sandy land, China. *Front. Plant Sci.* **2021**, *12*, 655517. [CrossRef]
77. Guarino, F.; Improta, G.; Triassi, M.; Cicatelli, A.; Castiglione, S. Effects of zinc pollution and compost amendment on the root microbiome of a metal tolerant poplar clone. *Front. Microbiol.* **2020**, *11*, 1677. [CrossRef]
78. Ferré, C.; Comolli, R.; Leip, A.; Seufert, G. Forest conversion to poplar plantation in a lombardy floodplain (italy): Effects on soil organic carbon stock. *Biogeosciences* **2014**, *11*, 6483–6493. [CrossRef]
79. Mureva, A.; Nyamugure, T.; Masona, C.; Mudyiwa, S.; Makumbe, P.; Muringayi, M.; Nyamadzawo, G. Community perceptions towards the establishment of an urban forest plantation: A case of dzivaresekwa, zimbabwe. *Int. J. Agric. Res. Innov. Technol.* **2014**, *4*, 16–23. [CrossRef]
80. Lüttge, U.; Buckeridge, M. Trees: Structure and function and the challenges of urbanization. *Trees* **2020**, *37*, 9–16. [CrossRef]
81. Wang, G.; Dong, Y.; Liu, X.; Yao, G.; Yu, X.; Yang, M. The current status and development of insect-resistant genetically engineered poplar in china. *Front. Plant Sci.* **2018**, *9*, 1408. [CrossRef] [PubMed]
82. Shahid, M.; Dumat, C.; Khalid, S.; Schreck, E.; Xiong, T.; Niazi, N. Foliar heavy metal uptake, toxicity and detoxification in plants: A comparison of foliar and root metal uptake. *J. Hazard. Mater.* **2017**, *325*, 36–58. [CrossRef] [PubMed]
83. Camarero, J.J.; De Andrés, E.G.; Colangelo, M.; De Jaime Loren, C. Growth history of pollarded black poplars in a continental Mediterranean region: A paradigm of vanishing landscapes. *For. Ecol. Manag.* **2022**, *517*, 120268. [CrossRef]
84. Hejmanowski, S.; Milewski, J.; Terpiński, Z. *Poradnik Zadrzewieniowca*; PWRiL: Warszawa, Poland, 1964.
85. Jakuszewski, T. Topola w Zadrzewieniu Kraju [w:] Topole (Populus L.). Nasze Drzewa Leśne, S. Białobok (red.). Monografie Popularnonaukowe, Tom XII s. 463-470, Zakład Dendrologii i Arboretum Kórnickie w Kórniku k. Poznania. PWN Publ.: Warszawa, Poland, 1973. Available online: https://rcin.org.pl/Content/187156/PDF/KOR001_154493_1973_Topola-w-zadrzewieni.pdf (accessed on 24 April 2024).
86. Zabielski Uprawa topoli w Polsce. In *Topole (Populus L.). Nasze Drzewa Leśne*; Białobok, S. (Ed.) Monografie Popularnonaukowe, 12, Zakład Dendrologii i Arboretum Kórnickie w Kórniku, PWN: Warszawa, Poland, 1973; pp. 413–462.
87. Fortuna-Antoszkiewicz, B.; Łukaszkiewicz, J.; Wiśniewski, P. Stan zachowania i walory krajobrazowe przywodnych zadrzewień topolowych Kanału Żerańskiego-metodologiczne studium przypadku. *MAZOWSZE Stud. Reg.* **2018**, *27*, 81–102. [CrossRef]

88. Çölkesen, İ.; Kavzoğlu, T.; Ateşoğlu, A.; Tonbul, H.; Öztürk, M.Y. Multi-seasonal evaluation of hybrid poplar (*P. Deltoides*) plantations using Worldview-3 imagery and State-Of-The-Art ensemble learning algorithms. *Adv. Space Res.* **2023**, *71*, 3022–3044. [CrossRef]
89. Li, J.; Gao, K.; Yang, X.; Guo, B.; Xue, Y.; Miao, D.; Huang, S.; An, X. Comprehensive analyses of four ptonfyc genes from populus tomentosa and impacts on flowering timing. *Int. J. Mol. Sci.* **2022**, *23*, 3116. [CrossRef] [PubMed]
90. Wilkaniec, A.; Borowiak-Sobkowiak, B.; Irzykowska, L.; Bres, W.; Świerk, D.; Pardela, Ł.; Durak, R.; Środulska-Wielgus, J.; Wielgus, K. Biotic and abiotic factors causing the collapse of robinia pseudoacacia l. veteran trees in urban environments. *PLoS ONE* **2021**, *16*, e0245398. [CrossRef] [PubMed]
91. Peng, Y.; Zhou, Z.; Zhang, Z.; Yu, X.; Zhang, X.; Du, K. Molecular and physiological responses in roots of two full-sib poplars uncover mechanisms that contribute to differences in partial submergence tolerance. *Sci. Rep.* **2018**, *8*, 12829. [CrossRef] [PubMed]
92. Hacquard, S.; Pêtre, B.; Frey, P.; Hecker, A.; Rouhier, N.; Duplessis, S. The poplar-poplar rust interaction: Insights from genomics and transcriptomics. *J. Pathog.* **2011**, *2011*, 716041. [CrossRef]
93. Seneta, W.; Dolatowski, J. *Dendrologia*; PWN: Warszawa, Poland, 2012.
94. Čížková, L. (Forestry and Game Management Research Institute, Jílové u Prahy, Czechia). *Personal communication*, 2021.
95. Piasecki, J.K. Tereny Zieleni Dzielnicy Warszawa-Ochota [Green Areas in Warsaw's Ochota District]. Master's Thesis, Manuscript in Collection of Landscape Architecture Department, WULS-SGGW, Warsaw, Poland, 1974. (In Polish)
96. Zielonko, A. Drzewostan przyuliczny w kolizji z infrastrukturą [Street tree stand colliding with infrastructure]. *J. Ogrod.* **1977**, *10*, 269–273. (In Polish)
97. Historical Photos of Rakowiecka Street in Warsaw-Photo A-H: FotoPolska.eu. Available online: https://warszawa.fotopolska.eu (accessed on 19 October 2019).
98. Fortuna-Antoszkiewicz, B.; Łukaszkiewicz, J.; Wisniewski, P. Przekształcenia kompozycji szaty roślinnej Parku Śląskiego w Chorzowie po 60 latach [The transformation of vegetation's composition 60 years after the establishment of Silesia Park]. *Urban. Archit. Files Pol. Acad. Sci. Kraków Branch* **2017**, *45*, 193–215.
99. Łukaszkiewicz, J.; Fortuna-Antoszkiewicz, B.; Wiśniewski, P. Silesia Park, Chorzów, Poland. In *Why Cities Need Large Parks*; Murray, R., Ed.; Routledge: London, UK; New York, NY, USA, 2021; pp. 264–273. ISBN 978-1-03-207293-7.
100. Rauf, H.A.; Wolff, E.; Hamel, P. Climate Resilience in Informal Settlements: The Role of Natural Infrastructure: A Focus on Climate Adaptation. In *The Palgrave Encyclopedia of Urban and Regional Futures*; Springer International Publishing: Cham, Switzerland, 2023.
101. Niemirski, W. Schlesischer Kulturpark in Kattowitz. Baumeister No7, Niemirski, W. Schlesicher Kulturpark in Kattowitz. *Zeitschrift für Architektur, Planung, Baumeister. Umwelt.* No. 7; Verlag Georg D. W. Callwey Publ.: Munich, Germany, 1971; pp. 826–827.
102. Knobelsdorf, W. (red.) Oaza pod rudym obłokiem: Śląski Park Kultury i Wypoczynku. Wyd. 1., Wyd. 1972, "Śląsk", Katowice.
103. Buchwald, A. Chapter 4.2. Określanie przyrostu drzewa i drzewostanu. In *Dentrometria [Dentrometry]*; Nasza Wiedza Publ.: Warszawa, Poland, 2021; pp. 170–176, ISBN-10. 6203227471, ISBN-13. 978-6203227475. (In Polish)
104. Ghezehei, S.B.; Nichols, E.G.; Maier, C.A.; Hazel, D.W. Adaptability of Populus to Physiography and Growing Conditions in the Southeastern USA. *Forests* **2019**, *10*, 118. [CrossRef]
105. Fang, S.; Liu, Y.; Yue, J.; Tian, Y.; Xu, X. Assessments of growth performance, crown structure, stem form and wood property of introduced poplar clones: Results from a long-term field experiment at a lowland site. *For. Ecol. Manag.* **2021**, *479*, 118586. [CrossRef]
106. Riccioli, F.; Guidi Nissim, W.; Masi, M.; Palm, E.; Mancuso, S.; Azzarello, E. Modeling the Ecosystem Services Related to Phytoextraction: Carbon Sequestration Potential Using Willow and Poplar. *Appl. Sci.* **2020**, *10*, 8011. [CrossRef]
107. Fang, S.; Xue, J.; Tang, L. Biomass production and carbon sequestration potential in poplar plantations with different management patterns. *J. Environ. Manag.* **2007**, *85*, 672–679. [CrossRef]
108. Aitchison, E.W.; Kelley, S.L.; Alvarez, P.J.; Schnoor, J.L. Phytoremediation of 1, 4-dioxane by hybrid poplar trees. *Water Environ. Res.* **2000**, *72*, 313–321. [CrossRef]
109. Goodarzi, G.R.; Ahmadloo, F. Assessment of quantitative and qualitative characteristics of traditional poplar plantations production in Markazi province. *For. Wood Prod.* **2022**, *75*, 201–216.
110. Coseo, P.; Hamstead, Z. Just, nature-based solutions as critical urban infrastructure for cooling and cleaning airsheds. *Nat.-Based Solut. Cities*, 2023; 106–146. [CrossRef]
111. Maouni, Y.; Boubekraoui, H.; Taaouati, M.; Maouni, A.; Draoui, M. Assessment of Urban Outdoor Comfort Variation in a Northwestern Moroccan City-Toward the Implementation of Effective Mitigation Strategies. *Ecol. Eng. Environ. Technol. (EEET)* **2023**, *24*, 1–18. [CrossRef] [PubMed]
112. D'Amato, G.; Girard, L.F.; Murena, F.; Nocca, F. 14. Climate change, air pollution and circular city model: A proposal for the urban sustainable development of the historical center of Naples (Italy). In *Reconnecting the City with Nature and History: Towards Circular Regeneration Strategies*; Torossa Publ.: New York, NY, USA, 2023; p. 367. Available online: https://www.torrossa.com/en/resources/an/5650226#page=369 (accessed on 24 April 2024).
113. Khurana, D.K. Mitigating Wood Shortages through a People Friendly Tree-Poplar. AGE, 7, 11. Available online: https://openknowledge.fao.org/home (accessed on 24 April 2024).
114. Konijnendijk, C.C.; Gauthier, M. Urban forestry for multifunctional urban land use. In *Cities Farming for the Future: Agriculture for Green and Productive Cities*; van Veenhuizen, R., Ed.; RUAF Foundation: Rome, Italy, 2006; pp. 414–416.

115. Miller, R.W.; Hauer, R.J.; Werner, L.P. *Urban Forestry: Planning and Managing Urban Greenspaces*; Waveland Press Inc.: Long Grove, IL, USA, 2015; ISBN 13: 978-1-4786-0637-6.
116. McCarthy, M.A.; Possingham, H.P. Active adaptive management for conservation. *Conserv. Biol.* **2007**, *21*, 956–963. [CrossRef]
117. Smith, I.A.; Dearborn, V.K.; Hutyra, L.R. Live fast, die young: Accelerated growth, mortality, and turnover in street trees. *PLoS ONE* **2019**, *14*, e0215846. [CrossRef]
118. Du, K.; Jiang, S.; Chen, H.; Xia, Y.; Guo, R.; Ling, A.; Liao, T.; Wu, W.; Kang, X. Spatiotemporal miRNA and transcriptomic network dynamically regulate the developmental and senescence processes of poplar leaves. *Hortic. Res.* **2023**, *10*, uhad186. [CrossRef] [PubMed]
119. Vornicu, L.; Okros, A.; Șmuleac, L.; Pascalau, R.; Petcov, A.; Zoican, Ș.; Jigău, R.; Zoican, C. Energetic poplars and their importance for the environment. *Res. J. Agric. Sci.* **2023**, *55*, 220.
120. Fuertes, A.; Oliveira, N.; Pérez-Cruzado, C.; Cañellas, I.; Sixto, H.; Rodríguez-Soalleiro, R. Adapting 3-PG foliar variables to deciduous trees in response to water restriction: Poplar short rotation plantations under Mediterranean conditions. *For. Int. J. For. Res.* **2024**, *97*, 107–119. [CrossRef]
121. Afzalan, N.; Muller, B. The role of social media in green infrastructure planning: A case study of neighborhood participation in park siting. *J. Urban Technol.* **2014**, *21*, 67–83. [CrossRef]

Disclaimer/Publisher's Note: The statements, opinions and data contained in all publications are solely those of the individual author(s) and contributor(s) and not of MDPI and/or the editor(s). MDPI and/or the editor(s) disclaim responsibility for any injury to people or property resulting from any ideas, methods, instructions or products referred to in the content.

MDPI AG
Grosspeteranlage 5
4052 Basel
Switzerland
Tel.: +41 61 683 77 34

Land Editorial Office
E-mail: land@mdpi.com
www.mdpi.com/journal/land

Disclaimer/Publisher's Note: The title and front matter of this reprint are at the discretion of the Guest Editors. The publisher is not responsible for their content or any associated concerns. The statements, opinions and data contained in all individual articles are solely those of the individual Editors and contributors and not of MDPI. MDPI disclaims responsibility for any injury to people or property resulting from any ideas, methods, instructions or products referred to in the content.